Insect Vectors and Plant Pathogens

N S Butter
Professor & Head
Department of Entomology (Retd.)
Punjab Agricultural University, Ludhiana, India

CRC Press is an imprint of the
Taylor & Francis Group, an **informa** business
A SCIENCE PUBLISHERS BOOK

CRC Press
Taylor & Francis Group
6000 Broken Sound Parkway NW, Suite 300
Boca Raton, FL 33487-2742

First issued in paperback 2020

ISBN-13: 978-1-138-58792-2 (hbk)
ISBN-13: 978-0-367-78084-5 (pbk)

Library of Congress Cataloging-in-Publication Data

Names: Butter, N. S. (Nachhattar Singh), author.
Title: Insect vectors and plant pathogens / N.S. Butter.
Description: Boca Raton, FL : CRC Press, Taylor & Francis Group, [2018] |
Includes bibliographical references and index.
Identifiers: LCCN 2018020884 | ISBN 9781138587922 (hardback)
Subjects: LCSH: Insects as carriers of plant disease. | Phytopathogenic
microorganisms.
Classification: LCC SB931 .B98 2018 | DDC 632/.7--dc23
LC record available at https://lccn.loc.gov/2018020884

Visit the Taylor & Francis Web site at
http://www.taylorandfrancis.com

and the CRC Press Web site at
http://www.crcpress.com

Dedicated
to
My Respected Father
Late Sardar Naginder Singh Butter
for
Planting an Education Tree
in The Family

Foreword

Plant diseases are a major yield and quality constraint for farmers throughout the world. The viral plant pathogens can infect crops systemically and at times lead to total crop failure, thereby jeopardizing the food and nutritional security of the population. The majority of plant viruses rely on vectors for plant-to-plant transmission in the field, vineyard or orchard. Numerous books have been written by authors from different countries, yet there is a general contention that scientists have been reticent of writing scientific books on the important subject of plant virus-vectors. It is, therefore, a matter of great satisfaction that an individual has come forward to take up this interesting but arduous task of writing a book on vector-borne viruses.

The present book on insect vectors and plant pathogens has come into existence as a means to cater to the syllabus requirements of postgraduate students majoring in Entomology and Plant Pathology disciplines. It is a distinctive book involving the invisible foes, plant viruses, and their association with arthropods, nematodes and fungal vectors. At the outset, the author while introducing the subject has made the contents of the book amply clear to the reader. The thematic sequence and the display pattern also add to the merit of this manuscript.

The diseases are inflicted by a variety of plant pathogens viz. viruses, bacteria, fungi, etc. Of these, the etiology of plant viruses has undergone a big transition especially in the recent past. The diseases which were previously considered to be of viral origin have now been attributed to other classes of pathogens. As a result, the new pathogens like Mycoplasma (MLOs), Phytoplasma (PLOs), Spiroplasma and Rickettsia Like Organisms (RLOs) have been discovered and the author has meticulously brought these pathogen classes into the light of scrutiny. Due to the dynamic complexity of biotic problems, especially in the wake of climatic change, extreme weather events, and global warming, etc., it becomes all the more imperative to create such a useful and accessible compilation of knowledge. The author has tried to justify it from all angles.

This is not the end of the story. To start with, the identification and biology of vectors, their feeding mechanisms and the transmission of viruses have been carefully explained in an interesting manner, so as to make these aspects easily comprehensible to the reader. Viral structural proteins are required for virion retention at specific sites in vectors. The transmission of certain viruses requires non-structural proteins to link the virion and the insect. Various circulative and replicative viruses require non-structural proteins for dissemination in the vector itself. Further changes in the plant pathogen scenario are expected to occur over time, but this issue has been well addressed in the book. To aid the reader in their understanding of the text, the technical nomenclature has been provided in the text along with corresponding viruses. It is therefore mandatory to impart the required information on taxonomy and the cryptogamic categorization of plant viruses.

The earlier concept of viral transmission through pin-pricking or use of carborandum does not hold up well now. Instead, the concept of specificity being afforded by transmission determinants has been highlighted. The emphasis on this topic is expected to pave the way for better understanding of the subject and for the development of strategy based on alteration of transmission determinants through mutation. It would provide new vistas for research in future and would be of immense value to the researchers.

We are rather fortunate that plant viruses at present do not infect human beings, but looking at the rate of change and new developments in the field, no possibility can be definitively ruled out. If we look at the effects (both positive and negative) of plant viruses on their vectors as depicted in the text, we learn that anything can happen. The management of viral diseases is in itself a tricky problem, but the strategy suggested, which is based on an integrated approach involving cross protection, biotechnology and modern techniques of alteration in the virus cycle of vectors through mutation will be invaluable in the containment of plant viruses. Fundamental knowledge of the virus–vector relationship will lead to new bio-rational control strategies.

I am confident that this reader-friendly, comprehensive book containing information on the latest developments in the field of insect vectors will be a great source of knowledge for the teachers involved in post graduate teaching, for researchers aiming to modify or reorient their priorities and for extension specialists who seek a new perspective with which to identify/contain these crop maladies. Above all, the extensive compendium of recent discoveries and innovations will prove greatly beneficial to students and research scholars the world over.

I am immensely delighted to congratulate the author for his strenuous efforts and sincere commitment to this noble academic endeavor.

I wish him success in all walks of life.

Ludhiana
Dated: 15 January, 2018

(Dr N S Malhi)
Ex Vice-Chancellor
Guru Kashi University Talwandi Sabo
Bathinda, Punjab, India

Preface

I had the opportunity to deliver lectures on the subject of insect vectors as a resource person in summer schools organized by different State Agricultural Universities in India. On the basis of the feedback that I received from the scientist participants, it was apparent that the trainees' knowledge in the field of insect vectors was notably limited. I had to redraft my lectures to discuss the preliminaries of the subject in some situations. These occasions gave me the opportunity to explore the reasons why this subject is neglected. The most important being the unavailability of trained manpower, I realized that the inclusion of courses on the topic in the syllabi of almost all the universities remained a notion, but was not actually introduced. The subject of insect vectors remained sandwiched between plant pathology and entomology disciplines. Ours being an Institute with adequate faculty and resources, I was able to obtain postgraduate training in this important discipline. By virtue of my training in insect vectors, I was associated with the teaching of courses to postgraduates for more than two decades. The subject matter pertaining to pathogens of crop plants and insect vectors was addressed, but in isolation under two disciplines. Based on my training, experience, and interest in the subject, I decided to author a publication on insect vectors of plant pathogens which hitherto had remained squeezed between the two other important disciplines.

The science of insect vectors has undergone a big change. New diseases are being identified and added regularly to the list of pathogens and, to our dismay and surprise, out of the newly discovered plant pathogens forty seven per cent are diseases of viral origin and vectored by arthropods. Furthermore, being a Professor of Plant Protection, I have had sufficient exposure to the subject through teaching and research. Above all, my interest persuaded me to compile a document that will be a strong tool for scientists to teach with, and provide guidelines for the future.

The book is divided into fifteen chapters to cover the subject matter from all angles. The opening chapter is brief as the matter stands already compiled by eminent scientists from both entomology and plant pathology streams. The sole objective of this chapter is to introduce the subject. In the following chapters, the important details pertaining to the history of pathogens, vectors, symptoms and the economic importance of vectors/viruses have been dealt with as per their relevance in the compilation. While highlighting the role of arthropods, nematodes, and fungi, other agents of the transmission of plant pathogens have also been elaborated. The relevant aspects pertaining to insect vectors, i.e., vector identification, biology, feeding apparatus, mechanism of transmission and control of pathogens through vectors find a comprehensive discussion in this write-up. The vast majority of plant pathogens are transmitted by the order Hemiptera and out of these hemipterans, aphids are the major insect group acting as vectors of more than ninety per cent of all plant pathogens, therefore major stress has been laid on hemipterous insects. The principal focus has been on transmission determinants, affording specificity under different categories of the transmission mechanism. The role of transmission determinants like coat protein and helper component has been discussed A brief description of new diseases, at least one from each genus of plant viruses, has been included in this compendium to elucidate the interaction of vector and virus. Phytoplasmal etiology pathogens have been detailed separately from plant viruses on account of their importance. The transmission of Phytoplasma and Spiroplasma has been taken up in detail. Another chapter has been contributed on the involvement of insects with biting and chewing mouthparts. The recent developments pertaining to this category of insects have been discussed in an exhaustive manner. The latest contribution made by scientists in the field of mites, nematodes, and fungi as vectors of plant viruses has been discussed. How the phytotoxemia is different from other crop disorders, has been critically explained with support from suitable and common examples of crop disorders. The brief account on the classification of plant viruses has also been addressed in one of the chapters. Likewise, details on the electron microscope along with its uses have been included, so as to clarify the procedure of examining sub-microscopic entities. The latest developments in the management of plant pathogens through vector management have been discussed with special reference to the use of biotechnology, crop protection, and plant resistance.

Since the book contains the latest developments in the field of insect vectors with relevant examples, it will be of immense value to teachers who wish to impart knowledge, and to scientists engaged in research who wish to reorient their field of research. With the enrichment of scientific knowledge, it will be easier for extension workers to manage crop disorders with ease and positive results. This book will provide the latest literature on the subject to students and researchers of entomology, plant pathology, plant protection and virology.

(N S Butter)

Acknowledgements

The book on 'Insect Vectors and Plant Pathogens' was completed with the help of of several well-wishers of mine and various family members and scientists who contributed at different stages of the book's conception. I am thus highly indebted to them for their generous help. The idea of a book on this subject emerged during summer schools in which I acted as a resource person to deliver expert lectures. Subsequently, I received great help and encouragement from stalwarts of the subject of insect vectors. First of all, my deepest gratitude is due to my friend Dr. KR Kanaujia, Prof of Entomology and Director of Extension Education (Retired) GB Pant University of Agriculture and Technology who actually instilled in me the thought of a write-up on this topic. Secondly, Dr. MS Kang Vice Chancellor, Punjab Agricultural University, Ludhiana while visiting the Department of Entomology guided me and assured all the necessary help for the production of a book. For the last three to four years, I have been with my son in the USA and have had ample time to devote to this task. With the cooperation and help rendered by my son Amarinder Singh and my daughter-in-law Harprabhjit Kaur, I was able to initiate and successfully complete this project. Dr. Amit Sethi Entomologist, Pioneer DuPont, Des Moines, USA managed recent literature from the library of his organization and I feel highly obliged to him for this valuable help. Besides this Jaswinder Singh, my son-in-law and my daughter Raminder Kaur arranged access to St. Clair College Windsor (Canada) library and that proved extremely valuable to me. The suggestions of Dr. JS Kular Professor of Entomology (Retired) and Dr. RS Chandi Assistant Entomologist regarding topics to be included in the book were greatly appreciated. The high definition photographs arranged by Drs. Vijay Kumar Senior Entomologist, NV Krishanaiah, Senior Entomologist (Retired), Manmeet Bhuller Sr entomologist, Sandeep Singh Entomologist, Prof. Tarwinder Kaur and Prasad Burange Assistant professor of Entomology really added to the quality of the book, and I gratefully acknowledge their efforts. The constant help rendered by my wife Mrs. Shawinder Pal Kaur during the preparation of this book deserves heartiest thanks. The sincere wish and overall encouragement from the faculty of Entomology to bring about a publication on this very important subject actually worked as a motivational force to take up this missionary job. In a real sense, the constant mature advice and able guidance rendered by stalwart scientists Drs HS Rataul Professor and Head Entomology (Retired) and TH Singh, Dean Post Graduate Studies (Retired) from Punjab Agricultural University, Ludhiana remain fresh in my mind, compelling me to pursue this noble venture. At the last, I cannot restrain myself from extending my sincere thanks to my long-time friend Dr AK Dhawan Ex Additional Director of Research who was instrumental in providing all kinds of assistance at every stage of this project.

Contents

CHAPTER 1

Introduction

Historical Background, Pathogens, Symptoms, and Economic Importance

1.1 Historical Background

In the ancient era, man was a food gatherer. Little by little, he learnt the art of crop domestication and became quite familiar with the crop abnormalities that appeared in the field. After acquiring preliminary knowledge, the farmer began to develop tactics to guard the useful plants against the maladies and diseases that were prevalent during that time. The fossil evidence also indicates the presence of plant diseases 250 million years ago. The religious publications of that time also contained information related to crop ailments. The Bible, for example, made a mention of blight, blasts, rusts and mildews. In AD 752, Japanese empress Koken wrote a poem expressing the beauty of the yellow leaves of *Eupatorium chinensis* (infected with virus) without knowing its etiology. The illustrious playwright, William Shakespeare, also referred to mildews in his plays. Similarly, Scot (1574) described the nettle head condition of a diseased flowering plant, *Humulus lupulus* (commonly known as hop), in ignorance of its unfortunate illness. It is worthwhile to quote another event; in which, farmers preferred to cultivate variegated tulips instead of non-variegated ones, with a view to fetch a premium price in the market. This concept of "broken" tulips was advocated by Charles de Lie Clase (1576) in Holland. However, mankind's ancestors remained oblivious to these abnormalities and ailments for much of their history. With the discovery of plant viruses, the farmer was compelled to find ways to combat this new foe. To start with, the variegated tulip crop in The Netherlands was discontinued as the tulip variegation was a disease. It is thus rightly said that diseases are in occurrence since times immemorial, but their causes were poorly understood until the invention of the microscope in 1683 by A.V. Leeuwenhoek. Prior to this, Robert Hooke identified and illustrated a fungal disease named "rust of rose" for the first time in 1665. With the advancement of science, many more pathogenic diseases have been added to the long list of diseases. The diseases caused by fungi and nematodes were discovered by Monocern in 1735. Johan Needham (1743) identified wheat gall caused by a nematode. Subsequently, a need was felt to name the crop diseases. Soon, efforts were made to create a system of naming the diseases affecting cultivated plants. It was The Latin Binomial System of Nomenclature of Plants and Animals introduced by Linnaeus (1753) that led to the beginning of a new era of botany. In the beginning, the emphasis was on the adoption of monoculture of crops. This system led to an explosion of new and more devastating pathogens in the agriculture sector. An epidemic of blight completely destroyed the potato crop in Ireland, leading to a devastating famine (1845) and the mass exodus of the country. Over one million people are thought to have died of starvation, as the potato was a staple food of the Irish population. These vagaries of nature brought into focus the maladies and diseases of cultivated crops. During the year 1879, Robert Koch put forth the Germ Theory (Koch's postulates) to prove pathogenicity of disease and to further refine plant pathology science. Fire blight of apple and pear was identified as a disease of bacterial origin in 1885. Whilst serving in Wageningen (The Netherlands), a German chemist named Adolf Mayer (1886), demonstrated the transmission of Tobacco mosaic virus (TMV) disease from diseased to healthy plants and named the disease "Mosaikkrankheit". EF Smith (1890) described bacterial wilt in cucurbits. In the course of his work on tobacco mosaic disease, Dmitri Ivanowski (1892), a botanist from Russia, demonstrated that this disease is caused by organisms smaller than bacteria as the filtrate of a bacteria-proof filter was able to create disease in healthy tobacco plants. The findings of Ivanowsky were endorsed further by Dutch botanist Martinus Beijerinck (1898) and the term "virus" (poison) was coined to describe the causal agent, in conjunction a theory called *Contagium Vivum Fluidum* was put forth. It is relevant to mention that major strides were experienced in the 19th century. The graft transmission of Infectious variegation disease of Abutilon was first demonstrated by Bauer (1904). The involvement of leafhoppers in the spread of sugar beet curly top was reported for the first time by Smith and Boncquet (1915). The bacteriophages (viruses of bacteria) were reported by Twort and d'Herelle in 1915/1917. Protection of citrus fruits against the severe strain of Citrus tristeza with the introduction of a mild strain was acheived by HH McKinney in 1929, this method was later was labelled as cross protection. The thrips (Thysanoptera) as vectors of plant viruses were identified in the mid-nineteen-thirties, through the scientific studies made by Samuel and his team. In the 20th century, an American biochemist, Wendell M Stanley (1935) demonstrated the crystalline nature of plant viruses and won a Nobel Prize for this outstanding discovery. By further exploring his research, Frederick C, Bawden and NW Pirie (1936) reported on the nucleoprotein nature of plant viruses in the United Kingdom. Kaushe (1939) showed a virus for the first time under an electron microscope. In the same year, Holmes, FO (1939) developed the symptom-based binomial system of classifying plant viruses covered in the Vira phylum. There was a landmark development in science when Karl Maramorosch (1952) demonstrated the multiplication of Aster yellows virus (AYV) in leafhopper vector. Francis Crick and James Watson (1953) identified the double helix model of plant virus DNA. In addition to insects, the mite *Aceria tulipae* has been identified as a vector of Wheat streak mosaic virus by JT Slykhuis in 1955. Ribonucleic Acid (RNA) as the infective entity in plant viruses was discovered in 1956. Nematodes (*Xiphinema index*) were soon found to be associated with the spread of grapevine Fanleaf disease by a team led by WB Hewitt in 1958. Kassanis (1962) reported the phenomenon of satellite viruses, an occurence in which one virus becomes dependent on a second virus for the purpose of replication. In the same year, DS Teakle demonstrated the involvement of fungi (*Olpidium brassicae*) in the spread of Tobacco necrosis virus (TNV). Landmark innovation of Doi (1967) led to the identification of *Mycoplasma* as the cause of several plant diseases. The diseases earlier categorized as being of viral origin are now classified under *Mycoplasma* etiology. The term "viroid" (viruses without protein coat) was coined by TO Diener during 1971. In 1972, the *Rickettsia-Like Organism* (RLO) was discovered as a causal organism of the clover club leaf disease present in crucifers by IM Windsor and LM Black (1972). Till that time, all the viral diseases of plants had RNA type nucleic acid as an infective material. Cauliflower mosaic virus (CaMV) of plant origin was identified first as circular, double-stranded Deoxyribonucleic Acid (DNA) virus in 1980. The complete sequence of single stranded Tobacco mosaic virus (TMV) was identified in 1982. The idea of utilizing transgenic plants to manage TMV emerged in 1986. BB Sears and BC Kirkpatrick (1994) identified *Phytoplasma* as a causal organism in diseases and classified under Mollicutes.

1.2 Symptoms

Disease is derived from two words, namely "ease" and the reversing prefix "dis", which together mean that something is not at ease. Disease is thus defined as any abnormality in an organism due to biotic or abiotic stresses. The biotic agents known to inflict diseases in plants are viruses, viroid, bacteria (fastidious vascular bacteria and non-fastidious vascular bacteria), fungi, nematodes, *Phytoplasma (Mycoplasma* and *Spiroplasma*), *Rickettsia-Like*

Organisms, *Trypanosomes* (Protozoa), algae and some parasitic plants. Insect injury is also categorized as a disease. Environmental stresses, like frost, flood and drought damage, deficiency of nutrients, salt concentration, strong winds, wildfire, etc., are also considered as agents of disease, as is human involvement in soil compaction and the presence of fertilizers, pesticides and irrigation, industrial smoke, etc. The diseases in the field are recognized by observing symptoms which are generally divided into external and internal symptoms. The most common symptoms exhibited by compromised crop plants are enations, vein clearing, puckering, mosaic, stripes, streaks, stunting, chlorosis, vein banding, rosette, curling, cupping, rolling, witches broom, wilting, discoloration, necrosis, blight, canker, dieback, damping off, mummy, rot, scorch, scald, pitting, spots, scab, mildew, gummosis, knots, galls, yellows, etc.

- Enation is an outgrowth (leaf like structre) of leaf vein on the undersurface of the leaf without vascular tissues (Cotton leaf curl disease or CLCuV).
- Vein clearing is the initial symptom in virus infected leaves and the leaf veins look transparent due to lack of green color (Cucumber mosaic disease or CMV).
- In the mosaic pattern, the alternate patches of light yellow and green color become conspicuous in the leaf, or it can also present itself as light and dark patches of green color in the leaf (Chili mosaic virus or ChMV).
- The leaflets become narrow/fern-like/straplike in shoestring (CMV).
- The raised inter-venal leaf lamina with sunken veins in Tobacco leaf curl virus (TbLCV) is responsible for rough and leathry leaves called "rugose".
- A rough and uneven leaf lamina is commonly known as "puckering" (Tomato leaf curl disease).
- The long narrow discolored line differentiated by color or texure is called a "streak" (Pea streak).
- The elongated discoloration of parts of monocotyledonous plants are known as "stripes" (Barley stripe).
- Specking/stippling is discoloration in the form of small spots on leaf lamina.
- In ring spotting, the severely and moderately diseased zones appear alternately in the lamina of the leaf.
- Proliferation is the rapid development of sprouts or growth of cells by multiplication from the axils of floral organs or other parts.
- "Stunting" is dwarfing of a plant due to shortening of internodes (Chili leaf curl virus).
- The general yellowing of a plant due to chlorophyll deficiency is named "chlorosis" (Aster yellows).

- Plant leaves showing yellow veins with green inter-veinal areas are a sign of vein banding (Citrus greening).
- The compact appearance of plants with short internodes and excessive axillary branches is called "bunchy appearance of plants" (Groundnut rosette).
- The upward-rolling of leaf margins is generally referred to as "curling" (Chili leaf curl virus).
- The upward curling, coupled with raised leaf lamina is called "cupping" (Tomato leaf curl virus-ToLCuV) while the thickened leaf lamina is called "crinkling".
- Hyperplasia is an abnormal increase in the number of cells.
- Atrophy is complete arrestation in the development of cells or organs.
- Hypertrophy is an abnormal increase in the size of cells.
- The outgrowth on plant parts due to hyperplasia or hypertrophy in plants is called gall formation.
- Thickened leaves with upward folding of margins are classified as "leaf rolling" (Potato leaf roll virus-PLRV).
- Broom-shaped plants with small leaves, short internodes, and excessive axillary branches are known as witch's broom (Potato witches broom).
- The absence of chlorophyll in the leaves is known as "discoloration" (Aster yellows).
- The dead area of a plant tissue is referred to as "necrosis", while yellowing is "chlorosis". The progressive decrease in plant vigour is called "decline" as in "citrus decline".
- Cork formation is present as normal cells activate to divide and form cork cells.
- Bark scaling is a typical scab or psora.
- The presence of a dark melanin like substance in necrotic lesions is known as "browning" or blackening.
- The complete absence of green color from the leaves caused by sunlight or chemicals the condition is called "bleaching" or "blanching".
- The burning of leaf margins is known as "leaf scorch".
- The symptom of leaves showing a burnt appearance is generally called blight (Potato blights).
- The rough outgrowth on the surface of plant tissues is known as "canker" (Citrus canker).
- Dieback is the death of twigs from the tip down (Chili dieback).
- The mortality of young seedlings in the nursery is known as the "Damping off" of seedlings (damping-off of seedlings of chili).
- The misshapen and shriveled fruits/plant parts are referred to as "mummies" (misshapen fruits of tomato).
- Rot is the decomposed and putrefied part of the plant (Root rot of cotton).

- Scald is the bleaching of plant/plant parts through excess sunlight (Sun scald of tomato fruits).
- The pit is groove formation in plant tissues (groves of woody plants).
- Spots are necrotic areas on parts of the plant (Tikka disease of groundnut).
- Scab is crust formation or depressions and sunken spots on plant parts (Apple scab).
- Galls are outgrowths on plant body caused by insect feeding or fungal attack (Mango galls).
- Malformation is the conversion of floral parts into leafy structures (Mango malformation).
- Knots are the formation of nodules on roots as a result of nematode attack (Root-knot of cucurbits).
- Mildew is the covering of leaf surface with an off-white, powdery mass (Powdery mildew of cucurbits).
- Gummosis is the oozing of sap or gum-like material that is found solidified on stems or leaves (Gummosis of citrus).
- Withering is the shrinking of leaves due to a deficiency of water.
- Wilting is a symptom apparent in plants as a result of the loss of turgidity due to thirst.
- The discoloration of veins in the leaf is called "vein chlorosis". The dark colored area between the leaf veins is called "vein banding".
- The irregular discoloration of the complete lamina is "variegation".
- Sterility is the conversion of flowers into leaf-like structures, as in Pigeon pea sterility mosaic (PPSMV).

Based on symptoms, the diseases can be categorized into different categories depending on the agency involved in inflicting symptoms.

1.3 Pathogens

To make the reader familiar with the subject, the disease-causing pathogens are briefly described in this chapter. Viruses are sub-microscopic, infective entities containing nucleic acid enclosed in a protein coat. Being obligate parasites, viruses multiply in living cells, make use of ribosomes of host cells and attack plants, animals (invertebrates and vertebrates), bacteria, fungi, algae, protozoa, and archae (Tobacco mosaic virus). A virus particle is called a virion. It contains one type of nucleic acid, either RNA or DNA, in its core. It is surrounded by layers of protein coat (1/2) and an outer lipid layer (envelope). The majority of plant viruses contain solely RNA, but there are those that do contain DNA as well. So far around 5000 viruses are known. Overall, seventy-five per cent of plant viruses consist of single-stranded RNA (ssRNA), while only seventeen per cent have single-stranded DNA (ssDNA). Of the RNA-containing viruses, sixty-five per cent have positive single-stranded RNA, while only ten per cent have negative ssRNA. The unique feature of plant pathogenic viruses is that, since these pathogens require living plant cells to multiply, they do not attack human beings. Based on morphology, viruses are grouped into five categories, i.e., isometric, rod-shaped, filamentous, geometric and bacilliform. TNV (*Necrovirus*), at twenty-six nanometres in diameter, is in the isometric category. Viruses in the rod-shaped (flexuous/rigid) category are generally around twenty to twenty-five nanometres in diameter, with a length between 300 to 500 nanometres. Tobacco mosaic virus (TMV), belonging to the genus *Tobamovirus*, is a member of the rod-shaped virus category. Filamentous viruses are twelve nanometres in diameter with a length of 1000 nanometres. Potato virus Y (PVY) of *Potyvirus* genus is 740 nanometres long and falls into this category. The geometric categories are twinned isometric (thirty by eighteen nanometres). Maize streak virus (MSV) (*Mastrevirus*; Geminiviridae) is a typical example of this group. The fifth group is comprised of bacilliform, round rods thirty nanometres in width and 300 nanometres in length (e.g., Swollen shoot of cocoa of *Badnavirus* genus). Whereas, the diameter of isometric shaped viruses is between twenty-five to fifty nanometres. The International Committee on Taxonomy of Viruses (ICTV) collaborated with David Baltimore (Baltimore, 1971) to divide viruses into seven categories containing five orders (Caudovirales, Herpesvirales, Picomavirales, Mononegavirales, and Nidovirales), eighty-two families, 307 genera and 2083 species. The Baltimore system of classification is based on viral messenger RNA (mRNA) synthesis and comprises seven groups viz.

i) dsDNA type (*Adenovirus, Herpesvirus* and *Poxvirus*),
ii) ssDNA type (+) sense DNA (*Parvovirus*),
iii) dsRNA type (*Reovirus*),
iv) (+) ssRNA type (*Picornavirus* and *Togavirus*),
v) (–) ssRNA type (*Orthomyxovirus* and *Rhabdovirus*),
vi) ssRNA–RT type (+) sense RNA with DNA intermediate (*Retrovirus*),
vii) dsDNA-RT type (Hepadnavirus)

These viruses measure between 10–300 nanometres in length (some may have a length of 1400 nanometres and diameter of eighty nanometres). The virion is a virus particle consisting of Nucleic Acid (NA) and protein coat (capsid). The science which deals with the study of viruses is known as virology. The plant viruses containing DNA are *Caulimovirus*-Cauliflower mosaic virus (CaMV), *Badnavirus*-Rice tungro bacilliform virus (RTBV) (dsDNA) and *Geminivirus*-Tobacco leaf curl virus-(TbLCV) (Geminiviridae) (ssDNA). The plant infecting viruses belong to Reoviridae are *Phytoreovirus*-Wound tumor virus (WTV), *Fijivirus*-Maize dwarf virus (MDV), *Oryzavirus*-Rice ragged stunt virus-RRSV) and Partitiviridae includes *Alphacryptovirus*-White clover cryptic virus-1 (WCCV-1), *Betacryptovirus*-White clover cryptic virus 2 (WCCV2) (dsRNA). In Rhabdoviridae, the genera are *Cytorhabdovirus*-Lettuce necrotic yellows virus (LNYV) and

Nucleorhabdovirus-Sowthistle yellow vein virus (SYVV) which have negative ssRNA. The positive ssRNA is in the family Bromoviridae (*Cucumovirus*-Cucumber mosaic virus-CMV, *Bromovirus*-Broad bean mottle virus-BBMV, *Ilarvirus*-Tobacco streak virus-TSV and *Alfamovirus*-Alfalfa mosaic virus-AMV), Comoviridae (*Nepovirus*-Tobacco ring spot virus-TRSV, *Comovirus*-Cowpea mosaic virus-CPMV, *Fabavirus,* Broad bean wilt virus-BBWV, *Tobamovirus*-Cucumber green mottle virus-CGMV, *Tobravirus*-Tobacco rattle virus-TRV, *Hordeivirus*-Barley stripe mosaic virus-BSMV, *Furovirus*-Potato mop top virus-PMTV, *Potexvirus*-Potato virus Y-PVY, *Capillovirus*-Apple stem grooving virus-ASGV, *Trichovirus*-Pea early browning virus-PEBV, *Carlavirus*-Carnation latent virus-CLV, *Potyvirus*-Lettuce mosaic virus LMV and *Closterovirus*-Beet yellows virus-BYV), Seqiviridae (*Sequivirus*-Parsnip yellow fleck virus-PYFV and *Waikavirus*-Maize chlorotic dwarf virus-MCDV) and Tombusviridae (*Dianthovirus*-Carnation ring spot virus-CRSV, *Luteovirus*-Barley yellow dwarf virus-BYDV, *Machlomovirus*-Maize chlorotic mottle virus-MCMV, *Marafivirus*-Maize rayado fino virus-MRFV, *Necrovirus*-Tobacco necrosis virus-TNV, *Sobemovirus*-Southern bean mosaic virus-SBMV, *Tymovirus*-Turnip yellow mosaic virus-TuYMV, *Tombusvirus*-Tomato bushy stunt virus-TBSV, *Umbravirus*-Coconut rosette virus-CRV and *Enamovirus*-Pea enation mosaic virus-PEMV) and Bunyaviridae (*Tospovirus*-Tomato spotted wilt virus-TSWV). *Idaeovirus* genus has not been assigned to any family so far. The rod-shaped viruses constitute fifty per cent of the total viruses attacking vascular plants. With respect to transmission, aphids (*Potyvirus, Cucumovirus* and luteoviruses), leafhoppers (Rhabdoviridae, Reoviridae), whitefly (begomoviruses), thrips (*Tospovirus*), beetles (*Comovirus* and *Sobemovirus*), nematodes (*Nepovirus, Tobravirus*), plasmodiophorids (*Bunyavirus, Bymovirus, Furovirus, Pecluravirus* and *Pomovirus*) and mites (*Rymovirus* and *Tritimovirus*) are acting as vectors (Table 1.1). The information contained in the table indicates the involvement of all the insect vectors, mites, nematodes, and fungi. The virus genera transmissible through insects, mites, fungi, and nematodes are sixty-four, ten, six and three, respectively. The maximum number of virus genera is transmissible by aphids (twenty), followed by planthoppers (seven), beetles (seven), whiteflies (six), leafhoppers (five), thrips (five), true bugs (four), mealy bugs (three) and others (seven). The opening chapter thus gives an idea of plant virus vectors to the reader. All categories of vectors have been analysed for a detailed study on identification, biology, feeding mechanisms, transmission of pathogens and effects of pathogens on their vectors under different chapters. Besides viruses, the details pertaining to the transmission of bacteria, fungi, nematodes, *Phytoplasma* and *Rickettsia* through insects have also been given (Table 1.2). These viruses, being obligate parasites, replicate in the plant system. Initially, there is an attachment of virus capsid proteins with receptors present on the outer cell membrane. The virus moves into the host cell through plasmodesmata injecting NA inside and leaving

Table 1.1 Plant viruses (genera) transmitted by different categories of vectors with mechanism of transmission.

Sl. No.	Genus	Virus	Taxon	Mechanism	Reference(s)
1	*Carlavirus*	Potato virus S (PVS)	Aphid	Non-persistent—stylet borne	Burrows and Zitter, 2005
2	*Alfamovirus*	Alfalfa mosaic virus (AMV)	Aphid	Non-persistent—stylet borne	Garran and Gibbs, 1982
3	*Carlavirus*	Lily symptomless virus (LSV)	Aphid	Non-persistent—stylet borne	Asjes, 2000
4	*Cucumovirus*	Cucumber mosaic virus (CMV)	Aphid	Non-persistent—stylet borne	Ullman et al., 1991
5	*Fabavirus*	Broad bean wilt virus-1 (BBWV-1)	Aphid	Non-persistent—stylet borne	Ferriol et al., 2012
6	*Macluravirus*	Alpinia mosaic virus (AlpMV)	Aphid	Non-persistent—stylet borne	Liou et al., 2003
7	*Potyvirus*	Potato virus Y (PVY)	Aphid	Non-persistent—stylet borne	Van Hoff, 1980
8	*Badnavirus*	Citrus yellow mosaic badnavirus virus (CYMBV)	Aphid (Needs Confirmation)	Non-persistent—foregut borne	Ghosh et al., 2014
9	*Closterovirus*	Citrus tristeza virus (CTV)	Aphid	Non-persistent—foregut borne	Roy and Brlansky, 2009
10	*Sequivirus*	Parsnip yellows fleck virus (PYFV)	Aphid	Non-persistent—foregut-borne	Elnagar and Murant, 1976
11	*Waikavirus*	Anthriscus yellows virus (AnYV)	Aphid	Non-persistent—foregut-borne	Elnagar and Murant, 1976
12	*Caulimovirus*	Cauliflower mosaic virus (CaMV)	Aphid	Non-persistent—foregut-borne	Hohn, 2007
13	*Luteovirus*	Barley yellow dwarf virus (BYDV)	Aphid	Persistent—circulative	Gray and Gildow, 2003
14	*Nanovirus*	Fababean necrotic stunt virus (FBNSV)	Aphid	Persistent—circulative	Sicard et al., 2015
15	*Polerovirus*	Cereal yellow dwarf virus (CYDV)	Aphid	Persistent—circulative	Gray and Gildow, 2003
16	*Umbravirus*	Carrot mottle virus (CMoV)	Aphid	Persistent—circulative	Stubbs, 1952

Table 1.1 contd. ...

...Table 1.1 contd.

Sl. No.	Genus	Virus	Taxon	Mechanism	Reference(s)
17	*Babuvirus*	Banana bunchy tops virus (BBTV)	Aphid	Persistent—circulative	Bressan and Watanabe, 2011
18	*Luteovirus*	Potato leaf roll virus(PLRV)	Aphids	Persistent—circulative	Sylvester, 1956
19	*Enamovirus*	Pea enation mosaic virus-1 (PEMV-1)	Aphids	Persistent—circulative	Salgueiro and Hull, 1999; Shikata et al., 1966
20	*Nanovirus*	Subterranean clover stunt virus (SCSV)	Aphid	Persistent—circulative	Johnstone and Mclean, 2008
21	*Cytorhabdovirus*	Lettuce necrotic yellows virus (LNYV)	Aphid	Persistent—propagative	Dietzgen et al., 2006
22	*Reovirus/ Rasalvirus (proposed)*	Raspberry latent virus (RpLV)	Aphid	Persistent—propagative	Quito-Avila et al., 2012
23	*Nucleorhabdovirus*	Strawberry crinkle latent ring spot virus (SCLRSV)	Aphid	Persistent—propagative	Krezal, 1982
24	*Waikavirus*	Rice tungro spherical virus (RTSV)	Leafhopper	Non-persistent—foregut-borne	Hibino, 1983
25	*Mastrevirus*	Maize streak virus (MSV)	Leafhopper	Persistent—circulative	Shepherd et al., 2010
26	*Curtovirus*	Sugarbeet curly top virus (SCTV)	Leafhopper	Persistent—circulative	Stanley, 2008
27	*Marafivirus*	Maize rayado fino virus (MRFV)	Leafhopper	Persistent—propagative	Edwards et al., 2016
28	*Phytoreovirus*	Rice dwarf virus (RDV)	Leafhopper	Persistent—propagation	Omura et al., 1998
29	*Oryzavirus*	Rice ragged stunt virus (RRSV)	Planthopper	Persistent—circulative	Gray and Banerjee, 1999
30	*Nanovirus*	Coconut foliar decay virus (CFDV)	Planthopper	Persistent—circulative	Wefels et al., 2015; Randles and Hanold, 1989
31	*Fijivirus*	Rice black-streaked dwarf virus (RBSDV)	Planthopper	Persistent—circulative	Lee and Kim, 1985

32	*Tenuivirus*	Rice stripe virus (RSV)	Planthopper	Persistent—circulative	Lee and Kim, 1985; Gray and Banerjee, 1999
33	*Cytorhabdovirus*	Barley yellow striate mosaic virus (BYSMV)	Planthopper	Persistent—propagative	Redinbaugh and Hogenhout, 2005
34	*Tenuivirus*	Rice grassy stunt virus (RGSV)	Planthopper	Persistent—propagative	Zheng et al., 2014
35	*Fijivirus*	Oat sterile virus (OSV)	Planthopper	Persistent—propagative	Ammar and Nault, 2002
36	*Nucleorhabdovirus*	Maize mosaic virus (MMV)	Planthopper	Persistent—propagative	McEwEn and Kawanishi, 1967
37	*Rhabdovirus*	Colocasia bobone disease virus (CBDV)	Planthopper	Transmission	Gollifer et al., 1977; Palomar, 1987
38	*Topocuvirus*	Tomato pseudo curly top virus (TPCTV)	Treehopper	Persistent—circulative	Tsai and McDaniel, 1990
39	*Torradovirus*	Tomato torrado virus (ToTV)	Whitefly	Non-persistent—stylet-borne	Verbeek et al., 2014
40	*Crinivirus*	Cucumber yellows virus (CYV)	Whitefly	Non-persistent—foregut-borne	Yamashita et al., 1979; Huang et al., 2010
41	*Closterovirus*	Tomato chlorosis virus (ToCV)	Whitefly	Non-persistent—foregut-borne	Navas-Castillo et al., 2000
42	*Crinivirus*	Cucurbit yellow stunting disorder virus (CYSDV)	Whitefly	Non-persistent—foregut-borne	Celix et al., 1996
43	*Crinivirus*	Cucurbit chlorotic yellows virus (CCYV)	Whitefly	Non-persistent—foregut-borne	Huang et al., 2010
44	*Carlavirus*	Cowpea mild mottle virus (CMMV)	Whitefly	Non-Persistent—stylet-borne/ foregut-borne	Iwaki et al., 1982; Menzel et al., 2011
45	*Ipomovirus*	Sweet potato mild mottle virus (SPMMV)	Whitefly	Persistent—circulative	Dombrovsky et al., 2014
46	*Begomovirus*	Tomato yellow leaf curl virus (TYLCV)	Whitefly	Persistent—propagative	Ghasin et al., 1998

Table 1.1 contd. …

...Table 1.1 contd.

Sl. No.	Genus	Virus	Taxon	Mechanism	Reference(s)
74	*Bromovirus*	Beet necrotic yellow vein virus (BNYVV)	Fungus	Transmission	Verchot-Lubicz et al., 2007
75	*Carmovirus*	Melon necrotic spot virus (MNSV)	Fungus	Transmission	Wada et al., 2008
76	*Pomovirus*	Potato mop top virus (PMTV)	Fungus	Transmission	Merz, 2008
77	*Potexvirus*	Pepino mild mottle virus (PepMMoV)	Fungus	Transmission	King et al., 2012
78	*Ophiovirus*	Lettuce big vein virus (LBVV)	Fungus	Transmission	Sasaya et al., 2008
79	*Bymovirus*	Barley mild mosaic virus (BMMV)	Fungus	Transmission	Shukla et al., 1998
80	*Potyvirus*	Potato virus Y (PVY)	Mite	Non-persistent—stylet-borne	Orlob, 1968
81	*Cilevirus/ Rhabdovirus*	Citrus leprosis virus-C (CiLV-C)	Mite	Transmission	Melzer et al., 2012
82	*Emaravirus*	Rose rosette virus (RRV)	Mite	Transmission	Laney et al., 2011
83	*Tenuivirus*	Maize red leaf stripe virus (MRLStV)	Mite	Transmission	Skare et al., 2006
84	*Trichovirus*	Cherry mottle leaf virus (CMoLV)	Mite	Transmission	Ma et al., 2014
85	*Dichorhavirus*	Orchid fleck virus (OFV)	Mite	Transmission	Kondo et al., 2003
86	*Rymovirus*	Ryegrass mosaic virus (RGMV)	Mite	Transmission	Gamliel-Atmsky et al., 2009
87	*Tritimovirus*	Wheat streak mosaic virus (WStMV)	Mite	Transmission	Paliwal, 1980
88	*Nepovirus*	Black currant reversion virus (BCRV)	Mite	Transmission	Susi, 2004
89	*Ilarvirus*	Plum latent virus (PLV)/Prunus necrotic ring spot virus (PNRSV)	Mite	Transmission	Proeseler, 1968

90	*Nepovirus*	Fan leaf grapevine virus(GFLV)	Nematode	Transmission	Jones et al., 2013; Hewitt et al., 1958
91	*Tobravirus*	Tobacco rattle virius(TRV)	Nematode	Transmission	Jones et al., 2013
92	*Tobravirus*	Pea early browning (PEBV)	Nematode	Transmission	Gibbs and Harrison, 1964
93	*Cheravirus*	Cherry rasp leaf virus (CRLV)	Nematode	Transmission	Hansen et al., 1974

Table 1.2 Transmission of bacteria, fungi, nematodes, *Phytoplasma* and *Rickettsia-like organisms* through insect vectors.

Sl. No.	Disease	Causal organism	Vector (Order)	Insect species	Source(s)
1	Bacterial central rot of onion	Bacterium	Thrip (Thysanoptera)	*Frankliniella fusca*	Dutta et al., 2015
2	Fire blight of apple/ pear	Bacterium (*Erwinia amylovora*)	Honey bees (Hymenoptera)	*Apis malleifera*	Vanneste, 1996
3	Stewarts wilt	Bacterium (*Pantoea stewartii*)	Pea aphid (Hemiptera)	*Acyrthosiphon pisum*	Stavrinides et al., 2010
4	Bacterial rot of apples	Bacterium (*Pseudomonas melophthora*)	Apple maggot (Diptera)	*Rhagoletis pomonella*	Martin, 1959
5	Olive knot disease	Bacterium (*Pseudomonas savastanoi*)	Olive fly (Diptera)	*Bactrocera oleae*	Young, 2004
6	Laurel wilt	Bacterium (*Raffaelea lauricola*)	Red bay Ambrosia beetle (Coleoptera)	*Xylebrcis glabratus*	Fraedrich et al. 2008
7	Mummy berries of blueberry	Fungus (*Monilina vaccinia-corymbosi*)	Ants, Bees, Wasps (Hymenoptera)	Pollinators	Batra and Batra, 1985
8	Brown rot of apple	Fungus (*Sclerotinia fructigena*)	Earwigs (Dermaptera)	*Forficula auriculidae*	Croxall et al., 2008
9	Endosepsis of fig	Fungus (*Fusarium moniliforme var fici*)	Fig wasp (Hymenoptera)	*Blastophaga psenes*	Michailides and Morgan, 1998
10	Boll rot of cotton	Fungus (*Fusarium* sp. and others)	Bollworms and cotton stainers (Lepidoptera and Hemiptera)	*Pectinophora gossypiella* and *Helicoverpa armigera* and *Dysdercus koenigii*	Chinthagunta et al., 2009
11	Rust	Fungus (*Puccinia punctiformis*)	Aphids (Hemiptera)	*Aphis fabae* sp. *Cirsiiacanthoidis*	Kluth et al., 2002
12	Brown rot of plum/ peach/cherry	Fungus (*Sclerotinia fructigena*)	True flies (Diptera)	*Drosophila subobscura*	Lack, 1989

13	Pine wilt	Nematode (*Bursaphelenchus xylophilus*)	Beetles (Coleoptera)	*Monochamus alternatus* and *M Scutellus*	Takasu, 2009
14	Citrus greening	*Phytoplasma*	Citrus psylla (Hemiptera)	*Diaphorina citri*	Manjunath et al., 2008
15	Pear decline	*Phytoplasma*	Citrus psylla (Hemiptera)	*Cacopsylla pyri*	Carraro et al., 2001; Vereijssen and Scott, 2013
16	Apple proliferation	*Phytoplasma*	Psylla (Hemiptera)	*Cactopsylla picta*	Tedeschi et al., 2006
17	Lethal yellowing of coconut	*Phytoplasma*	American palm cixidid planthopper (Hemiptera)	*Haplaxius crudus*	Gurr et al., 2015
18	Papaya bunchy tops	*Rickettsia*	Leafhopper (Hemiptera)	*Empoasca papayae and E stevensi*	Haque and Parasram, 1973
19	Carrot proliferation	*Rickettsia*	Leafhopper (unconfirmed report) (Hemiptera)	Not known	Franova et al., 2008

the protein coat outside the cell membrane. The viral enzymes degrade the virus capsid and expose the genome to the host cell. The virus replicates with the synthesis of messenger RNA and viral protein and is followed by the assembly of replicated genome material. This process causes lysis in the cell and results in necrotic symptoms on the plant parts as the cell walls disintegrate. TMV is a well-worked, rod-shaped virus, wherein the protein and nucleic acid are arranged in a helix fashion. The particle length is taken as the length of RNA. In spherical shaped virus particles, the RNA/DNA is in the center and protein units are arranged around it so as to provide protective cover as in CaMV, except that it has DNA as an infective entity. Bacilliform viruses resemble bacillus bacteria, and AMV is representative of this category. The viroids are the smallest infectious plant viruses that have a single-stranded circular arrangement of RNA but lack a protein coat/capsid. Unlike viruses which parasitize the cellular region, viroids make use of cellular transcription instead. The potato spindle tuber virus disease is the first disease caused by viroid, of the 30 disorders of viroid etiology. Subsequently, Citrus exocortis, Grapevine yellow speckle, Tomato apical stunt, Coconut cadang-cadang, Coconut tinangaja, Apple scar skin, Apple hop stunt, Chrysanthemum stunt, and Avocado sunblotch have been discovered. These pathogens are generally spread via vegetative propagation, pollen, and seed. The nomenclature and classification of viruses are deliberated upon by The International Committee on Taxonomy of Viruses (ICTV). So far, the committee has identified and named 3600 viruses. Of these, 900 are plant viruses. Accordingly, the virus classification into taxonomic levels is as follows; order (virales), family (viridae), subfamily (virinae), genus (virus) and species (virus). The naming has been effectuated, taking into account the host and symptom of the first pathogen. Mollicutes is a class of prokaryotes (without a cell wall). These organisms are also plant pathogens. The diseases caused by these organisms are known to produce typical symptoms like stunting, yellowing, witches broom, and phyllody. These Mollicutes include *Phytoplasma* and *Spiroplasma. Phytoplasma* show sensitivity towards the tetracycline antibiotics group, possess a cell membrane without a cell wall and are vectored by insects. Coconut root wilt disease is of Phytoplasmic etiology, and is transmitted by *Stephanitis typica*, a bug native to the state of Kerala (Southern India). *Spiroplasma* is small spiral shaped bacteria without cell walls, it is classed under Mollicute and moves in a corkscrew fashion. Paulownia witches broom is due to *Spiroplasma* and is transmissible via stink bug. Bacteria are unicellular micro-organisms without cell walls, organelles with a definite nucleus that multiply through binary fission. Under bacteria, there are two categories viz. fastidious and non-fastidious vascular bacteria. Of the fastidious xylem-inhabiting ones, *Xylella fastidiosa*, the causal organism of Pierce disease of grapevines is vectored by xylem feeders, i.e., spittlebugs and sharpshooters.

Whereas, phloem feeders like leafhoppers, psyllids, and heteropterous bugs are vectors of phloem-borne pathogens (Beet latent rosette virus-BLRV–transmissible via piesmid bugs). Fungi are spore-producing, eukaryotic, single-celled or multinucleate organisms lacking chlorophyll and vascular tissues. They grow on organic matter that is decomposed by them (Stigmatomycosis of citrus, cotton, and coffee are bug-borne). *Rickettsia* is pleomorphic, unicellular, non-motile and does not form spores. Gram-negative organisms resembling both bacteria and viruses, reproduce by binary fission in host cells being an obligate parasite and are sensitive to antibiotics. Both the Beet leaf curl disease in Europe, and Beet latent rosette disease are caused by *Rickettsia-Like Organisms* (RLO) and transmitted by tinged bug *Piesma quadratum*. Nematodes are cylindrical unsegmented worms of phylum Nematoda. They are narrow at both ends and parasitize both plants and animals (Pine wilt nematode is beetle-borne). Protozoa are mainly single-celled eukaryotic flagellates, from animal kingdom Protista. They are free-living parasites, that latch onto insects, plants, birds and mammals, etc., ingest food particles, and live in water (Pine wilt disease caused by Pine wilt nematode is transmissible through long-horned beetles). Algae are unicellular or multicellular organisms containing chlorophyll and pigments. They are found in freshwater and saltwater or moist soil, but lack roots, stems and leaves. Parasitic plants, namely mistletoe and dodder, are known to transmit viruses or virus-like organisms which are systemic in nature. The pathogens are known to cause diseases in human beings, plants, and bacteria. Of these pathogens, viruses are discussed in detail in this write-up, as the devastating diseases are inflicted on plants. The viruses also cause diseases which do not fall under the purview of this book. Similarly, the viruses are known to attack bacteria (Bacteriophages) in nature, but this too is not featured in this compilation.

1.4 Economic Importance

Plants are sessile and do not play any role in the spread of plant pathogens. However,the plants do spread pathogens through seed, pollen and human involvement. Therefore, the plant viruses are dependent on insects for their spread and survival. The disease development is dependent on the favorable interaction of pathogen, host and environment. The disease losses inflicted on perennial crops are enormous due to their long span in the field. In addition, losses are also experienced in annual crops, as is evident from many virus epidemics reported throughout the globe. Indirect damage is also apparent in situations where huge investment is essential to maintain a good crop through the vegetative periods. Over the last forty years, the human population has increased by ninety per cent, whereas grain production has only increased by twenty-five per cent. To achieve food security, the grain production should be thirty-nine per cent more by 2020, to feed an additional population of one point five billion, according to recent estimates (Shamim et al., 2013). The various pathogens are known to destroy heavy amount of produce both in storage and field. According to one estimate the plant pathogens are taking a toll of fifteen per cent on production, as is evident from the outbreaks experienced all over the world (Oerke and Dehne, 2004). In the recent past, plant disease outbreaks posed a serious threat to crop production and many failures have been experienced on almost all continents. The late blight of potato outbreak in Ireland in 1845 caused a famine. As a result of this famine, more than one million people, i.e., 1/8 of the total population perished due to starvation. On account of the food shortage, 1.5 million people emigrated to other countries and continents, North America in particular. Besides, the epidemics of powdery and downy mildew of grapes in France (1851–1878), coffee rust in Ceylon (1870s), southern bacterial wilt of tobacco in North America (1900s), sigatoka leaf spot and Panama disease of banana in Central America (1900–1965), black rust in wheat in Central America (1916, 1935, 1953–54) and southern corn leaf blight of maize in USA (1970) were experienced, and these noticeable diseases were responsible for large-scale destruction of crops. Fortunately, these diseases were not insect-borne for their secondary spread. The insect-borne pathogens are more complicated to manage as four agencies viz. plant pathogen, vector and environment are involved. It is, therefore essential to have a chemical that takes all four agencies into account. The suitable management strategy is not available to tackle such complex disorders inflicting huge losses in crop production. In all, seventy-three genera (forty-nine families) of viruses and thirty viroids are known to cause the world an annual loss to the tune of sixty billion USD. The viruses/viroids are the second largest agency after fungi, known to take major a toll on production. Of these viral diseases, Tobacco mosaic, Tomato spotted wilt, Tomato yellow leaf curl, Cucumber mosaic, Potato virus Y, Cauliflower mosaic, African cassava mosaic, Plum pox virus disease, Brome mosaic and Potato virus X are ten diseases caused by viruses that are economically important throughout the world. These are known to cause extensive damage to a number of different crops. According to one estimate, a loss to the tune of one billion dollars due to one viral disease (tomato spotted wilt disease) collectively in tomato, pepper peanut and tobacco crops in whole world has been calculated (Prins and Goldbach, 1998). Cassava is the third important source of calories after rice and maize in Africa, Asia and Latin America. Africa alone contributes about fifty per cent of the cassava roots produced globally. African cassava mosaic virus, transmissible by whitefly, is a limiting factor in the successful cultivation of cassava. In 1990, Cassava mosaic virus caused epidemics in Uganda, Tanzania, Sudan, Kenya and DRC, forcing the farmers to abandon its cultivation. The total loss in yield recorded in Africa alone is fifty million metric tons per year due to this dreaded disease alone. This comes out to be roughly two billion dollars per year. Besides this disease, Swollen shoot of cocoa (CSSV) is another serious disease in West Africa, responsible for enormous losses in cocoa. To prevent the spread of this disease, 100 million trees have been cut since 1945 in Ghana alone. The Sugar beet yellows (SYV), a disease of viral origin, is known to cause great losses in yield of sugar beet in Europe and North America. Banana bunchy tops virus (BBTV) is another serious problem throughout the world especially in Hawaii and Australia. The disease was first detected in 1989 in Bahu (Hawaii). The disease spread rapidly to major hawaiian islands and to the whole of Australia by 2002. It totally destroyed the banana industry in the entire regions where banana cultivation was taking place. Seed borne diseases of cucurbits are known to destroy entire crops in certain regions. Geminiviruses transmissible through whiteflies are known to cause huge losses in the yield of tomatoes, chili, cassava, okra and cotton. Crop failures of cucurbits in India have been experienced in the recent years due to these diseases. The outbreaks of Citrus tristeza virus have been seen throughout the globe over the last few decades wherever there is any cultivation of citrus. South Africa alone has suffered great losses to the tristeza viral disease since 1930. Cotton leaf curl virus (CLCuV) disease has invaded Punjab recently and become a limiting factor in the successful cultivation of cotton in the Indian Punjab. The loss to the tune of sixty-nine per cent due to CLCuV (*Begomovirus*) has been estimated in India (Dasgupta et al., 2003). The state of Punjab had been cultivating three crops of tomato during the year but had to abandon the monsoon crop on account of heavy infestation of tomato leaf curl virus-TLCuV. It is still being regarded as a restraining factor in the successful cultivation of crops infected with whitefly-borne virus. Crop losses in India have also been attributed to cassava mosaic virus (*Begomovirus*) in cassava (18–25%), Bud necrosis virus (*Tospovirus*) in groundnut (> 80%), Bean yellow mosaic virus (*Begomovirus*) in mung bean/black gram/soybean (21–90%), Pigeon pea sterility mosaic virus (*Tenuivirus*) in pigeon pea (> 80%) Potato mosaic–Potato virus Y (*Potyvirus*) in potato (85%), Rice tungro viruses (*Badnavirus* and *Waikavirus*) in rice (10%), Sunflower necrosis virus (*Ilarvirus*) in sunflower (12–17%) and Tomato leaf curl virus (*Begomovirus*) in tomatoes (40–100%) (Dasgupta et al., 2003). Humans have been aware of plant viruses for over 100 years, but the interference of man has further aggravated the situation in terms of the introduction and identification of more serious diseases in our agroecosystem. With every passing day many new diseases are threatening human survival. The potato virus Y alone is responsible for inflicting potato crop losses to the tune of forty to forty-five per cent in India. The Tungro virus disease is caused by two viruses (Rice tungro spherical and Rice tungro bacilliform viruses). In India it is responsible, in some cases, for up to a seventy per cent loss in rice production. Many other diseases caused by viruses have also been described as economically important in crops (Hull, 2002), cereals (Plumb, 2002), potatoes (Brunt, 2001; Salazer, 1996; De Bokx and Van der Want, 1987) and sugarbeets (Stevens et al., 2005). In cereal crops such as wheat and barley, the losses in yield due to Barley yellow dwarf virus alone are in the range of five to thirty per cent.

References

Allen C (2010). Virus transmission in orchids through the feeding damage of Australian cockroach, *Periplaneta australassiae.* Acta Horticulturae, 878: 375–379.

Ammar ED and Nault LR (2002). Virus transmission by Leafhoppers, Planthoppers and Treehoppers (Auchenorrhyncha: Homoptera). Advances in Botanical Research, 36: 141–167.

Asjes CJ (2000). Control of aphid-borne Lily symptomless virus and Lily mottle virus in Lilium in the Netherlands. Virus Research, 71: 23–32.

Baltimore D (1971). Expression of animal viruses. Bacteriology Reviews, 35: 235–241.

Batra LR and Batra SWT (1985). Floral mimicry induced by mummy berry fungus exploits hosts pollinators as vectors. Science, 228: 1011–1013.

Boylan-Pett W, Ramsdell DC, Hoopingarner RA and Hancock, JF (1992). Honey bee foraging behavior and the transmission of the Pollen borne blue leaf mottle virus in highbush blueberry. Acta Horticulturae, 308: 99–108.

Bressan A and Watanabe S (2011). Immunofluorescence localization of Banana bunchy top virus (family; Nanoviridae) within the aphid vector, *Pentalonia nigronervosa* suggests a virus tropism indistinct from aphid transmitted luteoviruses. Virus Research, 155: 520–525.

Brunt AA, Crabtree K, Dallwitz MJ, Gibbs AJ, Watson L and Zurcher FJ (1996). Viruses of Plants: Descriptions and lists from the VIDE database CAB International, Wallingford, UK, 1484 pp.

Brunt AA (2001). The main viruses infecting potato crops. In: Loebenstein G, Berger PH, Brunt AA and Lawson RH (eds). Virus and Virus-like Diseases of Potatoes and Production of Seed Potatoes. Kluwer Academic Publishers, Dordrecht, 65–67.

Burrows ME and Zitter TA (2005). Virus problems of potatoes. USDA-ARS and Dept of Plant Pathology, Cornell University, Ithaca, NY, 14853.

Cabanas D, Watanabe S, Higashi CHV and Bressan A (2013). Dissecting the mode of Maize chlorotic mottle virus transmission (Tombusviridae: *Machlomovirus*) by *Frankliniella williamsi* (Thysanoptera; Thripidae). Journal of Economic Entomolology, 106 (1): 15–24.

Carraro L, Loi N and Ermacora P (2001). The life cycle of Pear decline *Phytoplasma* in the vector, *Cacopsylla pyri* (*Pyrus communis* L.). Journal of Plant Pathology, 81: 87–90.

Celix A, Lopez-sese A, Almarza N, Gomez-Guillamon ML and Rodrigues-Cerezo E (1996). Characterization of Cucurbit yellow stunting disorder virus, a *Bemisia tabaci* transmitted, *Closterovirus*. Phytopathology, 86: 1370–1376.

Chinthagunta L, Verma OP, Devraj P, Naik SL and Krishna A (2009). Incidence of boll rot, boll and locule damage in different Bt cotton crop. International Journal of Plant Protection, 2: 171–175.

Croxall HE, Collingwood CA and Jenkins, JEE (2008). Observations on brown rot (*Sclerotinia fructigena*) of apples in relation to injury caused by earwigs (*Forficula auricularidae*). Annals of Applied Biology, 38 (6): 833–843.

Dasgupta I, Malathi VG and Mukherjee SK (2003). Genetic engineering for virus resistance. Current Science, 84: 340–354.

De Bokx JA and Van der Want JPH (1987). Viruses of potatoes and seed production. 2nd edition, Wageningen, 259 pp.

Dietzgen RG, Callaghan B, Wetzel T and Dale JL (2006). Completion of the genome sequence of Lettuce necrotic yellows virus, type species of the genus *Cytorhabdovirus.* Virus Research, 118: 16–22.

Dombrovsky A, Reingold V and Antignus Y (2014). *Ipomovirus*-a typical genus in the family Potyviridae transmitted by whiteflies. Plant Management Science, 70 (10): 1553–1567.

Dutta B, Burman BK, Srinivasan R, Avci U, Ullman DE, Langston DB and Gitaitus RD (2015). Transmission of *Pantoea ananatis* and *P. agglomomerans*, the causal agent of Center rot of onion, *Allium cepa* by onion thrips (*Thrips tabaci*) through feces. Phytopathology, 04: 812–819.

Edwards MC, Weiland JJ, Todd J, Stewart LR and Lu S (2016). ORF-43 of Maize rayado fino virus is dispensable for systemic infection of maize and transmission by Leafhoppers. Virus Genes, pp: 1–5.

Elnagar S and Murant AF (1976). The role of the helper virus, Anthriscus yellows in the transmission of Parsnip yellow fleck virus by the aphid, *Cavariella aegopodii*. Annals of Applied Biology, 84: 169–181.

Ferriol, I, Rubio L, Perez-Panades J, Corbonell EA, Davino S and Belliure B (2012). Transmissibility of Broad bean wilt virus by aphids: Influence of virus accumolation in plants, virus genotypesand aphid species. Annals of Applied Biology, 162: 71–79.

Fraedrich SW, Harrington TC, Rabaglia RJ, Mayfield AE, Eickwort JM and Miller DR (2008). A fungal symbiont of the redbay Ambrosia beetle causes a lethal Wilt in Redbay and other Lauraceae in the Southern United States. Plant Disease, 92: 215–224.

Franova J, Karesova R, Navratil M and Nebesarova J (2008). A Carrot proliferation disease associated with *Rickettsia-Like Organism* in the Czech Republic. Journal of Phytopathology, 148: 53–55.

Gaborjanyi R and Szabolcs J (1987). Brome mosaic virus transmission by cereal leaf beetle, *Oulema melanopus*, Coleoptera, Chrysomelidae. Cereal Research Communications, 15: 259–264

Gaddam SA, Kotakadi VS, Reddy MN and Saigopal DVR (2012). Antigenic relationships of Citrus yellow mosaic virus by immunological methods. Pelagia Research Library-Asian Journal of Plant Science and Research, 2: 566–569.

Gamliel-Atmsky E, Freeman S, Sztejnberg A, Maymon M, Ochoa R, Belausov E and Palevsky E (2009). Interaction of the mite, *Aceria mangiferae* with *Fusarium mangiferae,* the causal agent of Mango malformation disease. Phytopathology, 99: 152–159.

Garran J and Gibbs AJ (1982). Studies on Alfalfa mosaic virus and alfalfa aphids. Australian Journal of Agricultural Research, 33: 657–664.

Ghasin M, Morin S, Zeidan M and Czosnek H (1998). Evidence for transovarial transmission of Tomato yellow leaf curl virus by its vector whitefly, *Bemisia tabaci*. Virology, 240: 295–303.

Ghosh DK, Bhose S, Mukherjee K, Aglave B, Warghane AJ, Motghare M, Baranwal VK and Dhar AK (2014). Molecular characterization of Citrus yellow mosaic *Badnavirus* (CYMB) isolates revealed the presence of two distinct strains infecting citrus in India. Phytoparasitica, 42: 681–689.

Gibb KS and Randles JW (1991). Transmission of Velvet tobacco mottle virus and related viruses by mired, *Cyrtopeltis nicotianae.* Advances in Disease. Vector Research, 7: 1–17.

Gibbs AJ and Harrison BD (1964). A form of Pea early browning virus found in Britain. Annals of Applied Biology, 54: 1–11.

Goldbach R and Peters D (1996). Molecular and biological aspects of *Tospovirus*. In: Elliot RM (ed). The Bunyaviridae, Springer Science Business, Media, New York.

Gollifer DE, Jackson G, Dabek AJ and May YY (1977). The occurrence and transmission of viruses of edible Aroids in the Solomon Islands and Southwest Pacific. Tropical Pest Management, 23: 171–177.

Gray S and Gildow FE (2003). *Luteovirus*-aphid interactions. Annual Reviews of Phytopathology, 41: 539–66.

Gray SM and Banerjee N (1999). Mechanisms of arthropod transmission of plant and animal viruses. Microbiology and Molecular Biology Reviews, 63: 128–148.

Gurr GM, Bertaccini A, Gopurenko D, Kruger RR, Alhudaib KA, Liu J and Fletcher MJ (2015). *Phytoplasma* and their insect vectors; Implications for date palm. pp. 287–314. In: W Wakil, JR Faleiiro TA and Miller FA (eds). Sustainable Pest Management in Date

Palm: Current Status and Emerging Challenges. Springer Science Plus Business Media Dordrecht, The Netherlands.
Hansen AJ, Nyland G, McElroy FD and Stace–Smith R (1974). Origin, cause, host range and spread of Cherry rasp leaf disease in North America. Phytopathology, 64: 721–727.
Hardy VG and Teakle DS (1992). Transmission of Sowbane mosaic virus by *Thrips tabaci* in the presence and absence of virus carrying pollen. Annals of Applied Biology, 121: 215–320.
Haque SQ and Parasram S (1973). *Empoasca stevensi*, a new vector of bunchy top disease of papaya. Plant Disease Reporter, 57: 412–413.
Hewitt WB, Raski DJ and Goheen ACI (1958). Nematode vector of the Soil borne fanleaf virus of grapevines. Phytopathology, 48: 586–595.
Hibino H (1983). Transmission of two rice tungro associated viruses and rice waika virus from doubly or singly infected source plants by leafhopper vectors. Plant Disease, 67: 774–777.
Hohn T (2007). Plant virus transmission from insect point of view. Proceedings of National Academy of Sciences, USA, 104: 17905–17906.
Huang LH, Tseng HH, Li JT and Chen TC (2010). First report of Cucurbit chlorotic yellows virus infecting cucurbits in Taiwan. Plant Disease, 94: 1682–1682
Hull R (2002). Matthews Plant Virology. Academic Press, New York, NY, USA
Iwaki M, Thongmeearkum P, Prommin M, Honda Y and Hibi T (1982). Whitefly transmission of some properties of Cowpea mild mottle virus on soybean in Thailand. Plant Disease, 66: 365–368.
Jensen SG (1985). Laboratory transmission of Maize chlorotic mottle virus by three species of corn rootworms. Plant Disease, 69: 864–868.
Johnstone GR and Mclean GD (2008). Virus disease of subterranean clover. Annals of Applied Biology, 110: 421–440.
Jones JT, Haegeman AW, Danchin GJE, Hari SG, Helder J, Jones MGK, Kikuchi T, Rosa ML Juan EPR, Wesemael WML and Perry RN (2013). Top ten plant parasitic nematodes in molecular plant pathology. Molecular Plant Pathology, 4: 946–961.
King AMQ, Adams MJ, Carstens EB, Lefkowitz EJ (eds) (2012). *Potexvirus* pp. 912–915. In: Virus Taxonomy, Classification and Nomenclature of Viruses, 9th Report of ICTV (P1327), London, UK; Elsevier Academic Press.
Kluth S, Kruess A and Tscharntke T (2002). Insect vectors of plant pathogens: Mutualistic and antagonistic interactions. Oecologia, 133: 193–199.
Kondo A, Malda T and Tamada T (2003). Orchid fleck virus, *Brevipalpus californicus* mite transmission, biological properties and genome structure. Experimental Applied Acarology, 30: 215–223.
Koudamiloro A, Nwilene FE, Togola, A and Martin A (2015). Insect vectors of Rice yellow mottle virus. Journal of Insects, 20 (5): 1–12.
Krezal H (1982). Investigation on the biology of strawberry aphid (*Chaetosiphon fragaefolii*), the most important vector of strawberry viruses in West Germany. Acta Horticulturae, 129: 63–68.
Lack KJ (1989). The spread of apple brown rot (*Monolinia fructigena*) by insects. Annals of Applied Biology, 115: 221–227.
Laney A, Keller K, Martin R and Tzanetakis J (2011). Rose rosette virus: a discovery seventy years in the making characterization of Rose rosette virus. Journal of General Virology, 92: 1727–1732.
Lecoq H and Katis N (2014). Chapter five: Control of cucurbit viruses. Advances in Virus Research, 90: 255–296.
Lee KW and Kim SK (1985). The effects of Black-streaked dwarf virus on the vector (*Laodelphax striatellus* U.) and host plants. Korean Journal of Plant Pathology, 1 (3): 190–194.
LeMaguet J, Fuchs JJ, Chadoeuf J, Beuve M, Herrbach E and Limaire O (2012). The role of mealybug, *Phenacoccus aceris* in the spread of Grapevine leaf roll associated virus, (GLRaV-1) in two French vineyards. European Journal of Plant Pathology, 135: 415–427.
Liou RF, Yan HZ and Hong JL (2003). Molecular evidence that aphid transmitted Alpinia mosaic virus is a tentative member of genus *Macluravirus*. Archives of Virology, 148: 1211–1218.

Lukanda M, Owati A, Ogunsanya P, Volimzigha K, Katsongo K, Ndemere H and Kumar PL (2014). First report of Maize chlorotic mottle virus infecting maize in the Democratic Republic of Congo. Plant Disease, 98 (10) pages 1448 L//PDx.doi.org/10. 1094 psis -05-14-0484-PDN.

Ma YX, Li JJ, Li GF and Zhu SF (2014). First report of Cherry mottle leaf virus in China. Plant Disease, 98 (6): 161.3.

Manjunath KC, Hilbert SE, Ramadugu C, Webb S and Lee RF (2008). Detection of *Candidatus liberibacter asiaticus* in *Diaphorina citri* and its importance in the management of citrus huanglongbing in Florida. Phytopathology, 98: 387–397.

Michailides TM and Morgan DP (1998). Spread of endosepsis. In: Calimyrna Fig Orchards. Phytopathology, 88: 637–647.

Markham R and Smith KM (1949). Studies on the virus of Turnip yellow mosaic. Parasitology, 39: 330–342.

Martelli GP (2014). Grapevine leaf roll. Journal of Plant Pathology, 96: 551–570.

Martin C (1959). Bacterial rotting of apple fruit. Annals of Applied Biology, 47: 601–611.

McEwEn FL and Kawanishi CV (1967). Insect transmission of Corn mosaic: laboratory studies in Hawaii. Journal of Economic Entomology, 60: 1413–1417.

Melzer MJ, Sether DM, Borth WB and Hu JS (2012). Characterization of viruses infecting citrus Volkamieriana with leprosis like symptoms. Phytopathology, 102: 122–127.

Menzel W, Abang MM and Winter S (2011). Characterization of Cucumber vein clearing virus, a whitefly (*Bemisia tabaci* G.) transmitted *Carlavirus*. Archives of Virology, 56: 2309–23011.

Merz U (2008). Powdery scab of potato, occurrence, life cycle, and epidemiology. American Journal of Research, 85: 241–246.

Mohamed NA and Mossop DW (1981). Synosurus and cocksfoot mottle viruses; a comparison. Journal of General Virology, 55: 63–74.

Nagaich BB, Upreti GC and Verma KD (1972). Studies on transmission of Potato virus X and S by *Epilachna* and *Coccinella* beetles. Science and Culture, 38: 27–28.

Navas-Castillo J, Camero R, Buena M and Morioness E (2000). Severe yellowing outbreaks in tomato in Spain associated with infections of Tomato chlorosis virus. Plant Disease, 84: 835–837.

Notte PLA, Buzkan N, Chueiri E, Minafra A and Martelli GP (1997). Acquisition and transmission fo Grapevine virus A by the mealybug, *Pseudococcus longispinus*. Journal of Plant Pathology, 78: 79–85.

Oerke EC and Dehne HW (2004). Safeguarding production: Losses in major crops and the role of crop protection. Crop Protection, 23: 275–285.

Omura T, Yan J, Zhong B, Wada M, Zhu Y, Tomaru M, Maruyama W, Kikuchi A, Watanabe Y, Kimura I and Hibino H (1998). The P2 protein of Rice dwarf *Phytoreovirus* is required for adsorption of the virus to cells of the insect vector. Journal Virology, 72: 9370–9373.

Orlob GB (1968). Relationship between tetranychus urticae koch and some plant viruses. Virology, 35: 121–133.

Paliwal YC (1980). Relationship of Wheat streak mosaic virus and Barley stripe virus to vector and nonvector eriophyid mites. Archives of Virology, 63: 406–414.

Palomar MK (1987). Relationship between taro feathery mosaic disease and its vector, *Tarophagus proserpina* Kirk. Annals of Tropical Research, 9: 68–74.

Plumb RI (2002). Viruses of poaceae: a case history in plant pathology. Plant Pathology, 51: 673–683.

Prins M and Goldbach R (1998). The emerging problems of *Tospovirus* infection and non conventional methods of control. Trends in Microbiology, 31(6): 31–35.

Proeseler G (1968). Ubertragungsversuche mit dem latenten prunus-virus und der gallmilbe vasates fockeui Nal. Phytopathologica Zeitschrift, 63: 1–9.

Proeseler G (1978). Akquisitions und Zirculationszeit des Rubenkrausel Virus. ARCH Phytopathology Pflanzenschutz, Berlin, 14: 95–98.

Proeseler G (1980). Piesmids, pp 97–113. In: Harris KF and Maramorosch K (eds). Vectors of plant pathogens, New York, Academic Press, 467 p.

Quainoo AK, Wetten AC and Allainguillaume J (2009). Transmission of Cocoa swollen shoot virus by seeds. Journal of Virological Methods, 150: 45–49.

Quito-Avila DF, Lightle D, Lee J and Martin RR (2012). Transmission biology of Raspberry latent virus, the first aphid-borne reovirus. Phytopathology, 192: 547–553.

Raccah B and Fereres A (2009). Plant virus transmission by insects. In: Encyclopedia of Life Sciences (ELS) John Wiley and Sons Ltd, Chichester. Doi:10 1002/997804 7001 590 02 A 002155 a 0000760 pub 2.

Randles JW and Hanold D (1989). Coconut foliar decay virus particles are twenty nm icosahedra. Intervirology, 30: 177–180.

Redinbaugh MG and Hogenhout SA (2005). Plant Rhabdoviruses. In: The Series, Current Topics in Microbiology and Immunology, 292: New York NY: Springer-Verlag, pp. 143–163.

Roy A and Brlansky RH (2009). Population dynamics of a Florida citrus Tristeza virus isolate and aphid transmitted sub-isolate: Identification of three genotypic groups and recombinants after aphid transmission. Phytopathology, 99: 1297–306.

Salazer LF (1996). Potato viruses and their control. International Potato Center, Lima, Peru, 214 pp.

Salgueiro S and Hull R (1999). Comparison of Pea enation mosaic virus (PEMV) isolate Sequence of coat protein and aphid transmission protein genes of European isolate. Journal of Phytopathology, 147: 391–396.

Sasaya T, Fujii H, Ishkawa K and Koganezawa H (2008). Further evidence of Mirafiori lettuce big vein virus but not of lettuce big vein associated virus with big vein disease in lettuce. Phytopathology, 98: 464–468.

Sdoodee R and Teakle DS (2007). Transmission of Tobacco streak virus by *Thrips tabaci*, a new method of virus transmission. Plant Pathology, 36 (3): 377–380.

Shamim MD, Pandey P, Singh A, Yadav P, Bhowmick PK, Srivastava D, Kumar D, Khan NK, Dwivedi DK and Singh KN (2013). Role of biotechnology in plant disease management: an overview. Journal of Genetic and Environmental Resources Conservation, 1: 215–220.

Shepherd DN, Martin DP, Van Derwalt E, Dent K, Varsani A and Rybicki EP (2010). Maize streak virus: an old and complex pathogen. Molecular Plant Pathology, 11: 1–12.

Shikata E, Maramorosch K and Granados RR (1966). Electron microscopy of Pea enation mosaic virus in plants and aphid vectors. Virology, 29: 426–436.

Shukla, DD, Ward CW, Brunt, AA and Berger, PD (1998). Phytoviridae family. AAB Descriptions of Plant Viruses No 366.

Sicard A, Zeddam JL, Yvon M and Blanc S (2015). Circulative non-propagative aphid transmission of *Nanoviruses*: an over simplified view. Journal of Virology, 89: DOI 10.1128/JVI 00780-15.

Skare JM, Wijkamp I, Denham I, Rezende JA, Kitajima EW, Park JW, Desvoyes B, Rush CM, Michels G, Schlthof KB and Scholthof HB (2006). A new eriophyid mite borne membrane –enveloped virus like complex isolated from plants. Virology, 347: 343–353.

Stavrinides J, No A and Ochman H (2010). A single genetic locus within phytopathogen *Pantoea stewarti* gut colonization and pathogenicity in an insect host. Environmental Microbiology, 12: 147–155.

Stanley J (2008). Beet curly top virus. Encyclopedia of Virology (third edition), pp. 301–307.

Stevens M, Freeman B, Liu HY, Herrbach E and Limaire O (2005). Beet *Poleroviruses:* close friends or distant relatives? Molecular Plant Pathology, 6: 1–9.

Stubbs LL (1952). Further host range and transmission studies with a virus disease of carrots endemic in Australia. Australian Journal of Science Research, 5: 399–408.

Susi P (2004). Black currant reversion virus, a mite transmitted *Nepovirus.* Molecular Plant Pathology, 5: 167–173.

Sylvester ES (1956). Beet yellows virus transmission by green peach aphid. Journal Economic Entomology, 49: 782–800.

Takasu F (2009). Individual-based modeling of the spread of pine wilt disease location, dispersal and the allele effect. Population Ecology, 51: 399–409.

Tsai CW, Rowhani A, Golino DA, Daane KM and Almeida RPP (2010). Mealybug transmission of Grapevine leaf roll viruses: An analysis of virus vector specificity. Phytopathology, 100: 830–834.

Tsai JH and McDaniel LL (1990). Characterization and serological analysis of Pseudo curly top virus. Plant Disease, 74: 17–21.

Tedeschi R, Ferrato V, Rossi J and Ahmad A (2006). Possible *Phytoplasma* transovarial transmission in the psyllids *Cacopsylla melanoneura* and *Cacopsylla pruni*. Plant Pathology, 55: 18–24.

Turka I (1978). Lygus rugulipennis Popp (Heteroptera: Miridae) vector of potato viruses. Trudy ladv Lauksaimn Akad, 164: 65–73.

Ullman DE, Cho JJ and German TG (1991). Occurrence and distribution of Cucurbit viruses in the Hawaiian Islands. Plant Disease, 75: 367–375.

Van Velsen RJ and Crowley NC (1961). Centrosema mosaic: a plant virus disease transmitted by both aphids and plant bugs. Nature, 189: 858.

Vanneste JL (1996). Honey bees and epiphytic bacteria to control fire blight, a bacterial disease of apple and pear. Biocontrol News and Information, 17: 67N–68N.

Van Hoof, HA (1980). Aphid vectors of Potato virus Y. Netherlands Journal of Plant Pathology, 86:159–162.

Verbeek M, Van Bekkum PJ, Dullemans PM and Vtugt VD (2014). Torradoviruses transmitted in a semi-persistent and stylet-borne manner by three whitefly vectors. Virus Research, 186: 55–60.

Verchot-Lubicz, J, Chang-Ming Y and Bamusinghe, D (2007). Molecular ecology of *Potexvirus*: recent advances. Journal of General Virology, 88: 1643–1655.

Vereijssen J and Scott I (2013). Psyllid can overwinter on non-crop host plants. New Zealand Grower, 68: 14–15.

Wada Y, Tanaka H, Yamashita E and Tsukihera T (2008). Structure of Melon necrotic spot virus determined at 2.8 A resolution. Acta Crystallographica Section F Structural Biology, Crystallization Communications, Feb, 2008, F64: 8–13, DOI: 10.1107/S51744309107066481.

Wefels E, Morin JP and Randles JW (2015). Molecular evidence for a persistent circulative association between Coconut foliar decay virus and its vector *Myndus taffini*. Australasian Plant Pathology, 44: 283–288.

Winter S, Hamacher A, Engelmann J and Lesemann DE (2006). Angelonia flower mottle, a new disease of *Angelonia augustifolia* caused by a hitherto unknown *Carmovirus*. Plant Pathology, 55 (6): 820.

Yamashita S, Yora K and Yoshino M (1979). Cucumber yellows virus, its transmission by greenhouse whitefly, *Trialeurodes vaporariorum* (Westwood) and the yellowing disease of cucumber and muskmelon caused by the virus. Annals of Phytopathological Society of Japan, 45: 4584–4596.

Young JM (2004). Olive knot and its pathogens. Australasian Plant Pathology, 33: 33–39.

Zheng L, Mao Q, Xie L and Wei T (2014). Infection route of Rice grassy stunt virus, a *Tenuivirus*, in the body of its brown planthopper vector, *Nilaparvata lugens* (Hemiptera: Delphacidae) after ingestion of virus. Virus Research, 188: 170–17.

QUESTIONS (EXERCISE)

Q 1. Why are the following scientists/personalities known in the field of insect vectors?

a) WM Stanley

b) Kaushe
c) Beijerinck
d) Carolus clusius
e) Doi
f) Robert Koch
g) Bawden and Pieri
h) D Ivanowski
i) A Mayer
j) AV Leeuwenhoek
k) EF Smith

Q 2. Name the pioneer scientists associated with the following discoveries.

a) Bacteriophages
b) Mosaikkrankheit
c) Virus multiplication of virus in leafhopper
d) First DNA virus
e) Viroid

Q 3. What do these stand for: (a) ICTV (b) TMV (c) DNA (d) MLO (e) RLO?

Q 4. Describe briefly the importance of plant viruses vectored by arthropods.

Q 5. Define the following terms:

a) Mango malformation
b) Mosaic
c) Vein clearing
d) Shoe string
e) *Phytoplasma*
f) Virus

CHAPTER 2

Modes of Spread of Plant Pathogens

The plant viruses, being obligate parasites, require an injury in order to get into the plant tissues. The spread of these organisms is altogether different. The transmission of plant pathogens takes place through two different mechanisms under natural conditions (vertical transmission and horizontal transmission). This topic is elaborated in detail in the paragraphs below. The spread of all the plant pathogens is through both living and non-living agents (Table 2.1; Fig. 2.1). The vertical transmission mode of spread involves vegetative propagation, such as grafting, mechanical means, seed/pollen transmission and parasitic plants. The other means viz. insects, mites, nematodes, fungi, protozoa, animals, birds, slugs and earthworms in the living category, and air, water and soil in the non-living, are known to spread via horizontal pattern.

2.1 Vertical Transmission

2.1.1 Vegetative propagation

Vegetative propagation is an important practice, used as a means to improve the quality of fruit and flowering plants. This process involves the use of plant parts, such as tubers, rhizomes, suckers, and corms, to plant crops. Viruses are systemic in nature; thus, all plant parts are known to carry plant viruses since the infected plant parts are often used as seed or for grafting purposes, without there being any knowledge of the viruses that they are spreading. The vegetative plant parts are taken to far-off places by various methods of human transportation. This method of propagation is extensively used, and is the most common way of spreading plant viruses in crop plants, particularly in horticultural crops. The vegetative propagation

Table 2.1 Transmission/spread of plant viruses through different means.

Sl. No.	Mode of spread	Part(s)	Virus	Mechanism	Source(s)/Remarks
1	Vegetative Propagation	Tubers	Potato virus X (PVX), Potato virus Y (PVY), Potato leaf roll virus (PLRV), etc.	Seed potato; contaminated implements; transportation of seed	Hollings, 1965
		Cuttings	Rose mosaic virus (RMV), Sugarcane mosaic virus (ScMV)	Cutting implements due to systemic nature of plant viruses	Hollings, 1965
		Rhizomes	Canna yellow mottle virus (CaYMV)	On account of systemic nature of plant viruses	Hollings, 1965
		Corms	White break mosaic virus (WBMV), Cucumber mosaic virus (CMV)	Viruses being systemic and all parts contain virus	Hollings, 1965
		Suckers	Banana bunchy tops virus (BBTV)	Viruses are systemic	Hollings, 1965
2	Mechanical means	Sap inoculation	Tobacco mosaic virus (TMV), potato virus X (PVX)	During sap inoculation; contaminated hands; contaminated implements	Gray and Banerjee, 1999
		Grafting	Citrus tristeza virus (CTV), Apple mosaic virus (ApMV)	All parts carry virus due to systemic nature of viruses	All viruses are transmissible
3	Vectors	Animals (cow, monkeys, grass rats)/ human activity; use of contaminated implements for tillage, etc.	Rice yellow mosaic virus (RYMoV), Tobacco Mosaic Virus (TMV), Seed borne viruses	Through injury while moving in field; transportation of seed; ingestion of contaminated seed pass out through faeces; human activity with agricultural implements	Sara and Peters, 2003

		Birds (sparrows, weaver bird, bulbul, and other birds)	Rice yellow mosaic virus (RYMoV)	Carry contaminated plant parts; ingest infected seed; contaminated body and appendages; pollination process	Peters et al., 2012
4	Non-living agents	Soil (through particles, soil inhabiting fungi and nematodes)	Lettuce big vein virus (LBVV), Wheat mosaic virus (WMV), Tobacco rattle virus (TRV), Tobacco ring spot virus (TRSV)	Adherence of viruses on soil particles and movement through tillage operations and through soil inhabiting organisms	Campbell, 1996
		Air (through injury of strong wind)	Tobacco mosaic virus (TMV)	Highly infectious viruses through strong wind injury	It causes injury for passive entry of the virus
		Water (soil borne pathogens)	Soil-borne fungi and nematodes	Transportation of soil borne viruses through the movement of soil organisms in the soil	Bacteria spreads Through Water
5	Parasitic plants	Cuscuta, mistletoe	Most viruses (viruses being systemic in nature)	Through human activity birds carry Cuscuta for nest building	Viruses Spreads Through penetration of haustoria
6	Natural spread	Seeds	Alfalfa mosaic virus (AMV), Arabis mosaic (ArMV), Cowpea mosaic virus (CPMV), Leaf crinkle virus (LCV), Bean common mosaic virus (BCMV), Cherry leaf roll virus (CLRV)	Use of infected seed in the case of Seed borne viruses	Latham and Jones, 2001; Sharma et al., 2007; Choi et al., 2006; Dinesh et al., 2007; Cooper et al., 1984
		Pollen	Prunus necrosis ring spot virus (PNRSV), Prune dwarf virus (PDV), Cucumber green mottle virus (CGMV)	Transportation of infected pollen through living agencies, especially through pollinating insects	Cooper et al., 1988; Liu et al., 2014

Table 2.1 contd. …

...Table 2.1 contd.

Sl. No.	Mode of spread	Part(s)	Virus	Mechanism	Source(s)/Remarks
7	Protozoa	*Phytomonas*	Hart rot of coconut, Phloem necrosis of coffee	Phloem inhabiting organisms are transmitted	Alves-Silva et al., 2013
8	Nematodes	Soil nematodes	Fan leaf of grapevine virus (GFLV), Arabis Mosaic virus (ArMV), Tobacco rattle virus (TRV) Raspberry ring spot virus (RaspRSV), Tomato ring spot virus (TRSV), etc.	Genera of nematodes viz *Xiphinema, Longidorus, Trichodorus, Paratrichodorus*	Jones et al., 2013
9	Fungi	Soil fungi	Lettuce big Vein virus (LBVV), Wheat Mosaic virus (WMV), Cucumber necrosis virus (CNV), Tobacco necrosis virus (TNV)	*Olpidium, Polymyxa, Spongospora, Synchytrium*	Campbell, 1996
10	Mites	Arachnids	Wheat streak mosaic virus (WStMV), Barley stripe virus (BSV), Brome mosaic virus, (BMV), etc.	Mites of families viz. Eriophyidae; Tetranychidae	Sarwar, 2015
11	Insects	Insects (Hemiptera; Thysanoptera; Coleoptera; Hymenoptera; Orthopteran; Dictyoptera; Dermaptera; Lepidoptera; Diptera)	Cucumber mosaic virus (CMV), Cauliflower mosaic virus (CaMV), Tomato spotted wilt virus (TSWV), Tomato yellow leaf curl virus (TYLCV), Turnip yellow mosaic virus (TYMV), etc.	Transmission mechanism viz. non-persistent–stylet borne, semi-persistent-foregut-borne, persistent circulative and persistent propagative	Nault, 1997

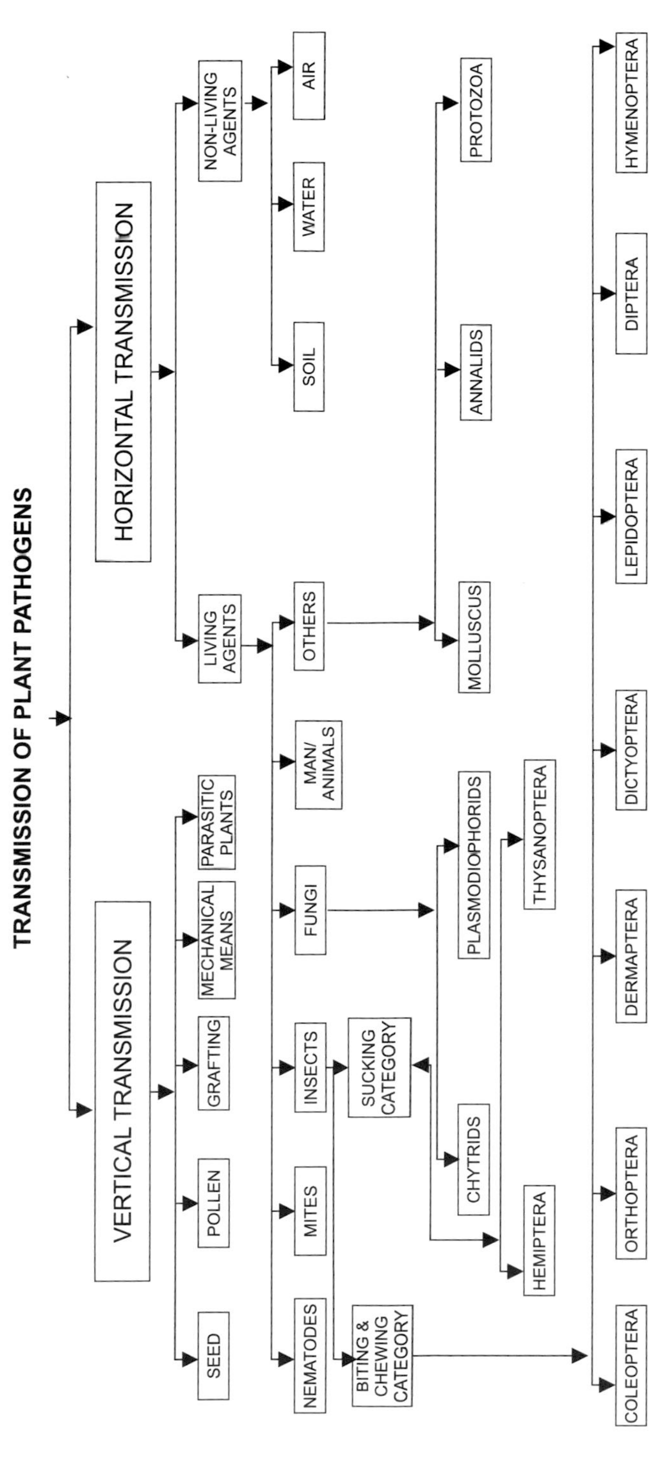

Fig. 2.1 Modes of Spread of Plant Pathogens.

through *Dalbulus maidis* (Delong and Wolcott) and *Circulifer tenellus* (Baker), respectively. Bacteria are ingested by insects while sucking the sap either from the vascular tissues of plants or feeding on infected parts/bacterial galls. Fire blight of apple and pear caused by *Erwinia amylovora* is spread by insects such as bees and wasps when these insects pick up the hibernating bacteria from ooze. Likewise, the spread of Bacterial wilt of cucurbits, caused by *E. tracheiphila,* takes place via the spotted cucumber beetle (*Diabrotica undecimpunctata howardi* Barber) and striped cucumber beetle (*Acalymma vittatum*). Pierce disease of grapes caused by *Xylella fastidiosa* is also spread by sharp shooters, namely *Graphocephala atropunctata, Draeculacephala mineria, Carnocephala fulgida* Nottingham and *Homalodisca coagulata* (Say). Citrus greening caused by *Liberibacter asiaticus* and *Liberibacter africanus* is transmitted by *Citrus psylla, Diaphorina citri* Kuwayama and *Trioza erytreae* Del Guercio. Another bacterium, *Peseudomonas savastanoi,* responsible for olive knot disease is transmitted by olive fly, *Dacus oleae* (Gmelin). The fungus causing Boll rot of cotton (*Fusarium moniliforme*) is spread by cotton bollworms (*Helicoverpa armigera* (Hubner), *Earias vitella* (Fab.), *E. insulana* (Boisduval), *Pectinophora gossypiella* (Saunders) and dusky cotton bug/ red cotton bug. The spread of Dutch elm disease (*Ceratocystis fimbriata*) occurs via nitidulid beetle (*Gleoprirus niger).* Similarly, Mango wilt (*Diplodia recifensis)* is beetle-borne (*Xyleborus affinis* Eichhoff), whereas the Perennial canker of apple (*Gloeosporium perennens*) is transmissible via woolly aphid (*Eriosoma*). The black stain disease of Douglas fir trees, caused by fungus *Leptographium wageniumis,* is spread by the bark beetle (*Hilastes nigrinus*) and the corn weevil (*Steremnius carinatus*). There are cases of spread of nematodes by insects. Pine wilt is caused by *Bursaphelenchus xylophilus* nematode and is transmissible by cerambycid beetles like *Monochamus alternatus* (Hope) and *M. carolinensis.* In most cases, these pathogens live in symbiosis and help the insect host in the digestion of food. Besides, pentatomid bugs are also transported and spread disease-causing protozoa. The details regarding how these diseases are spread, causing pathogens, is presented separately. The International Committee on Taxonomy of Viruses has been tasked with the classification of viruses. Till date, the committee has classified viruses into seven orders (Caudovirales (three families), Herpesvirales (three families), Ligmenvirales (two families), Mononegavirales (eight families), Nidovirales (four families), Picornavirales (five families), and Tymovirales (four families)), 111 families (including eighty-two families without orders), twenty-seven sub families and 609 genera. However, the detailed list of important plant viruses transmitted by insects (Table 2.2) and other means such as nematodes, mites, fungi, seed, etc. (Table 2.3) is presented.

2.2.1.2. Mites. Mites are members of class Arachnida under the phylum Arthropoda and the adults have four pairs of legs. These small creatures cause damage both by sucking sap directly from plants, and by acting as

Table 2.2 The presentation of acronyms, genera, family characteristics and insect transmission of important plant viruses.

Virus	Acronym	Genus	Family	Characteristics	Association of insects
African cassava mosaic virus	ACMV	*Begomovirus*	Geminiviridae	dsDNA, rod shaped, quasi-isometric, geminate	Whitefly
Alfalfa mosaic virus	AMU	*Alfamovirus*	Bromoviridae	(+)ssRNA, non-enveloped, isometric	Aphids, seed, dodder
Banana bunchy tops virus	BBTV	*Babuvirus*	Nanoviridae	ssDNA	Aphids
Banana streak GF virus	BSV-GF	*Badnavirus*	Caulimoviridae	dsDNA, bacilliform	Mealybug, sap, seed
Barley yellow dwarf virus	BYDV	*Luteovirus*	Luteoviridae	(+)ssRNA, isometric, non-enveloped	Aphid, cuscuta
Bean common mosaic virus	BCMV	*Potyvirus*	Potyviridae	RNA, flexuous filamentous	Aphids, seed, sap
Bean golden mosaic virus	BGMV	*Begomovirus*	Geminiviridae	ssDNA, isometric, non-enveloped,	Whitefly
Bean yellow mosaic virus	BYMV	*Potyvirus*	Potyviridae	RNA	Aphids, seed
Bean pod mottle virus	BPMV	*Comovirus*	Comoviridae	(+)ssRNA	Bean leaf beetles
Beet curly top virus	BCTV	*Curtovirus*	Geminiviridae	ssDNA, isometric, non-enveloped	Leafhoppers, dodder
Beet pseudo yellows virus	BPYV	*Crinivirus*	Closteroviridae	(+)dsRNA, flexuous, filamentous, non-enveloped	Whitefly
Beet yellows virus	BYV	*Closterovirus*	Closteroviridae	(+)ssRNA, filamentous, non-enveloped	Aphid, dodder whitefly

Table 2.2 contd. …

...Table 2.2 contd.

Virus	Acronym	Genus	Family	Characteristics	Association of insects
Broadbean wilt virus-2	BBWV	*Fabavirus*	Comoviridae	(+)ssRNA, isometric, non-enveloped	Aphids
Cocoa swollen shoot virus	CSSV	*Badnavirus*	Caulimoviridae	dsDNA, circular genome, bacilliform	Mealybugs
Carnation latent virus	CLV	*Carlavirus*	Betflexiviridae	(+)ssRNA, filamentous, non-enveloped	Aphids
Carnation mottle virus	CarMV	*Carmovirus*	Tombusviridae	(+)ssRNA, isometric, non-enveloped	Aphids
Carrot mottle virus	CMoV	*Umbravirus*	Not known	(+)ssRNA, isometric, enveloped	Aphid
Cauliflower mosaic virus	CaMV	*Caulimovirus*	Caulimoviridae	dsDNA, filamentous, Non-enveloped	Aphids
Chlorosis striate mosaic virus	CSMV	*Mastrevirus*	Geminiviridae	ssDNA	Leafhopper
Citrus tristeza virus	CTV	*Closterovirus*	Closteroviridae	(+)ssRNA, flexuous rods, enclosed in protein coat	Aphids
Coconut foliar decay virus	CFDV	*Nanovirus*	Nanoviridae	dsDNA, icosahedral	Leafhoppers, cixiid planthopper
Commelina yellow mottle virus	ComYMV	*Badnavirus*	Caulimoviridae	dsDNA, bacilliform, non-enveloped	Mealybugs
Cowpea chlorotic mottle virus	CCMV	*Bromovirus*	Bromoviridae	(+)ssRNA	Bean flea beetles
Cowpea mosaic virus	CPMV	*Comovirus*	Comoviridae	(+)ssRNA, isometric, non-enveloped	Beetles, Seed

Cucumber green mottle mosaic virus	CGMMV	*Tobamovirus*	Virgaviridae	(+)ssRNA, rod shaped	Sap, aphids, beetles, seed
Cucumber mosaic virus	CMV	*Cucumovirus*	Bromoviridae	(+)ssRNA, isometric non-enveloped	Aphids, seed, dodder
Cucumber vein yellowing virus	CVYV	*Ipomovirus*	Potyviridae	dsDNA	Whitefly
Cymbidium ring spot virus	CymRSV	*Tombusvirus*	Tombusviridae	ssRNA	Cockroaches, sap, aphids
Fiji disease virus	FDV	*Fijivirus*	Reoviridae	dsRNA, isometric, non-enveloped	Planthoppers
Groundnut rosette virus	GRV	*Umbravirus*	Not known	(+)ssRNA	Aphids
Johanson grass mosaic virus	JGMV	*Potyvirus*	Potyviridae	(+)ssRNA, flexuous, filamentous	Aphids
Lettuce infectious yellows virus	LIYV	*Crinivirus*	Closteroviridae	ssRNA, long flexuous particles	Whitefly
Lettuce mosaic virus	LMV	*Potyvirus*	Potyviridae	RNA	Seed, aphids
Lettuce necrotic yellows virus	LNYV	*Cytorhabdovirus*	Rhabdoviridae	(–)ssRNA, bacilliform, envelope	Aphid
Maclura mosaic virus	MacMV	*Macluravirus*	Potyviridae	(+)ssRNA	Aphids
Maize chlorotic mottle virus	MCMV	*Machlomovirus*	Tombusviridae	(+)ssRNA, isometric, non-enveloped	Beetles, mechanical
Maize dwarf mosaic virus	MDMV	*Potyvirus*	Potyviridae	ssRNA	Aphids

Table 2.2 contd. …

...Table 2.2 contd.

Virus	Acronym	Genus	Family	Characteristics	Association of insects
Maize mosaic virus	MMV	*Nuceorhabdovirus*	Rhabdoviridae	(–)ssRNA, bullet shaped, envelope	Leafhoppers
Maize rayado fino virus	MRFV	*Marafivirus*	Tymoviridae	(+)ssRNA, Isometric, Non-enveloped	Leafhoppers
Maize streak virus	MSV	*Mastrevirus*	Geminiviridae	ssDNA, geminate shape	Leafhoppers
Maize rough dwarf virus	MRDV	*Fijivirus*	Reoviridae	RNA, isometric	Planthopper
Onion yellow dwarf virus	OYDV	*Potyvirus*	Potyviridae	RNA	Aphids, seed
Pangola stunt virus	PaSV	*Fijivirus*	Reoviridae	dsRNA, double capsid, polyhedral shape	Planthoppers
Papaya ring spot virus	PRSV	*Potyvirus*	Potyviridae	RNA	Aphids, mechanical, leafminers
Parsnip yellow fleck virus	PYFV	*Sequivirus*	Sequiviridae	(+)ssRNA, isometric, non-enveloped	Aphid
Pea enation mosaic virus-1	PEMV	*Enamovirus*	Luteoviridae	(+)ssRNA, isometric, non-enveloped	Aphids, seed
Pea seed borne mosaic virus	PSbMV	*Potyvirus*	Potyviridae	RNA	Seed, aphids, sap
Peanut mottle virus	PeMoV	*potyvirus*	Potyviridae	RNA, flexuous	Aphid, seed, mechanical
Peanut stunt virus	PSV	*Cucumovirus*	Bromoviridae	(+)ssRNA, isometric	Aphids, sap
Potato virus Y	PVY	*Potyvirus*	Potyviridae	(+)ssRNA, filamentous, non-enveloped	Mites, aphids, contact

Potato leaf roll virus	PLRV	*Polerovirus*	Luteoviridae	(+)ssRNA	Aphid
Potato virus-A	PVA	*Potyvirus*	Potyviridae	RNA	Aphids
Potato virus-S	PVS	*Carlavirus*	Betaflexiviridae	(+)ssRNA	Aphids
Potato yellow dwarf virus	PYDV	*Nucleorhabdovirus*	Rhabdoviridae	(–)ssRNA bacilliform, enveloped	Planthoppers, leafhoppers
Rice ragged stunt virus	RRSV	*Oryzavrus*	Reoviridae	dsRNA, isometric, non-enveloped	Planthoppers
Rice stripe virus	RSV	*Tenuivirus*	Not known	(+) and (–)ssRNA, spherical thin filamentous, non-enveloped	Planthoppers
Rice tungro spherical virus	RTSV	*Waikavirus*	Sequiviridae	(+)ssRNA isometric, non-enveloped	Leafhoppers
Sorghum mosaic virus	SrMV	*Potyvirus*	Potyviridae	RNA, flexuous filamentous	Aphids
Southern bean mosaic virus	SBMV	*Sobemovirus*	Not known	(+)ssRNA, isometric, non-enveloped	Bean flea beetles, soil, Seed
Sowthistle yellow vein virus	STYVV	*Nucleorhabdovirus*	Rhabdoviridae	(–)ssRNA, Bacilliform, enveloped	Aphid
Soybean dwarf virus	SbDV	*Luteovirus*	Luteoviridae	(+)ssRNA, Isometric	Aphids
Soybean mosaic virus	SMV	*Potyvirus*	Potyviridae	RNA, flexuous filamentous	Aphids, seed
Strawberry crinkle virus	SCV	*Cytorhabdovirus*	Rhabdoviridae	RNA, bacilliform, enveloped	Aphids
Subterranean clover stunt virus	SCSV	*Nanovirus*	Nanoviridae	ssDNA, small isometric, non-enveloped	Aphids
Sugarcane mosaic virus	ScMV	*Potyvirus*	Potyviridae	RNA	Aphids

Table 2.2 contd....

...Table 2.2 contd.

Virus	**Acronym**	**Genus**	**Family**	**Characteristics**	**Association of insects**
Sugarcane yellow leaf virus	ScYLV	*Polerovirus*	Luteoviridae	(+)ssRNA	Aphids
Sweet potato mild mottle virus	SPMMV	*Ipomovirus*	Potyviridae	(+)ssRNA, filamentous, non-enveloped	Whitefly, dodder
Tobacco leaf curl virus	TbLCV	*Begomovirus*	Geminiviridae	ssDNA	Whitefly
Tobacco mosaic virus	TMV	*Tobamovirus*	Virgaviridae	(+)ssRNA, rod shaped, non-enveloped	Contact, chewing insects, dodder
Tobacco ring spot virus	TRSV	*Nepovirus*	Comoviridae	ssRNA, isometric	Nematodes, grasshoppers, thrips, flea beetles, seed
Tobacco streak virus	TSV	*Ilarvirus*	Bromoviridae	(+)ssRNA, isometric, non-enveloped	Pollen, seed, thrips
Tomato aspermy virus	TAV	*Cucumovirus*	Bromoviridae	RNA, isometric	Aphids
Tomato mosaic virus	ToMV	*Tobamovirus*	Virgaviridae	(+)ssRNA	Aphids, grasshopper, Seed, dodder
Tomato pseudo curly top virus	TPCTV	*Topacuvirus*	Geminiviridae	ssDNA, isometric, non-enveloped, geminate	Treehoppers
Tomato spotted wilt virus	TSWV	*Tospovirus*	Bunyaviridae	(–)ssRNA, large isometric, enveloped	Thrips
Tomato yellow leaf curl virus	TYLCV	*Begomovirus*	Geminiviridae	ssDNA	Whitefly
Turnip yellow mosaic virus	TYMV	*Tymovirus*	Tymoviridae	(+)ssRNA, icosahedral, non-enveloped	Beetles

Table 2.3 The presentation of acronyms, genera, families characteristics and modes of transmission of important plant viruses other than insects.

Virus	Acronym	Genus	Family	Characteristics	Transmission (Except insects)
Apple chlorotic leaf spot virus	ACLSV	*Trichovirus*	Betaflexiviridae	(+)ssRNA, filamentous, non-enveloped	No vector
Apple stem grooving virus	ASGV	*Capillovirus*	Betflexiviridae	(+)ssRNA, non-enveloped	Seed
Barley stripe mosaic virus	BSMV	*Hordeivirus*	Virgaviridae	(+)ssRNA, rod shaped, non-enveloped	Seed, pollen
Barley yellow striate mosaic virus	*BYMV*	*Bymovirus*	Potyviridae	(+)ssRNA, filamentous, non-enveloped	Fungi
Broadbean mottle virus	*BBMV*	*Bromovirus*	Bromoviridae	RNA	Sap
Brome mosaic virus	*BMV*	*Bromovirus*	Bromoviridae	(+)ssRNA, isometric, non-enveloped	Mechanical, nematodes
Carnation ring spot virus	*CRSV*	*Dianthovirus*	Tombusviridae	(+)ssRNA, isometric, non-enveloped	Nematodes
Citrus leaf rugose virus	CiLRV	*Ilarvirus*	Bromoviridae	RNA, isometric	Contaminated tools
Citrus variegation virus	CVV	*Ilarvirus*	Bromoviridae	(+)ssRNA, isometric	Sap, grafting, Seed
Garlic viruses A,B,C,D	GarV-A, GarV-B, GarV-C, GarV-D and GarV-E	*Allexivirus*	Flexiviridae	ssRNA, filamentous	Mites, bulbs

Table 2.3 contd....

...Table 2.3 contd.

Virus	Acronym	Genus	Family	Characteristics	Transmission (Except insects)
Grapevine fan leaf virus	GFLV	*Nepovirus*	Comoviridae	RNA	Nematodes, seed, grafting
Grapevine fleck virus	GFKV	*Maculavirus*	Tymoviridae	ssRNA	Vegetative propagation, grafting
Lettuce big vein associated virus	LBVaV	*Varicosavirus*	Not known	dsRNA, rod shaped, non-enveloped	Fungi
Maize white line mosaic virus	MWlMV	*Aureousvirus*	Tombusviridae	ssRNA, isometric	Mechanical, soil borne, seed borne
Odontoglossum ring spot virus	ORSV	*Tobamovirus*	Virgaviridae	RNA, rod shaped	No vector, mechanical
Olive latent virus-2	OLV-2	*Oleavirus*	Bromoviridae	ssRNA	Grafting
Ourmia mosaic virus	OurMV	*Ourmiavirus*	Not known	(+)ssRNA, bacilliform, non-enveloped	Mechanical
Papaya mosaic virus	PaMV	*Potexvirus*	Alphaflexiviridae	ssRNA	No vector
Pea early browning virus	PEBV	*Tobravirus*	Virgaviridae	RNA, rod shaped, tubular	Seed, nematodes
Peach rosette mosaic virus	PRMV	*Nepovirus*	Comoviridae	ssRNA, isometric	Nematodes, sap
Pelargonium zunate spot virus	PelarZSV	*Anulavirus*	Bromoviridae	ssRNA, isometric	Sap, seed, pollen
Plum pox virus	PPV	*Potyvirus*	Potyviridae	RNA, flexuous filmentous	Seed (unconfirmed report)

Potato virus X	PVX	*Potexvirus*	Flexiviridae	(+)ssRNA, filamentous, non-enveloped	Mechanical, contact
Potato mop top virus	PMTV	*Pomovirus*	Virgaviridae	(+)ssRNA	Fungi, sap
Prunus necrotic ring spot virus	PNRSV	*Ilarvirus*	Bromoviridae	RNA	Pollen, seed
Raspberry bushy dwarf virus	RBDV	*Idaeovirus*	Not known	(+)ssRNA, isometric, non-enveloped	Pollen, seed
Raspberry ring spot virus	RpRSV	*Nepovirus*	Comoviridae	RNA, isometric	Nematodes, seed
Ryegrass mosaic virus	*RGMV*	*Rymovirus*	Potyviridae	(+)ssRNA, filamentous, non-enveloped	Mite
Soil borne wheat mosaic virus	SBWMV	*Furovirus*	Virgaviridae	(+) ssRNA, hollow rigid rods	Fungi, sap
Sugarcane bacilliform virus	ScBV	*Badnavirus*	Caulimoviridae	dsDNA, bacilliform	Mechanical
Tobacco necrosis virus	*TNV*	*Necrovirus*	Tombusviridae	(+)ssRNA, isometric, non-enveloped	Fungi
Tobacco rattle virus	TRV	*Tobravirus*	Virgaviridae	(+)ssRNA, rod shaped, non-enveloped	Nematodes, seed and dodder
Tobacco stunt virus	TStV	*Varicosvirus*	Not known	dsRNA, rod shaped	Fungi, grafting, sap, dodder
Tomato bushy stunt virus	TBSV	*Tombusvirus*	Tombusviridae	(+)ssRNA, isometric, non-enveloped	Soil, mechanical, seed
Tomato ring spot virus	ToRSV	*Nepovirus*	Comoviridae	RNA, isometric	Nematodes, seed

Table 2.3 contd....

...Table 2.3 contd.

Virus	Acronym	Genus	Family	Characteristics	Transmission (except insects)
Wheat mosaic virus	WMV	*Furovirus*	Not known	(+)ssRNA, rod shaped, non-enveloped	Mites
Wheat streak mosaic virus	WSMV	*Tritimovirus*	Potyiviridae	(+)ssRNA, filamentous, non-enveloped	Mites, sap
White clover cryptic virus-2	WCCV2	*Betacryptovirus*	Partitiviridae	dsRNA, isometric, non-enveloped	Seed
White clover mosaic virus	WClMV	*Potexvirus*	Alphaflexiviridae	ssRNA, flexuous helical filamentous, elongated	Sap, seed, dodder

vectors of plant pathogens, especially of viruses. The eriophyid mites are also responsible for the transmission of Wheat streak mosaic virus (WStMV) (*Eriophyes tulipae* and *Abacarus hystrix*), Barley stripe mosaic virus (BSMV) and Brome grass mosaic virus (BgMV) (*E. tulipae*). In addition, *Phyllocoptes fructiphilus* is a vector of Rose rosette disease.

2.2.1.3 Nematodes. These organisms belong to phylum Nematoda and are considered as pathogens of plants, since they cause direct damage to crop plants and act as vectors of plant pathogens. Generally, they feed on roots but do also consume stems, leaves and seeds. Fan leaf of grapevine virus (GFLV) has been labelled as nematode-transmissible since 1958. At present, there are more than twenty diseases of viral etiology known to be vectored by nematodes which are either polyhedral or tubular (rods) in shape. Both adults and juveniles transmit viruses with their stylets. In all, five genera viz. *Xiphinema, Longidorus, Paralongidorus, Trichodorus* and *Paratrichodorus* are involved in the transmission of plant viruses. Raspberry ring spot virus (RpRSV) of cherry and Tomato black ring virus (TBRV) are vectored by *Longidorus elongatus* while Arabis mosaic virus (ArMV) in strawberry is carried by *Xiphinema diversicaudatum*. Another species, *X. californicum*, spreads Tomato ring spot virus (TomRSV). Whereas, *Trichodorus viruliferous* is instrumental in spreading Tobacco rattle virus (TRV) and Pea early browning virus (PEBV). *Paratrichodorus pachydermus* also transmits TRV. The infection of the nematodes facilitates the entry of many fungi, causingTobacco wilt, Carnation wilt and Raspberry crown galls. Nematodes also create favorable substratum for *Rhizoctonia* root rot infection in tomato through *Meloidogyne* spp. of nematodes.

2.2.1.4 Fungi. These organisms are responsible for causing very serious diseases in crop plants. Therefore, these are the number one enemy of man amongst all the pathogens; enormous losses in production are attributed to fungi. These pathogens are much more dangerous as soon as they start acting as vectors of plant viruses. *Olpidium brassicae* (Tobacco necrosis virus-TNV, Lettuce big vein virus-LBVV), *O. cucurbitacearum* (CNV), *Polymyxa graminis* (WMV), *Spongospora subterranea* (PMTV) are vectors of plant pathogens. Zoospore is the stage responsible for the spread of plant pathogens.

2.2.1.5 Protozoa (Trypanosomes). These animals also act as agents to spread phloem borne diseases. Sudden wilt of oil palm, Hart rot of coconut palm and Phloem necrosis of coffee are the important disorders caused by genera *Phytomonas*. The clogging of phloem is the most common mechanism involved in the production of wilt-like symptoms in infected hosts.

2.2.1.6 Earthworms. Club root disease of crucifers is caused by *Plasmodiophora brassicae*, a soil inhabiting fungus, the spread of which is facilitated by earthworms.

2.2.1.7 Slugs. These animals are from phylum Mollusca. The spread of many fungi (causing black rot of cabbage- *Phytomonas campestris*) is via slugs.

2.2.1.8 Animals/Man. The spread of the plant pathogens is both internal and external through human/animal activity. The spread of pathogens by animals is via contaminated mouthparts and the pathogen infested body. The mouthparts of animals become contaminated while feeding on diseased plants. The bodies of animals are infested externally with plant pathogens while roaming in infested fields. Rice yellow mottle virus (RYMV) is the first case in which the direct involvement of animals such as the domestic cow (*Bos* spp.), grass rats (*Arvicanthus niloticus*) and donkeys has been demonstrated (Sara and Peters, 2003). The parasitic plants mistletoe and dodder are known to transmit viruses or virus-like organisms owing to their systemic nature. The movement of externally infested animals in the field is also known to transmit pathogens. The spread of pathogens in such situations is entirely mechanical. The spores of many fungal pathogens cling to the body of animals and are carried over long distances. In addition, their movement in the field inflicts injury to plants that enables the entry of pathogens. The spread of highly infectious virus (Tobacco mosaic virus) is caused by the animal's movement in the contaminated field. Besides carrying the pathogens externally, the animals also ingest sporangia/ spores of numerous fungi which are subsequently transported to far-off places and spread via fecal matter. The spread of seed borne viral diseases is caused by animals who consume the contaminated seed. Man is another active agent instrumental in spreading plant pathogens by way of infected seed (CMV) transportation or by the transportation of planting material (PVY/PLRV) over long distances. Plant pathogens are also spread through use of contaminated implements (Banana bunchy tops virus—BBTV) and through mechanical injury (TMV) while roaming in the field. Mechanical injury triggers the passive transmission of plant viruses. Tillage is an operation regularly performed by human beings. The orchardists are also involved in pruning, cutting, budding and grafting in orchards. The use of contaminated implements is a major source of the spread of diseases such as Citrus tristeza virus (CTV) in citrus, and Sugarcane grassy stunt virus (ScGSV) in sugarcane. The transportation of seed consignments is inspected thoroughly and regularly in dedicated quarantine centres in order to control the spread of seed borne pathogens by human beings. Similar regulations are imposed in order to contain the spread of such diseases in planting material. Typical planting material includes tubers, rhizomes, suckers, cuttings of sugarcane/rose/chrysanthemum, and corms. Both domestic and international control centres are working to check the spread of pests and disease between regions and the countries, respectively. To prevent the sowing of virus-infected seed potato, thorough indexing is done. All potato viruses (PVX, PVY, PVS, PLRV, Potato rugose mosaic, etc.) are carried through potato tubers. Many diseases caused by bacteria (common scab, bacterial ring rot), viroids (Potato spindle tuber and Citrus exocortis) and fungi (late blight, powdery scab of potato, black scurf and silver scurf of potato) are also transmissible through diseased potato tubers. Banana bunchy top virus-BBTV (rhizomes), Sugarcane ratoon stunt virus-SRSV and Grassy shoot viruses and red rot fungus (cane cuttings) are spread through vegetative parts. Chrysanthemum stunt (viroid) and Chlorotic mottle virus in chrysanthemum, *Verticillium* wilt in dahlia and brown rot of potato (bacterium) are mechanically transmissible. The diseased seed is transported by man. The spread of seed borne diseases like Lettuce mosaic virus (LMV), Tobacco ring spot (TRSV) in soybean, Barley stripe mosaic virus (BSMV), Southern mosaic virus (all viruses), red rot of sugarcane, dieback of chili plant, Leaf stripe of barley, late blight of potato (fungi), etc., is mostly through human activity. Tillage operations carried by man also contribute to the spread of the bacterial leaf blight of cotton. Watering of rice fields is also known to spread the bacterial blight pathogen with human involvement.

2.2.1.9 Birds. Human disorders are already known to be carried by birds, but the information pertaining to the involvement of birds in the spread of plant pathogens is largely incomplete. There is evidence to suggest that banana bacterial wilt is spread by birds in and around the area of the African Sahara. Banana is a staple in the diet of people in the Saharan region. Birds are known to spread pathogens by carrying disease-infested branches for nest formation. Weaver bird (*Quelea quelea*), sparrows (*Passer domesticus*) and bulbul (*Pycnonotis barbatin*) have been found to be spreaders of Rice yellow mottle virus in Morocco (Peters et al., 2012). Furthermore, the appendages of birds can carry mildew-like fungi from one place to another. Mistletoe is also known to cause destructive disease in conifers and the spread of seed is through those birds which feed on berries/seeds and finally spread the plant matter from one place to other. The spread of chestnut blight is caused by birds as they root through conkers in search of insects, leaving their bodies coated with fungal spores.

2.2.2 Nonliving agents

2.2.2.1 Soil. A vital role is played in the spread of plant pathogens, both through the soil particles and the soil colonizing organisms (soil nematodes and soil fungi). The virus pathogens remain adhered to the charged soil

Table 3.1 Insect vectors of plant viruses from different insect orders.

Sl. No.	Taxon/order	Virus genus (Virus)
1	Aphid (Hemiptera)	*Alfamovirus* (Alfalfa mosaic virus) *Babuvirus* (Banana bunchy tops virus) *Badnavirus* (Citrus yellow mosaic badnavirus) (Needs investigations) *Carlavirus* (Lily symptomless carrier virus) *Caulimovirus* (Cauliflower mosaic virus) *Closterovirus* (Citrus tristeza) *Cucumovirus* (Cucumber mosaic virus) *Cytorhabdovirus* (Lettuce necrotic yellows virus) *Enamovirus* (Potato enation mosaic virus) *Fabavirus* (Broad bean wilt virus-1) *Luteovirus* (Onion yellow dwarf virus) *Macluravirus* (Alpinia mosaic virus) *Nanovirus* (Fababean necrotic stunt virus) *Nucleorhabdovirus* (Strawberry crinkle latent virus) *Polerovirus* (Cereal yellow dwarf virus) *Potyvirus* (Potato virus Y) *Reovirus*/*Rasalvirus*-proposed (Raspberry latent virus) *Sequivirus* (Parsnip yellows fleck virus) *Umbravirus* (Carrot mottle virus) *Waikavirus* (Anthriscus yellows virus)
2	Planthoppers (Hemiptera)	*Cytorhabdovirus* (Barley yellow striate virus) *Fijivirus* (Sugarcane fiji disease virus); *Nanovirus* (Coconut foliar decay virus) *Nucleorhabdovirus* (Maize mosaic virus) *Oryzavirus* (Rice ragged stunt virus) *Rhabdovirus* (Colocasia bobone disease virus) *Tenuivirus* (Rice stripe virus)
3	Leafhoppers (Hemiptera)	*Curtovirus* (Beet curly top virus) *Marafivirus* (Maize rayado fino virus) *Mastrevirus* (Maize striate virus) *Phytoreovirus* (Wound tumor virus) *Waikavirus* (Rice tungro spherical virus)
4	Treehopper (Hemiptera)	*Topocuvirus* (Tomato pseudo-curly top virus)
5	Scale insects (Hemiptera)	*Closterovirus* (Grapevine leafroll-associated virus-3)
6	Mealybugs Hemiptera	*Ampelovirus* (Grapevine leafroll-associated viruses); Badnavirus (Citrus yellow virus) *Trichovirus* (Grapevine trichovirus-A) *Vitivirus* (Grapevine virus A & B)

Table 3.1 contd. ...

...Table 3.1 contd.

Sl. No.	Taxon/order	Virus genus (Virus)
7	Whiteflies (Hemiptera)	*Begomovirus* (Tomato yellow leaf curl virus) *Carlavirus* (Tomato pale chlorosis virus) *Closterovirus* (Tomato chlorosis virus) *Crinivirus* (Cucurbit yellow stunting disorder virus) *Ipomovirus* (Sweet potato mild mottle virus) *Torradovirus* (Tomato torrado virus) Unassigned genera (Citrus chlorotic dwarf virus)
8	True bugs (Hemiptera)	*Carlavirus* (Potato mosaic virus-M) *Polerovirus* (Potato leaf roll virus) *Potyvirus* (Centrosema mosaic virus) *Rhabdovirus* (Beet curly top virus) *Sobemovirus* (Velvet tobacco mottle virus)
9	Thrips (Thysanoptera)	*Carmovirus* (Angelonia mottle virus) *Ilarvirus* (Tobacco streak virus) *Machlomovirus* (Maize chlorotic mottle virus) *Sobemovirus* (Sowbane mosaic virus) *Tospovirus* (Tomato spotted wilt virus)
10	Flies (Diptera)	*Sobemovirus* (Rice yellow mottle virus)
11	Earwigs/ (Dermaptera)	*Tymovirus* (Turnip yellow mosaic virus)
12	Cockroaches (Dictyoptera)	*Potexvirus* (Cymbidium mosaic virus)
13	Grasshoppers/ (Orthoptera)	*Sobemovirus* (Rice yellow mottle virus)
14	Honey bees/ (Hymenoptera)	*Ilarvirus* (Blueberry shock virus)
15	Beetles (Coleoptera)	*Bromovirus* (Brome mosaic virus) *Carmovirus* (Black gram mottle virus) *Comovirus* (Cowpea severe mosaic virus) *Machlomovirus* (Maize chlorotic mottle virus) *Potexvirus* (Potato virus X) *Sobemovirus* (Cocksfoot mild mosaic virus) *Tobamovirus* (Cucumber chlorotic mottle virus) *Tymovirus* (Turnip yellow mosaic virus) Unassigned (Urdbean crinkle virus)

further divided into two sub-orders: Homoptera and Heteroptera. The sub-order Homoptera has two divisions viz. Auchenorrhyncha (leafhoppers, planthoppers and froghoppers/spittlebugs) and Sternorrhyncha (aphids, psyllids, whiteflies, scale insects and mealybugs). Auchenorrhynchans are identified by the origin of stylets being from the ventral side of the face, one to three segmented bristle-like an antenna, three segmented tarsi, and a process is initiated, the insect continues to release saliva at an intermittent rate, which ultimately ends up in the salivary sheath. The insect continues to ingest and release the sap at twenty second intervals.

3.2.1.3 Planthoppers. Planthoppers are from the superfamily fulgoroidea (Auchenorrhyncha) and can be differentiated from other members due to the presence of a Y-shaped, thickened anal vein in the forewing, 3-segmented antennae with bulbous pedicel, and a fine filamentous arista (Plate 3.2 Planthopper vector). The hind tibia is provided with the broad and moveable apical spur. Nymphs have a biological gear meant to connect the hind legs during jumping. This biological gear disappears in adults. The female lays about 300–350 eggs in leaf sheaths covered with plugs. Hatching takes place after about six days. The nymphs are cottony white and undergo five nymph instars in order to become adults in approximately ten to eighteen days. The adult longevity is twelve to seventeen days. The feeding mechanism is similar to other members of Auchenorrhyncha. The leafhoppers, being strong fliers, possess the ability to fly for longer periods of time (seven

Plate 3.2 Planthopper Vector.

hours in *Nephotettix virescens* Ishihara). The planthoppers make use of visual cues and are attracted to the colour yellow. The host-oriented flights of leafhoppers could also be up to 560 nm light wavelength, as in *Dalbulus maidis* (De Long and Wolcott). Like other hemipterans, the leafhoppers, after landing on the plant surface, will select a suitable site for feeding on the abaxial side of the leaf.

3.2.1.4 Whiteflies. Whiteflies are tiny, weak fliers with four wings. The wings are equal in size, opaque with fewer veins and remain covered with mealy powder (Plate 3.3 Whitefly vector). These insects, being hemimetabolous (unlike other hemipterans), contain four developmental stages viz. egg, nymph, pseudo pupa and adult stage. The eggs are stalked and laid near the bases of trichomes on the lower leaf surface. The hatch within a week and release nymphs which pass through four nymph instars. The first instar

Plate 3.3 Cotton Whitefly Vector.

(tiny crawler) is mobile, the other three instars are immobile. Nymphs turn into adults in three to four weeks. The glandular trichome extends shelter to the sessile nymphs. Most crawlers prefer to settle on the abaxial side of the leaf, but they can move and cover a maximum distance of up to twenty cm (Summers et al., 1996; 2004). The adults are present throughout the year.

3.2.1.5 Psyllids. These insects are commonly referred to as jumping lice and fall in the family Psyllidae under the order Hemiptera. Two segmented tarsi with two claws and five to ten segmented antennae are unique to this family, and thus facilitate the identification. Psyllidae forewings are thicker than the hind wings. They have piercing and sucking mouthparts and cause damage by injecting toxins and transmitting plant pathogens while sucking sap. Eggs are laid in depressions in the plant tissues/fresh buds/ folds of leaves. On hatching, nymphs are released which then pass through five instars. These creatures are almost sessile, move a small distance and produce waxy filaments. Nymphs resemble adults except for the wings that are absent in nymphs. After feeding, they turn into adults and sit with raised abdomen (Plate 3.4 Psylla vector). The margins of forewings are brownish, which again facilitates their identification. The life cycle is completed in about fifteen to forty-seven days. The feeding mechanism is similar to that of other homopterous insects.

Plate 3.4 Psylla Vector.

3.2.1.6 Mealybugs. Mealybugs are from the family Pseudococcidae, in the order Hemiptera. Mealybugs are generally observed as being sexually dimorphous. The males are winged while the females are wingless. Males are born without a functional mouth and therefore eat very little. Females are pinkish in color, oval shaped, about 3 x 1.5 mm in size, and their body is always covered with a white, mealy powder. The wax glands, which are present on the dorsal side of the abdomen, allow the females to produce this mealy powder. Females have a small fringe of filamentous material around the body margins. A pair of long filaments at the end of the abdomen is peculiar in females. The legs are well developed. The female lays eggs in the ovisac attached to the ventral side of the body. Around 300 eggs are laid in about one or two weeks. These eggs hatch into crawlers within ten days. The crawlers, after moving for a day, fix themselves in one spot and become sessile, whereupon they proceed to feed and drain sap from the plant without moving. The crawlers are yellowish to blood red. The females die after laying eggs. The life cycle is normally completed in about one to two months. They have piercing and sucking mouthparts.

3.2.1.7 Scale insects. Scale insects are separated into two categories viz. soft scale and armoured scale. The soft scale insects have smooth scales, measuring ¼ of an inch, with a cottony/waxy surface, round and more convex than those of armoured scale insects. They belong to the Coccidae family. Armoured scale is flat, plate-like, measuring 1/8th of an inch, belonging to the family Diaspididae. Of these two categories of scale insects, the armored scale ones do not secrete honeydew, while secretion of honeydew is a common characteristic of soft scaled insects. Females are wingless while males are winged, as with mealybugs. Both sexes have compound eyes and ocelli. In armoured scale insects, the scale cover is independent of the body, legs are absent, beak is segmented and antennae are rudimentary. Whereas in soft scale insects, the females are flat with elongated oval bodies and a tough exoskeleton. The body covering is smooth or waxy, legs present, or absent and the antenna is also rudimentary or absent. Females lay eggs under the shell cover. The hatching takes place within one to three weeks and nymphs known as crawlers are released. The crawlers have legs to move. The movement takes place under the shell and these scale insects later settle in a suitable site. There are two nymph instars. Scale insects feed on leaves during the crawler stage, and devour twigs and branches during the second instar stage. Thus, their dispersal is only in the crawler stage. After settling, the crawlers lose their legs and continue to feed under the shell. The spread of scale insects also occurs via air, animals, and birds. The mechanism of feeding similar to that of the other hemipteran insects. These insects are also associated with the transmission of plant pathogens, but more detailed research is needed. Scale insects have been implicated in the spread of grapevine leafroll viruses. Besides the plant viruses, these are also labelled as transmitters of viroids. The disease called "Tinangaja" (caused by Viroid coconut tinangaja viroid) is prevalent among scale insects and mealybugs in nature. It can be identified with the presence of dwarf crown, small nuts without kernels and the stippling of leaves (Hodgson et al., 1998). A detailed investigation is still needed in order to confirm the involvement of scale insects in the transmission of this viroid.

3.2.1.8 Froghoppers/Spittlebugs. Spittlebugs are also known as froghoppers, due to their resemblance to frogs. This grouping is like the leafhoppers but can be differentiated by the presence of one or two stout spines on the hind tibia and a circle of stout spines at the apex. Spittlebugs have short and conical hind coxae, while the number of stout spines on the hind tibia is large among leafhoppers. Likewise, the spine formation is an enlargement of scutellum in froghoppers but a stout thorn-like spine on the thorax is more typical of treehoppers. Spittlebugs also have the same mouthparts as that of the other hemipterans. The life cycle is simple. The female lays eggs under the bark of trees in October and remains in an overwintering phase until March. The hatching of eggs takes place in March and April. The newly hatched nymphs develop into adults after passing through five instars. They continue to feed till October and complete the cycle in about forty-five to fifty days.

3.2.1.9 Treehoppers. Adults of treehoppers are identified by their enlarged pronotum, stout spines and a resemblance to leafhoppers. The pronotum, an enlarged prothorax, covers the head, thorax, and abdomen. The jumping hind tibia and two ocelli are important characteristics helpful in identification. Wings largely remain concealed under the pronotum. Eggs are covered with froth. After hatching, the nymphs undergo five molts to become adults. The adults live a gregarious life. Treehoppers have one or more generations in a year and are phloem feeders.

3.2.1.10 Heteropterous bugs (true bugs). These insects belong to the suborder Heteroptera and are identified by the presence of hemelytra (membranous forewings with a hardened consistency). Adults have trapezoidal pronotum and triangular or semi-elliptical scutellum covering half of the body. The antenna and tarsus are five and three-segmented, respectively. The legs are thin and smooth. The bugs have stink glands which emit a pungent fluid in order to deter their natural enemies. The metamorphosis is incomplete. The eggs are laid by bugs and, after hatching, give rise to nymphs. The nymphs reach maturity after passing through five instars. The first instar does not feed, but consumes the bacteria attached to the outer shell of the eggs. The eggs, while passing through the ovipositor, get smeared with bacteria and those bacteria serve as food for the first instar nymphs. These nymphs resemble their adult counterparts, but lack the wings that develop in the later stages of their life cycle. The development of wings starts from third instar nymphs. After becoming adults, they feed on vascular tissues by using both food and salivary canals. The adult longevity of males and females is between twenty-two to thirty-two and thirteen to forty-six days, respectively, depending on the temperature conditions. The total life cycle from egg to adult stage is completed in about twenty-three to forty-three days. The saliva of bugs is toxic which makes these insects less efficient vectors of plant pathogens. The toxic saliva is responsible for the necrosis of leaf lamina and the resulting dead tissues are unsuitable for the replication of obligate parasites like viruses.

3.3 Thysanoptera

3.3.1 Thrips

Thrips are small creatures with a body length of one to two millimetres. The wings have fringes of long hairs called bristle wings (Plate 3.5 Thrip vector). Thrips belong to the order Thysanoptera and have the rasping and sucking mouthparts. The mandibles are uneven, with the right-hand one being smaller in size. The antennae are six to nine-segmented. The adult movement is swift. These insects insert eggs in the plant tissues with the help of a serrated ovipositor. Eggs hatch in about three to four days and the nymph stage is ten to fourteen days in length, with a prepupal/pupal stage of seven days. The life cycle is completed within fifteen to thirty days, depending upon the species and the environmental conditions. They have both feeding and non-feeding stages, i.e., nymph stage and pre/pupal stage, respectively. Adult longevity is around four to five weeks.

3.4 Diptera

3.4.1 Dipterous flies

Four families, namely Agromyzidae, Anthomyidae, Diopsidae and Cecidomyiidae, contain vectors of plant pathogens. Metamorphosis in dipterous flies is complete. These insects have chewing mouthparts and they measure about two to three millimetres in length. Adults have one pair of wings (the hind pair being modified into halters), coloured frontal vitta on the head, oval compound eyes and hyaline wings. The maggots (larvae) are legless and taper at both ends. The eggs are laid on the under surface of

Plate 3.5 Thrip Vector.

leaves. Maggots also have biting and chewing mouthparts which are used to feed on leaves and make conspicuous mines in the leaf. Pupation takes place in the mines or in fallen leaves. The infant stages of these insects cause damage by making mines in the leaves and transmitting plant pathogens.

3.5 Coleoptera

3.5.1 Beetles

The families Durculionidae, Scolytidae, Melonidae and Chrysomelidae under the order, Coleoptera are associated with the spread of plant pathogens. Of these, Curculionidae, and Chrysomelidae are known to spread bacterial wilt of cucurbits in nature. Scolytidae and Melionidae beetles are vectors of fungi, viruses, bacteria and nematodes. The beetles are 3.5–12 mm long and have brightly coloured oval and convex shaped bodies. Their forewings are sclerotized to form a hardened cover for the hind wings. This protective shield is known as an elytron. The sheath-like elytra cover the soft hind wings over the dorsal surface of the abdomen. These are not used during flight. The body of beetles has two parts, separated by a moveable constriction. Larvae with large heads and prominent thoracic legs are called grubs. These insects also have biting and chewing mouthparts. Their metamorphosis is complete, and the life cycle comprises of four stages, i.e., egg, larval, pupal and adult stage. The beetles lay eggs in the galleries made in tree trunks, in the soil near a food source, or directly on a plant depending upon the species of beetle. In all, a total of sixty to 100 eggs are laid; they hatch within a week and newborn grubs emerge. The grubs pass through three to five instars to pupate. The pupal stage lasts for about seven to ten days, after which the adult beetle emerges. The total life cycle is completed in around two months. The communication among beetles for mating purposes is governed by pheromones, especially in bark beetles. These beetles feed on plants between tissues of parenchyma and vascular bundles. They chew the leaves and plant matter, leaving sizeable holes in their wake. Some beetles are scavengers, while others are predators that hunt a variety of insect species.

3.6 Hymenoptera

3.6.1 Ants, bees, and wasps

The order Hymenoptera contains two families responsible for the spread of plant pathogens. The Hymenoptera are identified by their having two wings, the forewings being larger than the hind wings. Their wings are netted with cross veins which form cells throughout the structure. The antenna is ten-segmented or more, and is longer than the head. The females have a conspicuous ovipositor which has been modified to hold a stinger.

Bees and ants are included in the sub-order apocrita, in which the thorax and abdomen are separated by a constriction. The families Apidae (bees) and Formicidae (ants) are include many of these social insects. Though they belong to the same order, they both possess different mouthparts. Ants have chewing mouthparts, while honey bees have specialised mouthparts meant for lapping and sucking the nectar from flowers. While visiting the flowers for the collection of pollen, the bees are completely smeared with pollen. The bees then proceed to visit several other flowers, irrespective of the status of these plants. If a plant visited by bees happens to be a diseased one, the bees can pick up the pathogens and transmit them to healthy plants. Ants are responsible for transportation of insects from one place to another. Like the bees, their bodies can become contaminated with fungal mass and ooze of many bacteria. These insects move between diseased and healthy plants in nature, thereby spreading any plant pathogens they come into contact with. Both categories of insects are social and live in colonies. Ants are 0.8 to five millimetres in length, with compound eyes, and their bodies are divided into head, thorax, and abdomen. Both fertilized and unfertilized eggs are laid by females. Females are produced from the fertilized eggs, while males are from the unfertilized eggs. Eggs hatch within two to three weeks. The hatchlings spend three weeks as legless larvae and then pupate. The pupation takes place in cocoons. After ten days the adult emerges, and the life cycle is completed in about six weeks to two months. Bee colonies have a queen, king, drones, and workers (females). The strength of the female workforce can be between twenty and eighty thousand bees. The queen's lifespan is around two to five years, during which she will lay up to 1500 eggs each day. Egg hatching takes place after three or four days, followed by the larval phase which lasts about nine days. Pupae grow into adults in about ten days. Drones longevity is around four to five weeks.

3.7 Orthoptera

3.7.1 Grasshoppers

Grasshoppers are members of the order Orthoptera. They have two pairs of wings; the forewings are narrow and of leathery consistency (tegmina) and the hind wings are large and membranous. Hind legs are powerful and the femur is hard and ridged. The inner ridges are sometimes provided with stridulatory pegs. The posterior edge of femur is lined with a double row of strong spines. The legs are modified for jumping. Grasshoppers have well-developed stridulatory organs. The metamorphosis is incomplete, with the insect having egg, nymph and adult stages. Clusters of eggs are laid in the soil and are glued together with froth in a pod near the roots of food plants. The eggs are laid in summer and overwinter during the egg stage. The hatching of eggs begins when suitable temperatures are reached as the weather becomes warmer in spring. The ideal temperature varies depending on the species of grasshopper. The nymphs that hatch resemble the adults, except for the wings which are absent in nymphs. In the place of wings, nymphs have wing pads. This nymph stage lasts for twenty-five to thirty days. After feeding on plants, the nymphs grow into adults and the adult longevity is approximately fifty days.

3.8 Lepidoptera

3.8.1 Butterflies and moths

These insects are from the order Lepidoptera. Their body and appendages are always covered with modified flattened hair known as scales. The larvae are called caterpillars and each have between one and eleven pairs of prolegs. They have long, coiled siphoning mouthparts meant for feeding on flowers. These adult butterflies are harmless pollen vectors; however, the larvae are responsible for massive economic damage since caterpillars have mandibulate mouthparts designed to chew on foliage. Metamorphosis is complete, with the insect having egg, larval (caterpillar), pupal (chrysalis) and adult stages. The adult butterflies copulate on emergence and then generally lay between 200 and 600 eggs on suitable plants. The eggs are small and round, with ridges/furrows on them and they hatch within two to three days. The caterpillars feed on foliage for about twenty-five days and then retreat into a chrysalis. In this stage, they do not feed, and pupae have sculptured integument that are not enclosed in a crown. The pupae are found hanging by threads from tree branches. Similarly, moths also lay eggs in clusters on young leaves, and the fecundity is as high as 1000 eggs in some insects such as *Helicoverpa armigera* (Hubner). The eggs hatch within three to five days and the neonates (first instar) feed by scraping the leaf lamina. They pass through five instars in about twenty days and pupate in soil or fallen debris. The pupal stage lasts for about ten days, after which, adults begin to emerge. Moths are active during the night, while butterflies remain active during the day. In both cases, the caterpillars are destructive due to their biting and chewing mouthparts.

3.9 Dermaptera

3.9.1 Earwigs

These insects are from the order Dermaptera, in which metamorphosis is incomplete. Earwigs are from this order and have chewing mouthparts. They measure around 3/4th of an inch in size and are identified by their reddish-brown, slender and flat bodies. The tip of the abdomen has a pair of forceps-like cerci (pincers) which are used to catch their prey. There are two pairs of wings, the forewings are tough, leathery and shorter than the hind wings. The hind wings are large, fan-like and plated. Adult earwigs overwinter in the colder months and then in February lay around twenty to fifty whitish eggs in the soil. Eggs hatch into nymphs within seven days. The nymphs become adults after passing through four to six nymph instars. The longevity of adults is up to one year. There is only one generation in a year. These insects are associated with the spread of plant pathogens.

3.10 Dictyoptera

3.10.1 Cockroaches

These insects belong to order Dictyoptera and possess chewing mouthparts and legs meant for running. These insects inhabit concealed places in houses and mainly feed on kitchen waste, but there are some species of cockroaches that also feed on plants. The Australian cockroach, *Periplaneta australassiae* fabricius has been identified as a pest of plants and is associated with the transmission of stylet-borne viruses among orchids (Allen, 2010). It has been identified as a vector of Cymbidium mosaic virus (CymMV) in orchids while feeding on flowers roots and fresh foliage. The life cycle, in general, is simple and metamorphosis is incomplete in this order. The eggs are laid in a sac-like structure (ootheca) that protrudes from the female abdomen and can hold around fifty eggs. The hatching takes place within fourteen to thirty-five days. Once detached from the abdomen the eggs are hatched within 24 hours. The nymphs become adults after passing through six to seven instars in about six to thirty-one weeks. The female longevity is around twenty to thirty weeks. The life cycle varies depending on the species and the weather conditions.

3.11 Feeding Mechanism of Hemipterans

Hemiptera is the largest order in terms of the number of categories of vectors. These insects are efficient vectors of plant pathogens. More than eighty per cent of these vectors of plant pathogens are in the sub-order Homoptera. It has two divisions as mentioned earlier. The members belonging to Auchenorrhyncha are feed in an intracellular manner while Sternorrhyncha feeding is intercellular. Of these vectors, forty per cent are grouped under Auchnorrhyncha and the remaining sixty per cent are classed as Sternorrhyncha. To understand the mechanism of transmission of plant pathogens, it is essential to have an understanding of the morphology, anatomy and other systems involved in the transmission of pathogens.

3.11.1 Sternorrhyncha

This group of insects includes aphids, whiteflies, psyllids, mealybugs and scale insects. Of these, the aphids are known to transmit the most plant pathogens. Therefore, the feeding mechanism of hemipteran (Sternorrhyncha) insects has been discussed, using aphids as an example. The mouthparts are an efficient way of virus acquisition and transmission of plant viruses. The mouthparts consist of a pair of mandibles, a pair of maxillae, labium, and labrum. Aphids have tactile hairs and chemoreceptors on the tip of the stylet, and can be alate or wingless. The alate aphids pass through different stages so as to successfully transmit the pathogen in nature. They fly from one place to another in search of food and different ecological conditions. For the alates to land on the substrate, their first step is to make use of visual and chemical cues in order to select a suitable site for feeding. They settle on the substratum only once they have identified suitable conditions. Wingless aphids are already on a plant and search within it for a suitable feeding site. These insects have to interlock maxillary stylets to form the salivary sheath for feeding. The interlocking ridges and grooves permit movement of the maxillary stylets over each other. The aphids always feed in the salivary sheath but before feeding, the saliva is secreted by the salivary glands for the formation of such a salivary sheath (Miles, 1972). Aphids have a pair each of principal and accessory glands associated with ducts and syringe. The contraction and relaxation of muscles control the salivary secretions. This guards the fragile stylets and avoids leakage of fluid while feeding. The aphids probe through these salivary sheaths. Probes continue until the stylets reach the target site (xylem, phloem, mesophyll cells, etc.) and feeding commences. During feeding, saliva continues to pour into check blockade of stylets. The contents of the phloem are rich in sugars; xylem sap, however, is extremely poor in nutrients. Since the sap being sucked by insects is imbalanced in terms of nutrients, the insects' anatomy is modified accordingly. In some insects, the midgut is provided with a modified filter chamber to separate and drain out the excess water from the sap. While in others, non-coiled midgut is present instead of a filter chamber. It encloses cells meant for the absorption of nutrients from the gut content, or harbors endosymbionts in mycetocytes which convert sap into nutrient-rich food. Hemocoel is found in aphids and the mycetomes are present in hemocoel, harboring the symbiotic organisms in special cells called mycetocytes.

Whiteflies feed on the phloem using their piercing and sucking mouthparts. The modified mouthparts penetrate through cuticle, epidermis and mesophyll and are able to reach the sieve tubes of phloem. The whiteflies also use tactile and gustatory cues to select suitable sites for feeding or oviposition; colour is also an important agent in the selection of plant species. After settling, but before feeding, a watery discharge is released on the surface of the leaf. The watery discharge helps to dissolve waxes on the leaf surface (Miles, 1999). The whiteflies also feed in the salivary sheath and the salivary sheath guides the movement of stylets in the leaf tissues while feeding. The interlocking of maxillary stylets forms the food and salivary canals. The whiteflies, unlike aphids, do not puncture the mesophyll; instead, they target the vacuoles and apoplast of cells (Kempema et al., 2007). On reaching the target, the feeding wounds are plugged with cellulose or proteins in order to halt the backflow of sap into the apoplast. They cause damage by sucking the sap and spreading plant pathogens.

3.11.2 Auchenorrhyncha

This order is further subdivided into two sub-orders viz. Auchenorrhyncha/ fulgoromorpha (planthoppers) and clypeorrhyncha/cicadomorpha (leafhoppers, treehoppers and sharpshooters (froghoppers/spittlebugs). The mouthparts of this group are also similar to the Sternorrhyncha group of vectors. The stylet is formed by two mandibles, two maxillae, cone-shaped labrum and three-segmented labia. The interlocking of maxillary stylets forms large food canals and small salivary canals. The feeding of these insects takes place in the salivary sheath which is formed by the salivary secretions released by salivary glands through the salivary canal. In Auchenorrhyncha hoppers, the distal end of the mandibles has five teeth on the external side. Each maxilla (2) and mandible (1) forms dendritic canals. The labial tip is a rosette in planthoppers (Dai et al., 2014) and has trichomes, uni-porous pegs, and basiconic sensilla along with two sub-apical labial sensory organs. The labial tip also has dorsal and ventral folds of sensory organs.

The mouth structure in Auchenorrhyncha is almost the same, but the structure of the mouth of spittlebugs in particular is slightly different from other members of the order. The maxillary stylets are smooth externally and form food and salivary canals, similar to other members. The mandibles of spittlebugs are carved in such a way so as to form a smooth dorsal region and a grooved ventral side (tooth region) near the tip of the convex part. This structure differentiates this category of insects from others. The various types of sensilla viz. trochoidal, basiconic, and multipeg are present in the labium tip region along with sensory folds. Ten small peg sensilla are arranged in 5 + 4 + 1 pattern with one large peg sensilla (Wang et al., 2015).

Like the aphids, leafhoppers have stout piercing and sucking mouthparts and feed intracellularly. Leafhoppers inject saliva into the leaf tissues and ingest fluid. Visual cues help the insect to select the crop to settle in, while chemical sensory organs present on the tip of labium help the insect to judge the chemical nature of sap. Before feeding begins, the salivary sheath is formed so as to avoid leakage of sap and watery saliva is secreted to prevent clogging in the opening of the stylet and to permit smooth movement of mandibles. The feeding lasts for a few seconds to many hours depending on a number of factors. The stylet reaches the target site within thirty seconds. These insects can withdraw their stylets quickly with slight disturbance without any damage to stylets. Leafhoppers are generally xylem or phloem feeders. The xylem feeders consume sap that is poor in nutrients and so to make the diet balanced one, these creatures are provided with the suitable structures. The filter chamber, coiled midgut, and symbiotic bacteria are essential for augmenting the richness of their diet. They have two mandibles and two maxillary stylets, and the interlocking of the latter is responsible for forming food and salivary canals.

Like hemipterans, thrips cause conspicuous feeding punctures (whitish/brownish marks on the plant parts) with their rasping and sucking mouthparts. The reduced and non-functional right mandible makes the mouth highly asymmetrical. While feeding, the adults rock the head up and down, two to six times, in order to insert their stylets. The food canal in the maxillary canal enters into the tissues through the intercellular wall. Saliva is injected, which is followed by the rasping of cellular content. Once the stylets are withdrawn, there is a leakage of material. The total destruction of the mesophyll and epidermal cells and distortion of the cuticle are important characteristics of thrip feeding.

3.12 Feeding Mechanism of Chewing Insects

3.12.1 Grasshoppers

Grasshoppers have chewing mouthparts, feed voraciously and inflict great injury on plants. With the help of their chewing mouthparts, the grasshoppers chew the food, which is mixed with salivary fluid in their buccal cavity in order to ease the digestion of the food/carbohydrates. The food taken into the crop meant for storage and the digestion continues. From crop, the storage organ, the food is passed on to the gizzard. It is a structure which has teeth-like plates to grind the food particles. After grinding, the food is passed on to the stomach where it is acted upon by the enzymes released by hepatic caeca to digest it properly. The tube-like structures at the junction of the midgut and hindgut (malpighian tubules) perform the excretory function; excretions such as urea, uric acid, and amino acids are eliminated here. The nutrients are absorbed in the ileum and undigested food is transferred to the colon. Water is absorbed and the solid is passed on to the rectum, to be excreted as fecal matter in the form of pellets. The insect families Acrididae, Gryllidae, Tettigoniidae, Tetrigidae, and Pyrgomorphidae are vectors of plant pathogens. These insects are regarded as poor vectors of plant pathogens. Rice yellow mottle virus is transmissible through grasshoppers.

Table 3.2 Categories of insect transmission mechanism of plant viruses (modified from Casteel and Falk, 2016; Katis et al., 2007).

Sl. No.	Character	Non-persistent stylet-borne	Non-persistent foregut-borne	Persistent circulative	Persistent propagative
1	Acquisition/inoculation access	Seconds to minutes	Minutes to hours	Hours to days	Hours to days
2	Tissues for acquisition	Epidermis/mesophyll	Epidermis/mesophyll	Xylem/phloem	Xylem/phloem
3	Tissues for inoculation	Parenchyma	Parenchyma/phloem	Phloem	Phloem
4	Effect of pre-acquisition fasting	Positive	Positive	Nil	Nil
5	Latent period	Nil	Nil	Hours to days	Weeks
6	Virus in vector hemolymph	Nil	Nil	Yes	Yes
7	Multiplication of virus in vector	Nil	Nil	Nil	Yes
8	Retention in vector body (Time)	Minutes (lost during molting)	Hours (lost during molting)	Days to weeks	Life
9	Retention (half-life)	Minutes	Hours	Days to weeks	Weeks to months (Life)
10	Vector taxon	Mostly aphids	Aphids	Whiteflies	Leafhoppers
11	Specificity	Low	Moderate	High	Very high
12	Transovarial transmission	Negative	Negative	Negative	Generally positive
13	Trans-stadial transmission	Nil	Nil	Yes	Yes
14	Kind of symptoms	Generally mosaic	Generally mosaic	Leaf curl type	Yellows/phyllody/witches broom
15	Hemiptera vectors (numbers)	168	41	141	33
16	Mechanical transmission	Positive	Positive (some)	Negative (except PEMV-2)	Nil
17	Seed transmission	Some	Nil	Nil (except TYLCV)	Nil

18	Virus genera	*Cucumovirus/Potyvirus/ Macluravirus/Luteovirus/ Closterovirus*	*Caulimovirus/Closterovirus/ Badnavirus/Tymovirus/ Tombusvirus/ Trichovirus/ Nepovirus*	*Luteovirus/ Geminivirus/ Nanovirus*	*Marafivirus/Tospovirus/ Tenuivirus/Reovirus/ Rhabdovirus*
19	Common (examples)	Alfalfa mosaic virus, Chili mosaic virus, Soybean mosaic virus, Cowpea mosaic virus	Parsnip yellows fleck virus, Beet yellows virus, Cauliflower mosaic virus	Carrot mottle virus, Banana bunchy tops virus, Citrus trristeza virus, Potato leafroll virus, Cotton leaf curl virus	Sowthistle yellow vein virus, Strawberry latent crinkle virus, Coriander feathery red vein virus

The salivary secretions, while passing through the salivary duct, carry the detached the virus from acrostyle. The virus is ejected through the stylet into test plants. The potyviruses transmissible through aphids encode the Helper component proteinase (HC-Pro) required for the transmission of the virus. In CaMV, P2 protein is encoded and acts as a bridge between aphid vector protein and the capsid protein of the virion of the virus. The virus-encoded proteins are responsible for detecting receptor sites suitable for the attachment of the virus in the stylet; one such protein has been identified (P2) in Cauliflower mosaic virus during infection. It interacts with a non-glycosylated protein found embedded under chitin in the stylet tip. In case of other non-persistent viruses (cucumoviruses), the attachment of the virus is directly within the cuticular linings of mouthparts and does not requires any helper component for mechanical transmission. The virus is carried on the distal tip and proximal region of maxillary stylets (Martin et al., 1990; Wang et al., 1996). It was detected through the formulation of a mutant. The mutation P2-GFP was created, in which the P2 protein is meant to bind the virion to the stylet of an aphid in CaMV (Blanc et al., 2014; Hoh et al., 2010; Plisson et al., 2005). This was tested using fluorescent techniques. The mutant P2-GFP highlighted the fact that the fluorescence was not randomly distributed in the vector species stylet, rather, it was found on the specific tiny region where the non-glycosylates proteinaceous receptors for binding were located in the stylet. That tiny spot which showed fluorescence was later identified and named "acrostyle". This binding was abolished by proteinase K but not by trypsin pronase E, n–hexane, chloroform-methanol or sodium metaperiodate.The above procedure demonstrated that the stylet-borne viruses did not enter into the blood and that the binding of the virus CaMV is undone under the action of enzymes present in the saliva and the virus is subsequently ejected into test plants. This process, therefore, strengthened the concept of ingestion–egestion mechanism (Harris, 1977; Harris et al., 1981). With further advances in science, another theory based on ingestion–salivation was put forth (Martin et al., 1997).

Non-persistent foregut-borne viruses: In this category, the pathogen is picked up within the stylets and carried in the foregut then ejected along with salivation into plants. There is no latent period and no multiplication of these viruses in vectors. However, these viruses are picked up from the source in a slightly longer acquisition process. Vector takes longer (minutes) to transmit the pathogen. The longer the acquisition access on the source, the more the transmission is efficient. The virus does not pass beyond the anterior region of the alimentary canal into hemolymph and salivary glands. It is inoculated by the vector through salivation into new healthy plants. According to this theory, the virus is acquired by ingestion and deposited on multiple sites in the anterior alimentary canal. The virus titer attached to the proximal tip of maxillary stylets is deposited back through salivation via fused food and salivary canals, eliminating regurgitation. In semi-persistent viruses, CP is an important factor in the transmission of whitefly-borne criniviruses (e.g., LIYV). In the case of Lettuce infectious yellows virus, minor coat protein (CPm) was identified using fluorescence techniques as essential for transmission of foregut-borne viruses transmitted by whiteflies. With respect to biotypes, it was also demonstrated that the fluorescence was present in vector biotype-A of *B tabaci* and absent in B (nonvector biotype). Out of the various proteins (CP, CPm, HSP70h, P59) tested, the CPm was identified as the only binding element present. On treating these proteins (CPm and CP) with antibodies, the transmission occurred in CP, whereas, the effect of transmission was neutralized under CPm and thus no transmission of LIYV occurred. This kind of testing highlighted the requirement of CPm in the transmission of this virus. LIYV is a CP mediated, foregut-borne virus, the occurrence of which was during regurgitation rather than in salvation (as in aphid vector of CaMV) (Ng and Zhou, 1990). The vector specificity is low in the non-persistent type of aphid-borne plant viruses, as SMV and CMV are transmissible through thirty-two and sixty species of aphids, respectively (Irwin and Goodman, 1981; Kennedy et al., 1962). The latent period is absent in the body of the vector. The virus is retained in the body of the vector for several hours, but the pathogen does not multiply in the vector. In the foregut-borne category, the virus concentration in the region of attachment of virus is still not known. In foregut-borne mechanism, the virus-containing material is attached to epicuticle linings of the anterior part of the alimentary canal. This mechanism was agreed upon (cucumoviruses-Cucumber mosaic virus) taking into account aphid, leafhopper and nematode vectors with piercing and sucking mouthparts. Later on, beetles were discovered as vectors of plant viruses with chewing mouthparts. Aphids, whiteflies, leafhoppers, and nematodes transmit pathogens from the non-persistent category. Prolonged acquisition access on virus source shreds the virus from the stylets. In potyviruses, Helper component (HC-pro) is essential for facilitation and retention in stylet by bridge formation between *Potyvirus* CP and aphid protein in stylet CMV 2b (Wang et al., 1996). The coleopteran beetles feed on plant parts and inflict wounds. Beetles place the biological pathogens on wounds, these pathogens get into the xylem tissues or invade cells in close proximity to wounds with regurgitation (Gergerich, 2001). Thus, beetles acquire and place the pathogen on the wound by regurgitation that most of the time gets inactivated, so there is no transmission.

Persistent circulative viruses: In circulative viruses (Luteoviridae, Geminiviradae, and Nanoviridae), the pathogen follows a definite path, i.e., stylet—food canal, alimentary canal (foregut, midgut, and hindgut), hemocoel—salivary glands, salivary canal, back to stylet during probing (Fig. 3.3). In luteoviruses, geminiviruses, and nanoviruses, the virion is taken

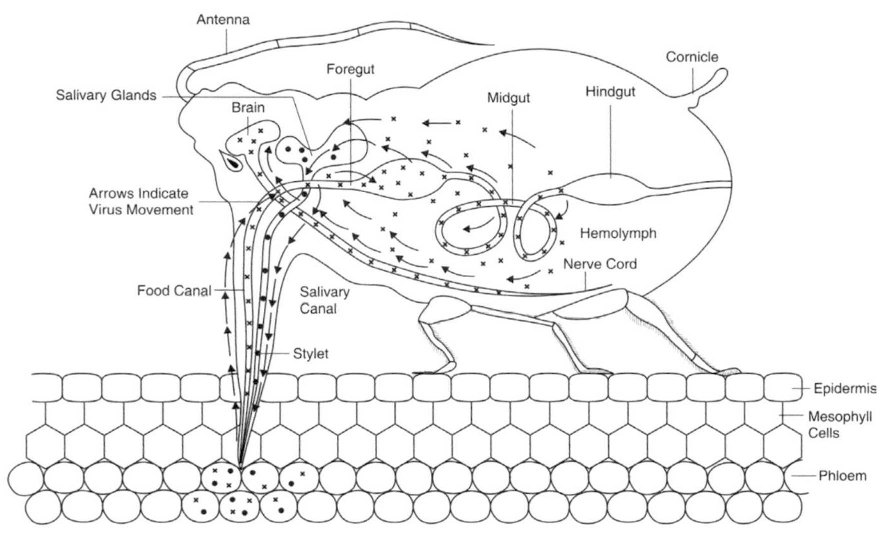

Red crosses show virus in sap; blue solid circles show virus with saliva
Path of persistent-circulative virus-stylet-alimentary canal-hemolymph-salivary glands and inoculations through stylets
Path of persistent-propagative virus is same as above in addition to passage through nerve cord

Fig. 3.3 Pathway of Persistent-circulative and Persistent propagative Viruses in Insect Vectors.

along the alimentary canal and adsorbed by the epithelial surface to make its way to the hindgut or midgut through receptor mediated endocytosis. After this process, the virions are delivered to hemocoel through exocytosis and ultimately get into the membranes of salivary glands (Gray et al., 2014; Reinbold et al., 2003). The plant viruses are responsible for physiological changes in their hosts, which ultimately brings about behavioural changes in the vector. CP and RTP are known to play a great role in their transmission through vectors. CP is meant to transcytose the virion in/out from gut to hemocoel and CP-RTP is needed for passing on the virion to accessory salivary gland membranes. Of these families, the viruses belonging to Luteoviridae and Nanoviridae are vectored by aphids while transmission of Geminiviridae viruses is via whiteflies or treehoppers (Topocuvirus-Tomato pseudo curly-top virus). In the body of the vector, the barriers such as midgut infection, dissemination in the body, salivary gland and transovarial barriers are encountered at different stages and crossed with the aid of transmission determinants in order to reach the salivary glands (Ammar, 1994; Hardy, 1988). From the salivary glands, the virus is passed on to healthy hosts through saliva in the salivary duct. The virus does not multiply in the body. It has a definite latent period before inoculation to new host plants. In this category, Banana bunchy tops (*Pentalonia nigronervosa* Coquerell), Groundnut rosette (*Aphis craccivora* Koch) and Potato leafroll viruses (*Myzus persicae* Sulzer) are aphid-transmissible viruses. In non-persistent circulative viruses, the acquisition and inoculation accesses happen in minutes/hours; retention of virus is for comparatively longer in the body of vector but without multiplication. The major component of the virion is identified as CP-Read through protein (CP–RTP). This CP is considered as a major determinant of luteoviruses and is capable of delivering the virus in hemocoel alone. The mutations in CP-RTP are known to neutralize the transmission. The begomoviruses (TYLCV-whitefly-borne) and *Mastrevirus* (MSV-leafhopper borne) are released in the gut wall, travel into hemocoel and reach the primary salivary glands. Symbionts do play a role in case CP is the sole determinant in geminiviruses. These geminiviruses require CP for transmission (Wang et al., 2014). In these genera, Heat shock proteins 70 (HSP 70) and GroEL.chaperone proteins are essential for transmission and stabilization of viruses, respectively (Rosen et al., 2015; Ghanim, 2014).

Persistent propagative viruses: In this group, the pathogen follows the same path as in circulative viruses, in addition to its multiplication and passage through eggs. Important plant viruses included in this category are BYMV, LNYV, and SCV. These are all aphid-borne in nature. CP and RT glycosylated proteins in Beet western yellows virus are known to prevent degradation of the virus by interacting with proteins produced by endosymbionts. In persistent propagative (Bunyaviridae; Rhabdoviridae), the pathogen multiplies inside the body of the vector. The virus genera in the propagative category which are transmissible through leafhoppers, planthoppers, and aphids include reoviruses, tenuiviruses, and marafiviruses. Tospoviruses (Bunyaviridae) are transmissible through thrips. The family Bunyaviridae has 150 virus species belonging to five genera. Of these genera, only *Tospovirus* contains plant pathogens and the most important member is thrip-borne Tomato spotted wilt virus. The virus virion has two glycoproteins embedded under the outer envelope. The pathogen acquisition takes place over several hours/days; the latent period lasts for weeks; virus retention is for life and transovarial transmission is positive. These viruses pass through gut and hemolymph and salivary tissues to reach the salivary glands for transmission (Fig. 3.3) (Bragard et al., 2013). Similarly, under propagative viruses, the transmission determinants (CP P2) are responsible for the movement and its successful transmission through clathrin coated vesicles as in the case of Rice dwarf virus. In addition, non-structural protein (Pns 10) also plays a role in the movement of viruses in plants. It is known to interact with actin and spread in the vector body of *Nephotettix cincticeps*. Another example (RBSDV), from the Reoviridae family, has been quoted to explain the role of transmission determinants through planthoppers, *Sogatella furcifera*. The role of non structural protein (NSP-P7-1), has been demonstrated. The plant viruses are dependent on arthropods for their spread. For this purpose, these must be retained in the body of the insect vector. To retain the virus at a specific site, structural proteins for virus placement in the body of the vector is a must. Besides, non-structural proteins are now also known to create a bridge between virion and insect vector (Whitfield et al., 2015). Such advancements in science resultantly paved the way for a virus to encode specific structural proteins on the surface of the virion for successful transmission. In addition, the binding of the virus at sites and its retention is governed by determinants of transmission. Glycoprotein (Gn/Gc) determinants are essential for the entry of a virus and its transmission through thrips (*Tospoviruses*) via binding through midgut and blocking entry of virus in the midgut (Sin et al., 2005; Whitfield et al., 2005; Garry and Garry, 2004). TSWV is known to generate non-structural movement proteins (NSm) for movement, but such proteins have not been recorded in thrip vector, *Frankliniella occidentalis* (Mann and Dietzgen, 2014; Lewandowski and Adkins, 2005; Storms et al., 1995). The completion of the virus cycle is dependent upon the transmission determinants, the details of CP and HC under categories/virus genera are presented. Tenuiviruses are non-enveloped and encode glycoproteins that aid in the uptake of the virus by serving as a helper component. Rice grassy stunt virus (RGSV) in *N lugens* invades the primary salivary glands and accessory salivary glands but not the ovarioles of the vector, while Rice stripe virus (RSV) is found in ovarioles and primary salivary glands in *Laodelphax striatellus*. These tenuiviruses encode glycoproteins that promote viral spread among vectors species. The transovarial transmission to offspring and spread within the vector is through non-structural proteins (NS) (Zheng et al., 2014; Wu et al., 2014; Deng et al., 2013). Such a protein, NS4 has been found to spread the virus, RSV in *Laodelphax striatellus*, but it requires interaction between host protein vitellogenin (Vg) and RSV major nucleocapsid protein Pc3 (Huo et al., 2014). Maize mosaic virus (Rhabdoviridae) was studied using immunofluorescence microscopy that demonstrated the spread of the virus from midgut to salivary gland via anterior diverticulum, esophagus, compound ganglion region, brain and compound eye cells. The glycoprotein (G) protrudes from the surface of the virion and interacts with the receptors of the midgut in order to further enter the epithelial cells by endocytosis. The receptors are still not known (Ammar et al., 2009; Jackson et al., 2005). In *Reovirus*, Rice dwarf virus (RDV) vectored by *Nephotettix cincticeps* uses P2 as CP for entry into gut cells through endocytosis via clathrin coated vesicles (Chen et al., 2011; Wei et al., 2007; Omura et al., 1998). P2 is also used for release, assembly, replication, and association. Another protein Pns 10 is for intra and inter-cell movement through tubular structures along chitin based tissues in the body, including salivary glands (Chen et al., 2015; Chen et al., 2012). Southern rice black-streaked dwarf virus (SRBSDV) initially escapes the infected midgut epithelial tissues using tubules to cross basal lamina barrier in the intestine to spread in *Sogatella furcifera* as it is composed of non-structural protein P 7-1 that interacts directly with Actin and uses the same tubule-mediated mechanism for movement in both vector and plants (Jia et al., 2014; Liu et al., 2011; Hoh et al., 2010; Leh et al., 1999). It speeds up the invasion process that shortened the latent period of SRBSDV to between six and nine days in *Sogatella furcifera* as against fourteen to twenty-one days of RDV in *Nephotettix cincticeps*. The non-structural proteins (NS) are for the rapid spread of the virus and one such protein NS4 has been identified for the rapid spread of RDV in the vector.

3.13.2 Specificity criterion

The vector is very important for the survival and spread of plant viruses. The viruses are obligate parasites and require injury for entry into the right cell for multiplication. The insects are the right agents for providing the required conditions for the entry of a virus into a cell through fine injury. Secondly, the spread of plant viruses is through vectors, though there are other methods of spread, such as vegetative propagation, seed, pollen and contaminated farm implements, but not all plant viruses are spread by these means. The specificity in the transmission of plant pathogens is in operation as all plant viruses do not spread through single insect species. There is a specificity that determines the transmission of the particular virus by a definite taxon. The specificity can be defined as the relationship between the virus and a particular vector that is responsible for its successful transmission. In simple terms it can be said that a virus vectored by whitefly (e.g., Cotton leaf curl virus-CLCuV) is not transmitted by thrips, and a thrip-borne virus (Tomato spotted wilt virus-TSWV) is not transmitted by whitefly. There are variations in the specificity level of different vectors as some are highly specific while others are not. The specificity was demonstrated in different taxa in insects. Bean common mosaic virus-CBMV is aphid-transmissible, but is not transmissible through leafhoppers, thrips, and whiteflies. Similarly, CLCuV is transmissible by whiteflies but not by aphids, leafhoppers, and thrips. Cucumber mosaic virus-CMV is transmissible through aphids but not through leafhoppers, thrips, and whiteflies. Maize mosaic virus (MMV) is transmissible through leafhoppers but not through other insects like aphids, thrips, and whiteflies. The leafhoppers have much higher specificity than aphids as a given virus is transmitted by single species of leafhopper, or two at the most; whereas, in aphids, there could be many more genera acting as a vector of a given virus. Thus, the insects with biting and chewing mouthparts are least specific in their vector ability. In highly specific vectors, there are variations in transmission due to the presence of determinants of transmission viz. coat protein or its derivatives, or non-structural proteins like a helper component (HC). The plant virus transmission is controlled through the transmission determinants, The role of determinants viz Coat Protein and Helper virus, there should be inclusion of bodies in different organs of the body of the vector; infectivity of the virus to be proven with a series of dilutions; assessment of virus concentration and length of incubation period in vectors. To demonstrate the multiplication of a given virus, there are definite criteria which are discussed in the proceeding paragraphs.

3.13.3.1 Serial passage technique. In this technique, a measured quantity of sap extract from a viruliferous insect (one ug) is taken and is injected equaly into ten non-viruliferous insects reared on non-hosts/healthy plants. These insects are allowed enough inoculation feeding access. After the optimum access, the insects are macerated, and again a measured quantity of sap extract is taken and injected into a counted number (10) of insects to make dilutions of sap. These insects are again allowed optimum feeding after which another sample of sap extract is taken from one of the ten insects and injected into another set of ten insects. Several passages are conducted, the virus being diluted at each passage. With the decreased virus concentration, the insect vectors are unable to cause infection. After passing through a series of passages, these are tested for infectivity. The viruses known to multiply in the body of the vector would continue to cause infection through the vectors. The transmission tests will be negative in case the viruses which do not multiply in their vectors. The serial passage technique was first demonstrated (Black and Brakke, 1952) in *Agallia constricta* (Van Duzee), a leafhopper vector of WTV.

3.13.3.2 Transovarial transmission. The process through which the viruses are carried by eggs from one generation to the other is known as transovarial transmission. It is one of the most important methods to test the multiplication of viruses in insect vectors. The procedure is explained by taking a hypothetical example of a leafhopper vector of the propagative virus. In this technique, the viruliferous female (say leafhopper) is taken and paired with a male which has developed on a healthy plant (non-host of the virus). After egg laying (fifty eggs), the adult female is killed. The eggs are allowed to develop on the non-host, when the eggs hatch, forty nymphs are removed from the host. Of these forty nymphs, twenty, being males, are discarded (assuming sex ratio is 50:50) and the remaining females are tested for infectivity on a daily basis. It is assumed that a large population of test plants is inoculated regularly for ten days. Of these inoculated plants, a virus infection to the tune of ninety per cent is obtained. As the virus concentration which was originally too small is further diluted by a factor of forty in nymphs of the first generation, it seems mandatory to accept that the virus multiplies in the vector. It becomes extremely difficult to explain how twenty insects could produce a ninety per cent infection rate in ten days. In the second generation also, the same procedure can be followed. When the females of subsequent offspring continue to infect plants without fresh acquisition access or recharge on the virus source, one is bound to accept that the multiplication of the virus occurs in the body of the vector. The testing can be done on any number of generations. It was in the work of Japanese worker (Fukushi, 1940) where Rice dwarf virus was demonstrated to be transovarially transmitted up to seven generations. The transovarial transmission of TYLCV through at least two generations has been reported in bemisia tabaci (Genn.) (Ghasin et al., 1998).

3.13.3.3 Long latent period. The time lapse after the acquisition of virus until inoculation of virus into plants is referred to as a "latent period". The latent period is required either for circulation of the virus in the body, or to reach the concentration level required to cause infection through multiplication. A latent period of forty-six days of Sow thistle yellow vein virus in *Amphorophora lactucae* (Linnaeius) (Duffus, 1963) was demonstrated with aphids as evidence of virus multiplication. Another case of a long latent period (nineteen days) is Strawberry crinkle virus (SCV). Likewise, the multiplication of Oat blue dwarf virus (OBDV) in aster leafhopper vector, *Macrosteles fascifrons* (Stal) was demonstrated (Banttari and Zeyen, 1976). When the initial virus titer is not sufficient to cause infection, the virus multiplies until it reaches the required concentration for causing infection in healthy plants. There are instances where the latent period is exceptionally long, and in other cases it is absolutely nil.

3.13.3.4 Inclusion bodies of virus particles in different parts of the body. The vector cells under electron microscope show aggregation of virus particles after the acquisition of virus, in addition to the presence of virus in the endoplasmic reticulum and plasma membranes of salivary glands. The wide spread of the virus on these sites could be attributed to the multiplication of the virus in the body of the vector. The virus particles were measured in the case of WTV in *Agallia constricta* (Van Duzee) after varying time intervals of inoculation and obtained virus concentration such as 10^6.6 (six days), 10^9.26 (thirty days), 10^8.7 (forty days) (Whitcomb and Black, 1961) after a lapse of time. In addition, intracellular virus aggregates of OBDV in *M fascifrons* (Stal) were recorded to prove the virus multiplication (Banttari and Zeyen, 1976). The characteristic features were identified on ultrathin sections in Rice dwarf virus (Fukushi et al., 1962) and located in cells of fat bodies, blood intestinal epithelium, salivary glands, malpighian tubules of vector *N impicticeps* Ishihara and the WTV (Shikata and Maramorosch, 1967) in these parts in A constricta as seen under electron microscope.

3.13.3.5 Detection of virus in vector through ELISA. The enzyme-linked immunosorbent assay (ELISA) was used to detect virus presence in different parts of the body of vector leafhopper, *Peregrinus maidis* Ashmead (Nault and Gordon, 1988; Falk et al., 1987). A purified version of MSV was fed

3.16 Fungi/Plasmodiophorids Protozoa

The chytrid fungi (*Olpidium* and *Synchytrium*) and plasmodiophorid protozoa (*Polymyxa* and *Spongospora*) are soil-inhabiting organisms which are known vectors of soil fungi. The vector fungi form resting spores; these spores are released into the soil on the disintegration of roots and produce zoospores. The zoospores enter through the roots of the plants and transmit the plant viruses. Plasmodiophorids transmit viruses in a similar fashion. The mass of protoplasm of plasmodium contains several nuclei but lacks a cell wall and produces resting spores which are again released from the roots. The zoospores, being flagellates, swim in water in the soil and infect the roots again, transmitting plant viruses. The details regarding the mechanism of transmission is separately presented in the book.

3.17 Mechanism of the Spread of Pathogens Other than Viruses (Fungi, Bacteria, Phytoplasma, Rickettsiae Like Organisms, Nematodes, and Protozoa) through Insects

Unlike viruses, these plant pathogens are spread in different ways. Amongst these, fungal plant pathogens are spread through internal and external means. In the internal spread, the spores of many fungi are carried either in the mouth or in the body, whereas in the external spread the insect body and its appendages get smeared with sticky spores and are carried from one place to another. Pollinating insects like bees and wasps also carry fungal pathogens externally. The insects are acting as predisposing factors for the entry of many fungi through feeding and ovipositional wounds. The gall-forming insects, such as the woolly apple aphid, are instrumental in the production of galls. These galls crack under the influence of cold winter weather and these cracks serve as entry points for fungi. Insects such as cotton strainers (red and dusky cotton bugs) make fine punctures in the developing cotton bolls and these feeding punctures serve as entry points for lint staining fungi. Fungi also live in the body of insects in a symbiotic relationship, in which both the fungus and the insect benefit. The spread of bacteria also takes place in a similar way to that of fungi. In external spreading, the spores of fungi become attached to the body and appendages of insects, while in the case of a bacterium, the insects are smeared with bacterial ooze containing bacteria. Many insect species, such as ants, feed on bacterial ooze and spread the bacteria. The Mollicutes (*Phytoplasma* and *Spiroplasma*) are also plant pathogenic and spread through insects in a persistent propagative manner. Aster yellow is an important disease of phytoplasmic etiology, transmitted by *Macrosteles fascifrons* Stal (leafhopper). Similarly, Corn stunt and Citrus stubborn are of spiroplasmic origin and transmissible through leafhoppers, *Dalbulus maidis* (De Long and Wolcott) and *Circulifer tenellus*, respectively. The relationship between the pathogen and the vector is persistent and propagative.

References

Allen C (2010). Virus transmission in orchids through the feeding damage of Australian cockroach, *Periplaneta australassiae*. Acta Horticulturae, 878: 375–379.

Ammar ED (1994). Progressive transmission of plant and animal viruses by insects. Factors affecting vector specificity of competence. Advances in Disease Vector Research, 10: 289–332.

Ammar ED and Nault LR (1991). Maize chlorotic dwarf virus-like particles associated with the foregut in vector and nonvector leafhopper species. Phytopathology, 81: 444–448.

Ammar ED, Jarlfors U and Pirone TP (1994). Association of *Potyvirus* helper component protein with virions and the cuticle lining the maxillary food canal and foregut of an aphid vector. Phytopathology, 84: 1054–1060.

Ammar ED, Tsai MG, Whitfield AE, Redinbaugh MG and Hogenhout SA (2009). Cellular and molecular aspects of *Rhabdovirus* interactions with insect and plant hosts. Annual Review of Entomology, 54: 442–468.

Banttari EE and Zeyen RJ (1976). Multiplication of oat blue dwarf virus in the aster leafhopper. Phytopathology, 66: 896–900.

Black LM and Brakke MK (1952). Multiplication of Wound tumor virus in insect vector. Phytopathology, 42: 269–273.

Blanc S, Drucker M and Uzest M (2014). Localizing viruses in their insect vectors. Annual Review of Phytopathology, 52: 403–425.

Blanc S, Schmidt I, Vantard M, Scholthof HB, Kuhl G, Esperanddieu P, Cerutti M and Louis C (1996). The aphid transmission factor of cauliflower mosaic virus forms a stable complex with microtubules in both insect and plant cells. Proceedings of National Academy of Sciences, USA, 93: 15158–15163.

Blanc S, Lopez-Moya JJ, Wang RY, Garcia-Lampasona S, Thornbury DW and Pirone TP (1997). A specific interaction between coat protein and helper component correlates with aphid transmission of a *Potyvirus*. Virology, 231: 141–147.

Blanc S, Schmidt I, Kuhl G, Esperandieu P, Lebeurier G, Hull R, Cerutti M and Louis C (1993). Paracrystalline structure of Cauliflower mosaic virus aphid transmission factor produced both in plants and in a heterologous system and relationship with a solubilized active form. Virology, 197: 283–292.

Bragard C, Caciagli P, Lemaire O, Lopez-Moya LL, McFarlanes S, Peters D, Susi P and Torrence L (2013). Status and prospects of plant virus control through interference with vector transmission. Annual Review of Phytopathology, 51: 177–201.

Brown DJF, Robertson WM and Trudgill DL (1995). Transmission of viruses by plant nematodes. Annual Review of Phytopathology, 33: 223–249.

Casteel CL and Falk BW (2016). Plant virus vector interactions: more than just for virus transmission. pp. 217–240. In: Wang A and Wang X (eds). Current Research Topics in Plant Virology. Publishers Springer International.

Chen B and Francki RIB (1990). Cucumovirus transmission by the aphid, Myzus-persicae is determined solely by the viral coat protein. Journal of General Virology, 71: 939–944.

Chen Q, Chen HY, Mao QZ, Liu QF, Shimzu T, Uehara-Ichiki T, Wu ZJ, Xie LH, Omura T and Wei TY (2012). Tubular structure induced by plant virus facilitates viral spread in its vector insect. PLoS Pathog, 2012, 8: e1003032.

Chen HY, Chen Q, Omura T, Uehara-Ichentialiki T and Wei T (2011). Sequential infection of Rice dwarf virus in the internal organs of its insect vector after ingestion of virus. Virus Research, 160: 389–394.

Chen Q, Wang HT, Ren TY, Xie LH, Wei T (2015). Interaction between non-structural protein Pns 10 of Rice dwarf virus and cytoplasmic actin of leafhoppers is correlated with insect vector specificity. Journal of General Virology, 96: 933–938.

Childress SA and Harris KF (1989). Localization of virus-like particles in the foreguts of viruliferous *Graminella nigrifrons* leafhoppers carrying the semi-persistent Maize chlorotic dwarf virus. Journal of General Virology, 70: 247–251.

Dai Wu, Pan L, Lu Y, Fin L and Zhang C (2014). External morphology of mouthparts of white backed planthopper *Sogatella furcifera* (Hemiptera: Delphacidae) with special reference to sensilla. Science Direct, 56: 8–16.

Deng JH, Li S, Hong J, Ji YH and Zhou YJ (2013). Investigation on subcellular localization of in its vector small brown planthopper by electron microscopy. Virology Journal, 10: 310.

Dietzen RG, Mann KS and Johnson KN (2016). Plant virus-insect vector interactions: Current and potential future research directions.Viruses, 8: 303. Doi:10.3390/v 8 110303.

Duffus JE (1963). Possible multiplication in the aphid vector of Sow thistle yellow vein virus; a virus with an extremely long latent period. Virology, 21: 194–202.

Falk BW, Tsai JH and Lommel SA (1987). Differences in levels of detection of Maize stripe virus capsid and major non-capsid proteins in plants and insect hosts. Journal of General Virology, 68: 1801–1811.

Fukushi T (1940). Further studies on the dwarf disease of the rice plant. Journal of Faculty of Agriculture, Hokkaido Imp University, 45: 83–154.

Fukushi T, Shikata E and Kimura I (1962). Some morphological characters of Rice dwarf virus. Virology, 18: 192–205.

Garret A, Kerlan C and Thomas D (1993). The intestine is a site of passage for Potato leafroll virus from the gut lumen into the hemocoel in the aphid vector *Myzus persicae*. Archives of Virology, 131: 377–392.

Gray S, Cilia M and Ghanim M (2014). Circulative non-propagative virus transmission: an orchestra of the virus, insect, and plant-derived instruments. Advances in Virus Research, 89: 141–199.

Garry and Garry RF (2004). Proteomics computational analyses suggest that the carboxyl-terminal glycoproteins of bunyaviruses are class II viral fusion protein (beta-penetrenes). Theoretical Biology and Medical Modelling, 1: 10.

Gergerich RC (2001). Mechanism of virus transmission by leaf-feeding beetles. pp. 133–142. In: Harris KF, Smith OF and Duffus JE (eds). Virus-Insect-Plant-Interactions. New York: Academic Press.

Ghanim M (2014). A review of the mechanisms and components that determine the transmission efficiency of Tomato yellow leaf curl virus (Geminiviridae, *Begomovirus*) by its whitefly vector. Virus Research, 186: 47–54.

Ghasin M, Morin S, Zadan M and Czosneck H (1998). Evidence for transovarial transmission of Tomato yellow leaf curl virus by its vector whitefly, *Bemisia tabaci*. Virology, 240: 295–303.

Govier DA and Kassanis B (1974). A virus-induced component of plant sap needed when aphids acquire Potato Virus Y from purified preparations. Virology, 61: 420–426.

Granados RR, Hirumi H and Maramorosch K (1967). Electron microscopic evidence for Wound tumor virus accumulation in various organs of an inefficient leafhopper vector, *Agalliopsis novella*. Journal of Invertebrate Pathology, 9: 147–159.

Gray SM (1996). Plant virus proteins involved in natural vector transmission. Trends in Microbiology, 4: 253–294.

Gray SM and Banerjee N (1999). Mechanisms of arthropod transmission of plant and animal viruses. Microbiology and Molecular Biology Reviews, 63: 128–148.

Gutierrez S, Michalakis Y, Van Monster MV and Blanc S (2013). Plant-microbe–interaction: plant feeding by insect vectors can affect life cycle population genetics and evolution of plant viruses. Functional Ecology, 27: 610–622.

Hardy JL (1988). Susceptibility and resistance of vector mosquitoes. pp. 87–126. In: TP Naresh (ed). The Arbovirus Epidemiology and Ecology. Boca Raton FC, CRC, Press.

Harris KF (1977). An ingestion-egestion hypothesis of noncirculative virus transmission. pp 165–220. In: Harris KF and Maramorosch K (eds). Aphids as Virus Vectors. New York, N.Y: Academic Press, Inc.

Harris KF, Treur B, Tsai I and Toler R (1981). Observations on leafhopper (Homoptera: Cicadellidae) ingestion–egestion behavior–its likely role in the transmission of non-persistent viruses and other plant pathogens. Journal of Economic Entomology, 74: 446–453.

Harris KF, Pesic-Van Esbroeck Z and Duffus JF (1996). Morphology of sweet potato whitefly, *Bemisia tabaci* (Hemiptera: Aleyrodidae) relative to virus transmission. Zoomorphology, 116: 143–156.

Hodgson RAJ, Wall GC and Randles JW (1998). Specific identification of Coconut tinangaja viroid for differential field diagnosis of viroids in coconut palm. Phytopathology, 88: 774–781.

Hoh F, Uzest M, Drucker M, Plisson-Chastang C, Bron, Blanc S and Dumas C (2010). Structural insights into the molecular mechanisms of cauliflower mosaic virus transmission by its insect vector. Journal of Virology, 84: 4706–4713.

Hull R (1994). Molecular biology of plant–virus-vector interactions. Advances in Disease Vector Research, 10: 361–386.

Hull R (2002). Matthews Plant Virology. Academic Press, New York, NY, USA.

Huo Y, Liu W, Zhang F, Chen X, Li L, Liu Q, Zhou Y, Wei T, Fang R and Wang X (2014). Transovarial transmission of a plant virus is mediated by vitellogenin of its insect vector. PLoS Pathogens, 10(4): e1004141. https/doi:org/10 1371/Journalppat 1004141/.

Irwin ME and Goodman RM (1981). Ecology and control of Soybean mosaic virus in soybeans. In: Maramorosch K and Harris KF (eds). Plant Diseases and Vectors, Ecology and Epidemiology, Academic Press, New York.

Jackson AO, Dietzgen RG, Goodin MM, Bragg JN and Deng M (2005). Biology of plant *Rhabdoviruses*. Annual Review of Phytopathology, 43: 623–660

Jia DS, Mao QZ, Chen HY, Wang AM, Liu YY, Wang HT, Xie LH and Wei TY (2014). Virus-induced tubule: a vehicle for the rapid spread of virions through basal lamina from midgut epithelium in the insect vector. Journal of Virology, 88: 10488–10500.

Katis NI, Tsitsipis JA, Stevens M and Powell G (2007). Transmission of plant viruses. pp. 353–390. In: Van Emden and Harrington R (eds). CAB International, 2007, Aphids as Crop Pests.

Kempema LA, Cui X, Holzer FM and Walling LL (2007). Arabidopsis transcriptome changes in response to phloem-feeding silver leaf whitefly nymphs. Similarities and distinctions in response to aphids. Plant Physiology, 143: 849–865.

Kennedy JS, Day MF and Eastop VF (1962). A Conspectus of Aphids as Vectors of Plant Viruses Commonwealth Institute of Entomology, London.

Kliot A and Ghanim M (2013). The role of bacterial chaperones in the circulative transmission of plant viruses by insect vectors. Virology, 5: 1516–1535.

Leh V, Jacquot E, Geldreich A, Hermann T, Leclerc D, Cerutti M, Yot P, Keller M and Blanc S (1999). Aphid transmission of Cauliflower mosaic virus requires the viral PIII protein. The European Molecular Biology Organization Journal, 18: 7077–7085.

Lewandowski DJ and Adkins S (2005). The tubule-forming NSm protein from tomato spotted wilt virus complements cell-to-cell and long-distance movement of tobacco mosaic virus hybrids. Virology, 34: 26–37.

Liu Y, Jia D, Chen H, Chen Q, Xie L, Wu Z and Wei T (2011). The P7-1 protein of Southern rice black-streaked dwarf virus, a *Fijivirus*, induces the formation of tubular structures in insect cells. Archives of Virology, 156: 1729–1736.

Mann KS and Dietzgen RG (2014). Plant *rhabdoviruses*: new insights and research needs in the interplay of negative-strand RNA viruses with plant and insect hosts. Archives of Virology, 159: 1889–1900.

by different plant species, irrespective of host, and the aphids normally colonize on fresh growth for probing and feeding. There are species which prefer different sites for feeding. Aphid *Myzus persicae* (Sulzer), a vector of Potato leafroll virus, prefers the lower leaf surface for settling (Calabrese and Edwards, 1976). *Macrosiphum euphorbiae* (Thomas) settles on older leaves while *Nasonovia ribisnigri* remain localized to the top of young growing tips (Nebreda et al., 2004). The most efficient vector of plant viruses, *Aphis gossypii* (Glover), prefers to move from one plant to another. Aphids inflict shallow probes first in the epidermis called exploratory probes (Pirone and Harris, 1977).

4.5 Mechanism of Virus Transmission

The transmission of the virus requires acquisition, retention, and inoculation for the completion of the process. In the case of viruses like Cucumber mosaic virus or Cauliflower mosaic virus which are carried on tip of stylets, acquisition of virus takes place via intracellular punctures. Under such situation, if the aphids are allowed access for more than ten minutes, the efficiency decreases. Finally, the aphid stylets on reaching the phloem tissues puncture the sieve elements of phloem, this is followed by salivation that continues for five to thirty minutes. During this period, watery saliva is discharged in order to eliminate the chances of sieve tissues becoming clogged with phloem proteins (Tjallingii, 2006). In phytophagous insect species, the watery saliva performs a physiological role during mechanical penetration of stylets. The composition of saliva varies depending on the insect species. The sap ingestion and the saliva release continue simultaneously during feeding (Powell, 2005; Pirone and Harris, 1977). The salivary secretions get mixed up with the sap. The viruses sucked up with the sap are retained in the body of vector. This retention is taken care of by virus-encoded proteins or helper component (Govier and Kassanis, 1974; Kassanis and Govier, 1971). The helper component binds the virion with the inner linings of foregut of the cuticle of aphids. Non-persistent viruses are retained on the common duct formed with the fusion of food and salivary canals or distal part of maxillary stylets (2–4 um length), or the epicuticle of maxillary food canal. Cauliflower mosaic virus is known to confine itself to the common duct made with a fusion of food canal and salivary duct (at the extremity of the salivary duct) in *Brevicoryne brassicae,* rather than on the stylet (Uzest et al., 2010; Uzest et al., 2007). Furthermore, the chemical nature of recapture in Cauliflower mosaic virus has been identified as a nonglycosylated protein found in chitin web at the tip of maxillary stylets. These results were supported in a study carried out subsequently, showing the retention of viruses on/in the common duct of food and salivary canals (Moreno et al., 2005b). Earlier there was no definite proof to pinpoint the location of non-persistent viruses transmissible by aphids, but now it has been established that the stylet-borne viruses are located at acrostyle while foregut-borne viruses are present in the anterior region of the foregut (Fig. 4.1).

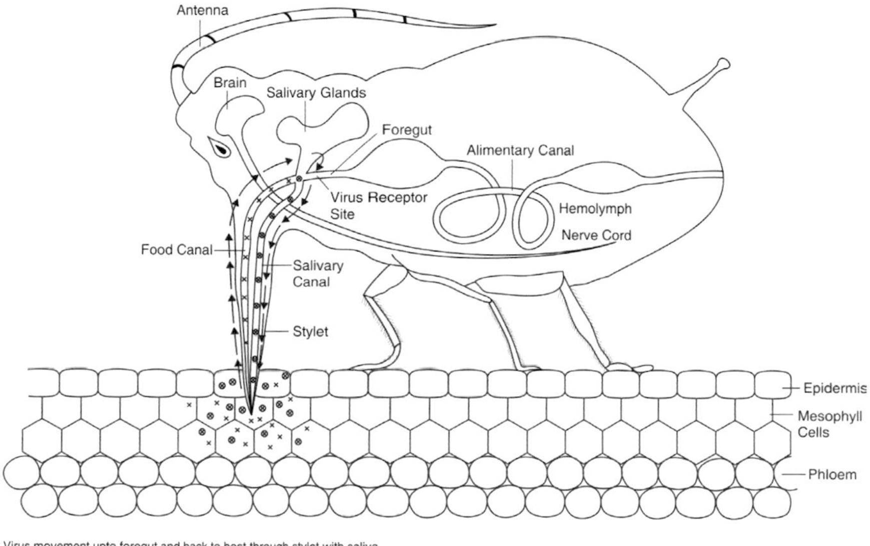

Fig. 4.1 Movement and Location of Non-persistent Foregut-borne Viruses in Insect Vectors.

4.6 Helper Component/Coat Protein

The transmission of viruses is under the action of encoded protein, this is the reason for which specificity is afforded in virus transmission. It was responsible for eroding the concept of pin pricking in the spread of plant viruses. Specificity is the ability of aphids to transmit a virus involving only one species of insect. It is afforded by certain determinants of virus transmission. Not all viruses are transmissible through all the insect species; this is due to specificity. The interaction of all four categories of plant viruses with the vectors has demonstrated that plant viruses encode structural proteins essential for transmission; they bind to specific sites in the vector for retention, and the viral determinants of vector transmission are responsible for disrupting transmission. The aphids have the distinction of transmitting all four categories of plant viruses, i.e., non-persistent stylet-borne, non-persistent foregut-borne, persistent circulative and persistent propagative (Stafford et al., 2012) (Table 4.1). Non-persistent stylet-borne viruses are from genera *Alfamovirus, Carlavirus, Cucumovirus, Fabavirus, Macluravirus* and *Potyvirus*, non-persistent/foregut-borne viruses are from genera *Caulimovirus, Closterovirus, Sequivirus Trichovirus*, and *Waikavirus*, persistent-circulative viruses are from genera *Enamovirus, Luteovirus, Nanovirus, Polerovirus* and *Umbravirus*, while persistent propagative viruses belong to genera *Cytorhabdovirus, Nucleorhabdovirus* and *Reovirus* (Quito-Avila et al., 2012; Link and Fuchs, 2005). It contains important genera involved in the spread of viruses by different transmission mechanisms. In the initial stage, a potential vector is fed on virus-infected plants and then transferred to healthy plants for inoculation, followed by disease development. While working on these lines, the scientists encountered an additional and much more complex problem. In potato viruses, PVC (C-strain of PVY) was not aphid-transmissible, while PVY was aphid-transmissible in nature. Of these, PVC also became aphid-transmissible when fed on plants of potato infected with both potato viruses. In another situation, the aphids were first fed on PVY infected potato plants followed by feeding on PVC infected plants, both these viruses (PVY and PVC) were also successfully transmitted through aphids. These circumstances led to the idea that aphids have picked up something important from PVY infected plants that triggered the acquisition of PVC. This observation led to great innovation, the result of which was subsequently named as Helper Component (HC) (Govier and Kassanis, 1974). The hetero encapsidation experiments conducted with CMV-M (non-aphid transmissible strain) and Tomato aspermy virus (TAV–aphid-transmissible) showed that coat protein (CP) of assembled TAV, not of CMV-M, led to transmission by aphids. This discovery led to recognition of the importance of coat protein in transmission through aerial vectors (Chen and Francki, 1990). The determinants of selective transmission include coat protein (CP) and its derivatives, like the read-through protein (RT) and minor coat protein (m CP); non-structural proteins are called helper component (HC). Plant viruses are enclosed within protein coat and react within the vector in the relevant sites (Blanc et al., 2014). In aphid-borne non-enveloped viruses, the interaction is direct. The coat protein is directly involved in non-circulative and circulative (Luteoviridae and Geminiviridae) viruses. As a result, some viruses are attached to inner linings of cuticle and get lost, along with molting material, while others circulate and multiply in insect vectors. The helper component (HC) is another determinant that contributes towards specificity in aphid-borne *Caulimovirus* and sequiviruses from the non-circulative category (Pirone and Blanc, 1996; Lung and Pirone, 1974). HC to bridge the virion within aphid vectors is demonstrated in potyviruses (Fifty kDa protein) and caulimoviruses (eighteen kDa protein) transmissible by aphids (Plisson et al., 2005; Guo et al., 2001; Blanc et al., 1993; Woolston et al., 1987). Besides the non-persistent viruses, the role of HC is also highlighted in circulative Faba bean necrotic yellows virus (FBNYV), a *Nanovirus* (Franz et al., 1999). The concept of specificity based on HC is explained taking into consideration two virus genera viz. potyviruses and cucumoviruses. Of the determinants of transmission, the involvement of HC in aphid vectors' specificity was first highlighted in 1984 (Sako et al., 1984). Under this concept, three species of aphids after the acquisition of virus were tested for homologous and heterogeneous HC virion combinations in potyviruses. Of these (*M. persicae* Sulzer, *A. craccivora* Koch and *Dactynotus gobonis* (Matsumura) all three of them transmitted Turnip mosaic virus (TuMV), while *M. persicae* Sulzer vectored only Watermelon mosaic virus 2 (WMV2). On using WMV2 as a source of HC, *M. persicae* Sulzer transmitted the purified Turnip mosaic virus; however, on using Turnip mosaic virus as HC source, none of the aphids transmitted WMV2. The inability of WMV2, HC to function in *D. gobonis* (Scopoli) or *A. craccivora* Koch could explain their failure to transmit. A similar study to explain the role of HC in the transmission of TuMV and TEV was carried out using four species of aphids. Similar results were obtained using purified virus acquired through membranes from homologous virion HC mixture (Suzucki et al., 2006). The differential transmission efficiency of vectors could be due to the difference in the epicuticle of food canal and chemical composition of saliva of vector species. In cucumoviruses, the specificity varies according to the change in certain amino acids in CP.

4.7 Non-persistent Stylet-borne Viruses

The non-persistent aphid-borne viruses belong to *Alfamovirus, Carlavirus, Cucumovirus, Fabavirus, Macluravirus and Potyvirus* genera. Of the four categories approved, the vectors of non-persistent/stylet-borne viruses are known to lose their ability in minutes or a few hours. Aphids are known to spread a large number of non-persistent/stylet-borne viruses on the distal tip of maxillary stylets (common duct formed by the fusion of food and salivary canal). To prove it, the stylets of *M. persicae* Sulzer were either dipped into formalin (0.03%) or exposed to ultraviolet radiation. The potato virus Y was brushed off from the stylets of aphid vector with this treatment. These non-persistent viruses have a narrow (potyviruses) to wide (cucumoviruses) range of possible hosts. The sources of spread of this category of viruses are from internal (infected seed and vegetative parts) and external sources (diseased cultivated crops and weeds). The spread of Cucumber mosaic virus in cucurbits, and Banana bunchy tops in banana, is via internal sources, i.e., infected seed and vegetative parts, respectively. As per the characteristics, these non-persistent type viruses are acquired by their aphid vectors though exploratory probes meant for locating suitable substratum, these probes last as little as five seconds; should the probing period last longer than five minutes, the aphid is unlikely to pick up the virus. Once these viruses in this category are acquired, they can be retained for a short period of up to 30 minutes, or they are lost during molting. Pre-acquisition fasting increases transmission efficiency in this category correlated to the presence of differential salivary inhibitors. Likewise, the non-transmissibility of TMV and PVX through aphids, and the successful transmission through insects with chewing mouthparts is demonstrated (Walters, 1952). The modified mechanical hypothesis is proved on account of short retention, nil latent period, vector specificity and loss of virus during molting (Day and Irzykiewicz, 1953). The earlier notion was that the protein coat was a simple cover over the nucleic acid, an infective entity. No doubt the nucleic acid is still the infective entity, but the secondary spread is decided by the protein coat component. Besides the inhibitors, the coat protein and helper component decide the transmissibility through the vector. One of the studies carried out on CMV (*Cucumovirus*) using *Aphis gossypii* Glover indicated efficient transmission of this virus by bringing about changes in coat protein of virus (Perry, 2001). Similar results were obtained using a different vector, *Myzus persicae* Sulzer. Likewise, on changing the amino acid sequence of coat protein from Asy-Ala-Gly (DAG) of an efficient strain of Tobacco vein mottling virus (TVMV) (*Potyvirus*) to Gly-Glu (DAG), an inefficient strain, it rendered the efficient strain as inefficient one through aphids. The outcome was due to the reaction that took place between the coat protein and the helper component (Moreno et al., 2012; Moreno et al., 2005a; Peng et al., 1998; Blanc et al., 1997). Though the highly infectious virus PVX (*Potexvirus*) is not aphid-transmissible in nature, with the introduction of a new sequence of an amino acid of coat protein from a Potato aucuba mosaic virus (aphid-transmissible) (PAMV) it also became aphid transmissible (Baulcombe et al., 1993). Lilly virus X (LVX) (*Potexvirus*), Tulip breaking virus (TuBV) (*Potyvirus*), Alfalfa mosaic virus (AMV) (*Alfamovirus*), Broad bean wilt virus (BBWV), Beet mosaic virus and Cucumber mild mosaic virus (CMMV) (*Fabavirus*), Lilly symptomless virus (LSV) (*Carlavirus*) and Cardamom chirke virus (CCV) (*Macluravirus*) are the important viruses of this category. All these viruses generally persist in their aphid vectors for a few hours only, except Beet mosaic virus, the retention of which is for twelve to seventy-two hours.

4.8 Non-persistent Foregut-borne Viruses

In foregut-borne viruses, the virus particles lie between both ends of the alimentary canal on the inner linings in the body of vector and are picked up from the phloem. During molting, the inner linings of the alimentary canal are shed along with the virus and there is no possibility of virus retention in the body of the vector. In other words, the viruses are carried between the stylet tip and the posterior part of the foregut. In this case the virion binding is in the anterior region of foregut and it is quite far away from the maxillary stylets and the salivary duct, therefore, salivation is unable to trigger the release of the virus, thus, such viruses circulate or replicate. In foregut-borne semi-persistent viruses (like the Cauliflower mosaic virus) three proteins are required. Viral capsid protein (VCP), loosely bound viral associated protein (VAP) and aphid transmission factor (ATF) are all required for successful transmission of virus (Hoh, 2007). Of these proteins, VAP encircles the inner shell of the virion from C-terminal and N-terminal facing outwards. These proteins are also required for cell-to-cell movement of the virion. Cauliflower mosaic virus (CaMV) (DNA) is located at a point of fusion of food and salivary canals (acrostyle) and is released with salivation for inoculation. It is very well explained, taking CaMV as an example. The virus is fifty-two nanometers in size, with icosahedral symmetry and seven capsids. P2 protein is essential for binding CaMV to the stylet tip. Aphids have nonglycosylated proteinaceous receptors in a special region of the stylet (Zhou et al., 2007). The site in which these receptors are impregnated within the chitin, in the swollen area at the bottom of common duct in the stylet, is known as the acrostyle (Uzest et al., 2010).

4.8.1 Helper component/coat protein

The mechanism of the helper component is also applicable in foregut-borne aphid-transmissible viruses. Aphid transmission factor in CaMV was identified as 18 kDa protein (Lung and Pirone, 1974) but subsequently, virus-encoded factors were investigated. In this context, DNA virus, CaMV, the CP proteins P2 (aphid factor) and P3 were identified as essential for transmission (Zhou et al., 2007). To be aphid-transmissible, P2 is acquired from the mesophyll, followed by P3 from the mesophyll/phloem cells. In this case, P3 bridges the P2 and the virion in the transmission process. Likewise, m CP is required on one end of the virion for successful transmission of Lettuce infectious yellows virus (LIYV). In this context, Willow carrot aphid *Cavariella aegopodii* (Scopoli) is a vector of Anthriscus yellows virus (*Waikavirus*) (AYV) in nature. Besides AYV, it also acquires and transmits Parsnip yellow fleck virus (*Sequivirus*) (PYFV) if the plant is infected with both viruses (Murant et al., 1976). However, PYFV alone is not transmissible through this aphid species. The further examination of viruliferous aphid vectors showed the presence of both viruses in the foregut region. In this category, one virus (AYV) produced a helper component which helped the PYFV to be aphid-transmissible. So AYV acted as helper component for the other in order to facilitate its transmission. Two other viruses from the *Potyvirus* genus are Potato aucuba mosaic virus and Potato virus Y (C-strain). Both these viruses have demonstrated the role of a helper component in aphid transmission (Blanc and Drucker, 2010; Blanc et al., 1998). Of the two viruses, C-strain of PVY is vectored by green potato aphid, *Myzus persicae* Sulzer, but the other virus is not transmissible through this aphid. The PVY acted as a bridge (helper component) to bind the coat protein of virus (Potato aucuba mosaic virus) to the inner lining of the foregut of the vector. As a result, both the viruses became aphid-transmissible (Blanc can be seen under an electron microscope after the required latent period has elapsed. If the virus is detected in the whole body of the vector, it is said to be multiplying in the vector. To test the multiplication based on latent period, the insect can be tested in serial experiments. In this context, the vector is used to inoculate test plants on a daily basis for several days or months, depending upon the longevity of the vector. The positive confirmation of transmission under serial experiments is essential to prove the multiplication of the virus in the body of the vector. The presence of infected plants would endorse the criterion of virus multiplication in the vector. Sow thistle yellow vein virus is a propagative virus in vector *Hyperomyzus lactucae* Linnaeus (*Amphorophora*) (Duffus, 1963). An aphid vector can acquire the virus within two hours of acquisition access. The latent period, however, is variable depending on the temperature; it is always sufficiently long. The latent period is eight days at 25°C and forty-six days at 5°C in the body of the aphid vector. The appearance of symptoms also changes with variations in temperature since it took between seven and fifty-five days to show symptoms of the virus. The aphid is able to acquire the virus five days before the appearance of symptoms. A viruliferous vector can inoculate a virus for up to fifty-two days after acquisition access on virus source. Another criterion was spread of virus in the body of vector was also met with by this virus as the virus particles found scattered in parenchyma and developing vascular tissues (xylem and phloem). This virus is always detected in the nucleus, only occasionally in the cytoplasm (Peters and Lea, 1972). Besides fulfilling the two criteria, the virus also passed from one generation to other generations through eggs of vector (Sylvester and Richardson, 1981). Injections of sap containing the virus were given into the hemolymph of *H. lactucae* Linnaeus and the vector continued to inoculate the virus up to three serial injections. It is therefore confirmed that this virus multiplies in the body of aphid vector. Another virus, (Strawberry crinkle virus-SCV), has been found distributed in the cytoplasm and nuclei of the cell in the body of the vector, *Chaetosiphon jacobi* (HRL) (Sylvester and Richardson, 1981). The multiplication has also been confirmed based on the conduct of three serial passages using injections in hemolymph. In addition to above two viruses, Coriander feathery red vein virus-CFRVV from *Rhabdovirus* genus is a bacilliform virus (size: 75 X 216 nm) and is vectored by *Hydaphis foeniculi passerini*. CFRVV is confirmed as being a virus from the propagative category. The vector can acquire the virus within two hours and can inoculate within fifteen to thirty minutes (Misari and Sylvester, 1983). The lifelong retention of the virus in vector is demonstrated. The virus successfully met the criteria of being transovarial (fifty-five per cent transmission) and serial passages identified for multiplication of virus in the body of vector. The virus was detected in annulated lamellae of phloem cells of plants (Hoefert and

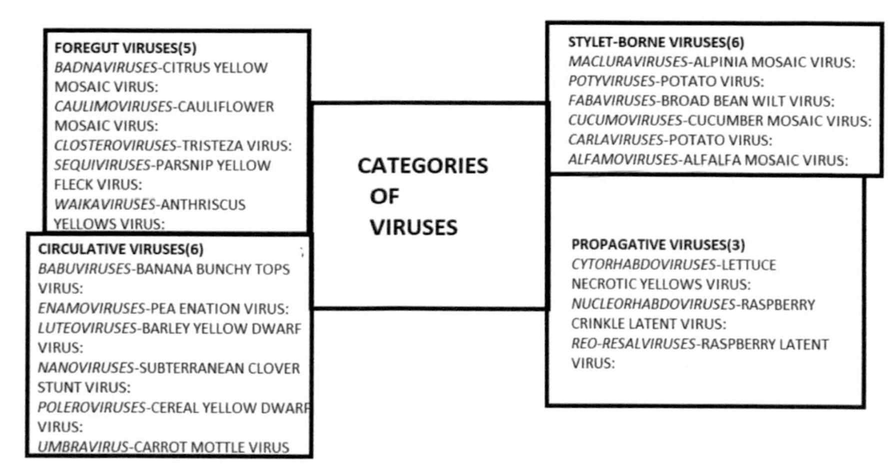

Fig. 4.2 Plant Diseases Caused by Aphid-Borne Virus Genera.

Steinkamp, 1977). Another virus from the rhabdovirus genus, vectored by *H. lactucae* Linnaeus, is Lettuce necrotic yellows virus (known in Australia since 1963). Based on gene sequence, it was found to be closely related to strawberry crinkle virus (Dietzgen et al., 2006). It has spherical subunits of hexagonal lattice (Sylvester and Richardson, 1981; Wolanski and Chambers, 1972). Further study was carried out on Sow thistle yellow vein virus and its vector in order to prove multiplication. The results were positive through serial passage. A plant virus, such as Lettuce necrotic yellows virus (LNYV), must multiply within a vector before it is transmitted by aphid vector. Such viruses will typically have a median or average latent period (LP) of days or even weeks. LNYV is also persistently transmitted by its aphid vectors after three to sixty days of latent period.

4.11 Diseases Caused by Aphid-borne Genera of Plant Viruses

Aphids are known to spread viruses from all the four categories of viruses, i.e., non-persistent, semi-persistent, circulative and propagative (Table 4.1; Fig. 4.2).

4.11.1 Non-persistent/stylet-borne viruses

The non-persistent aphid-borne virus genera are *Alfamovirus, Carlavirus, Cucumovirus, Fabavirus, Potyvirus* and *Macluravirus*. The detailed virus-vector relationship is presented using one typical virus from each category as an example.

4.11.1.1 *Cucumovirus* (Cucumber mosaic virus-CMV) is an isometric virus (diameter: 29–30 nm) with ssRNA and linear genomes (size 8.621 kb) including three other variable size genomes (3.389, 3.035, 2.197 kb). The virus belongs to the family Bromoviridae with a genome consisting of three RNA (RNA-1; RNA-2; RNA-3) each with 3250, 3050 and 2200 nucleotides, respectively. CMV has a protein coat and replicates in the cytoplasm. First detected in 1916, it is now known to attack 1200 species of plants belonging to 100 families. Economically, the disease is responsible for yield losses to the extent of about ten to twenty per cent. However, the cucurbits are more prone to acquiring this disease and infestation up to 100% has been a common feature in most fields. It is identified from the yellow-green mottling, yellow spots/streaks on leaves, yellow veins with reduced leaf size, variegated flowers, malformed/rough fruits and stunted size in different host plants. The severe epinasty with downward bending of leaf petioles is also encountered in cucurbits. It is transmitted through mechanical means, seed, parasitic plants and sixty to eighty species of aphids. The important species are *M. persicae, A. gossypii* Glover, *Acrythosiphon pisum* (Harris) and *Aphis glycines* (Matsumura). The virus is non-persistent, thus, it can be acquired within an acquisition access of sixty seconds. It is retained for a short period of a few minutes, and then lost in the normal feeding time of two minutes. This genus also includes peanut stunt virus and tomato aspermy viruses.

4.11.1.2 *Potyvirus* (Potato virus Y-PVY) is an important member of genus potyvirus which inflicts a loss in yield of between fifty and eighty per cent. It is non-enveloped, filamentous (680–900 x 11–15 nm) and covered with 200 copies of protein coat. The presence of protein coat in the form of inclusion body is an important feature that differentiates it from other viruses. It is a ssRNA-positive virus. In potato crop, it produces mild to severe mosaic, leaf crinkling, necrosis and leaf drooping in infected plants. The virus has different strains; therefore, symptoms differ slightly depending upon the strain. In some strains, stippling and yellowing are visible, while in others, the tubers show ring spots and internal necrosis. PVY is transmissible through sap, tubers and aphid vectors. The aphid *Myzus persicae* transmits it in a non-persistent manner. In addition, Bean common mosaic, Bean yellow mosaic, Lettuce mosaic, Peanut mottle, Peanut stripe and Soybean mosaic viruses are included in this genus.

4.11.1.3 *Fabavirus* (Broad bean wilt virus-BBWV) has isometric symmetry (25 nm), ssRNA and is a non-enveloped, bipartite virus. The virus replicates in the cytoplasm. On broad bean, it produces vein clearing, discoloration, mosaic, and ring spot symptoms on leaves, pods, and stems. In the advanced stages of disease development, the necrosis of terminal tip leaves and wilt symptoms become evident. There are two isolates of this virus viz. P123

(USA) and Ben (Spain); both are aphid-transmissible (Ferriol et al., 2012). Aphids *Myzus persicae* Sulzer and *Aphis gossypii* Glover are efficient vectors wherein transmission is non-persistent in nature since the aphid spp. do not retain the virus beyond twenty-four hours. The vector spp. acquire the virus with an acquisition access of ten to fifteen seconds.

4.11.1.4 *Alfamovirus* (Alfalfa mosaic virus-AMV) is icosahedral in symmetry (30–37 nm), non-enveloped and has positive single-stranded RNA. It has four types of particles (three bacilliform with 36, 48, 58 nm length and one spheroidal in shape with a diameter of 18 nm), three ssRNA (RNA-1, -2, -3) and subgenome RNA-4. Of these, RNA-1 is for protein encoding in replication, RNA-2 is essential for virion movement in the cells, and RNA-3 for capsid encoding. The length of the genome is 8274 nucleotides. RNA-1 (3644), -2 (2593), -3 (2037) and -4 (881) have variable nucleotides. During replication, the virus enters the cell and the particles disassemble. The coat protein interacts with initiation factor, which further initiates translation of RNA into P1 and P2 proteins. In a replication of the virus, RNA protein are synthesized to assemble the virus. The virus is known to infect 600 plant species from seventy different families. The important host plants include potato, pea, tobacco, tomato and blue beard. In alfalfa, it is identified by white flecks, ring spots, mottling, mosaic and malformation of leaves. Wilting of plants is visible in the advanced stage of infection. Fourteen species of aphids transmit the virus in a non-persistent manner, out of which *Myzus persicae* Sulzer is the most efficient one. It also spreads through cuscuta, mechanical means, pollen, and seed.

4.11.1.5 *Carlavirus* (Potato virus S-PVS) has ssRNA and is a flexuous filament-type virus measuring 620–690 x 12–15 nm. Until 1950, the virus did not gain much importance owing to latent infection. It became economically important as it inflicted loss to the extent of twenty per cent in yield. It can be identified by the mild chlorosis of leaves with roughness of the leaf lamina (puckering of leaf lamina). It is transmissible by aphids, mechanical means, and infected tubers. *Myzus persicae* Sulzer and *Aphis nasturtii* Kaltenbach are potential vectors which can transmit this virus in a non-persistent manner. With an acquisition access period of thirty seconds to two minutes, the transmission efficiency was 2.9 per cent in *M. persicae* Sulzer. An acquisition access of fifteen seconds gave transmission efficiency of eleven per cent in another species of aphid, *A. nasturtii* Kaltenbach. With an increased time period of acquisition access, the efficiency declined in the case of both spp. of aphids. This genus includes carnation latent virus as well.

4.11.1.6 *Macluravirus* (Alpinia mosaic virus-AlpMV) has been shifted from potyvirus genus. The virus belongs to a ginger family (Zingiberaceae) which also includes cardamom mosaic virus, Chirke and Foorkey viruses, Cucumber mosaic virus, and ginger chlorotic fleck virus. The particles are

via aphids (*Cavariella aegopodii* Scopoli) in a semi-persistent manner. It is a helper virus for Anthriscus yellow speck virus, which becomes transmissible through the species of aphid, *C. aegopodii*.

4.11.2.5 *Badnavirus* (Citrus yellow mosaic badnavirus-CYMBV/Citrus yellow mosaic virus-CYMV) is a non-persistent, dsDNA, non-enveloped bacilliform (30–130 nm) virus with a genome size of 7.1–76 kb that multiplies in the cytoplasm (Ahlawat et al., 1996). It was first recorded in Japan with characteristic symptoms of chlorosis and leathery texture of leaves with dwarfs of plants. Serologically, it bears a resemblance to satsuma dwarf nepovirus. It is now an important disease in India as losses in yield of between ten and seventy per cent have been recorded in Andhra Pradesh (in citrus, lemon, and grapefruit crops). The virus is known to exist in two isolates. It has a citrus yellow mosaic badnavirus as its synonym (Ghosh et al., 2014) The virus is transmissible through aphids (*Toxoptera citricida, Aphis craccivora, Myzus persicae*) in a semi-persistent manner (Needs Confirmation).

4.11.3 Persistent circulative aphid-borne viruses

This group includes virus genera namely *Luteovirus* (Barley yellow dwarf virus/Beet western yellow virus/Soybean dwarf virus), *Polerovirus* (Cereal yellow dwarf virus/Potato leafroll virus), *Enamovirus* (Pea enation mosaic virus), *Umbravirus* (Carrot mottle virus) and *Nanovirus/Babuvirus* (Banana bunchy tops virus/Subterranean clover stunt virus/Faba bean necrotic stunt virus (Sicard et al., 2015).

4.11.3.1 *Luteovirus* (Barley yellow dwarf virus-BYDV) has positive ssRNA, hexagonal symmetry and the virion is 25–28 nm in diameter but non-enveloped, and lacking lipid layer around the virion. It has two proteins viz. major protein and read-through protein. It is known to cause diseases in cereals like oat, wheat, rice, maize, grasses, etc. These are characterized by yellow and red thickened, leathery leaves, stunted plants and upward posture of leaves. The plants have a poorly developed root system. The virus is transmissible by aphids (*Schizaphis graminum* Rodem; *R. padi* Linnaeus; *Sitobion avenae* Fabricius; *Metopolophium dirhodum* Walker) in a persistent circulative manner. The acquisition access period is thirty minutes long. The vector generally acquires the virus within twelve to thirty hours of acquisition access. The virus circulates in the body of the vector and it requires one to four days of latency before it is transmitted.

4.11.3.2 *Polerovirus* (Cereal yellow dwarf virus-CYDV). It was a strain of Barley yellow dwarf virus earlier. Now it has been put under *Polerovirus* and is transmissible through *Rhopalosiphum padi* (Linnaeus). It is transmissible in ten species of aphid vectors. Of these, aphid *R. padi* Linnaeus is an efficient vector and transmits the virus in a non-persistent manner. The virus was first recorded in *Avena sativa, Hordeum vulgare* and *Triticum aestivum* from California (USA) in 1951. It causes chlorosis of leaves in wheat and oat, and blasting of floral parts and chlorosis and reddening of foliage in barley. It is an isometric, non-enveloped (25–30 nm in diameter), ssRNA virus with a genome size of 5.763 kb. It has linear symmetry.

4.11.3.3 *Umbravirus* (Carrot mottle virus-CMoV) is an ssRNA virus. The total genome (Unipartite) size is 4.75 kb (largest one). The virus was first recorded in *Dacus carota* in 1964 in England. It is now prevalent in Australia, New Zealand, Japan and Europe. In addition to the carrot, its natural hosts are anethum graveolus and *Anthriscus sylvestris*. In diseased plants, it produces yellowing and reddening. Carrot mottle virus has a particle of fifty to fifty-two nanometres in diameter but lacks capsid protein and open reading frames. The particle is surrounded by lipoprotein as a unit membrane. It is a lipid-containing persistent virus and it persists through molts in *Cavariella aegopodii* Scopoli aphid vectors. The vector can pick up the virus from the source if the host is infected with carrot red leaf virus. So, red leaf virus is a helper virus for Carrot mottle virus for the purpose of transmission through aphids. Aphids can acquire the virus in acquisition access of thirty minutes. With twenty-four hours of acquisition access, the aphid can transmit the virus in two minutes. The total time required for the vector to transmit it is nine hours. It has encapsulated helper protein and five genomes. Additionally, it has umbra viral ORF-3 host nuclear protein, fibrillarin. It causes yellowing and reddening of foliage in all members of the families viz. Amaranthaceae, Chenopodiaceae, Papillionacaeae, Solanaceae, and Umbelliferae. It is also sap transmissible.

4.11.3.4 *Babuvirus* (Banana bunchy tops virus-BBTV). The virus belongs to G-2 and is an ssDNA virus, first recorded in 1989 in Fiji (Family: Nanoviridae). In 1920, this virus disease destroyed the banana industry in Australia. It is characterized by the presence of bunchy tops at the top of the plant. In addition, the leaves show yellowing and are flattened. Besides, irregular streaks and spots on leaves are also found in affected plants. Also, the fruits are deformed. The virus is confined to the members of family Musaceae which includes banana, abaca, heliconia, and flowering ginger. The disease is now prevalent in South-East Asia, The Philippines, Taiwan, South Pacific Islands, India, and France. The virus has six circular ssDNA components, each pair being 1 kb in length. The viruses belonging to this genus are known to attack mostly mammals, birds and insects, in addition to plants. These viruses are divided into six to eight segments, each within the protein (Gronenborn, 2004). Banana bunchy tops virus is a typical member of this genus and is transmissible by *Pentalonia nigronervosa* Coquerel in a circulative non-persistent manner. The aphid becomes viruliferous with an acquisition access of four hours on virus source and the inoculation access is fifteen minutes. With an acquisition access of eighteen hours on virus source, the aphid vector can retain the virus for two weeks in the body. The latent period in the body of the vector is between twenty and twenty-eight hours. The aphid vector does not lose the virus during molting and transovarial transmission is negative.

4.11.3.5 *Enamovirus* (Pea enation mosaic virus-PEMV) was first recorded in 1914 and is a combination of PEMV-1 (*Enamovirus*) and PEMV-2 (Umbravirus) viruses. It is positive ssRNA. The encapsidated virus causes downward-rolling of trifoliate, followed by distinct mottling, crinkling and chlorosis of leaves and flecking of veins with enations (blister-like outgrowths). Stunted plant growth with the proliferation of basal branches is a common feature of this disease. It has a narrow host range and is limited to members of the Leguminosae family. It is transmissible by eight to ten species of aphids but *Acyrthosiphon pisum* Harris and *M. persicae* Sulzer are the most efficient vectors of this virus. The acquisition threshold is fifteen minutes and there is a definite latent period of five to seven hours. The first instar nymphs are more efficient vectors of this virus. The transmission efficiency was about ninety-nine per cent with an acquisition access of four hours. PEMV-2 is dependent on PEMV-1 for the purpose of aphid-borne transmission and encapsidation.

4.11.3.6 *Nanovirus* (Subterranean clover stunt virus-SCSV) is an ssDNA, non-enveloped, icosahedral virus with a diameter of eighteen to nineteen nanometres. The genome is multipartite, having six to eight components, circular, segmented and about 1 kb in length, with replication confined to the nucleus. The natural hosts from the family Fabaceae/legumes include *Trifolium pratense, T. glomeratum, T. repens, Medicago hispida, M. lupulina, M. minima* and *Wisteria sinensis.* The yellowing of margins in newly emerged leaves is peculiar in diseased plants. Besides, the leaves show complete yellowing coupled with puckering of leaf lamina that extends to old or mature leaves. Finally, reddening of leaves becomes conspicuous in mature leaves. The virus is transmissible through four species of aphids viz. *Aphis gossypii, A. craccivora, Myzus persicae* and *Macrosiphum euphorbiae* and the relationship is persistent and circulative. An efficient aphid, *A. craccivora*, can acquire and inoculate the virus within thirty minutes of feeding. The latent period is less than twenty-four hours in the vector. The vector retains the virus throughout its life, but the virus does not multiply in its body. All instars are vectors and can successfully transmit the virus. Helper component is required in transmission via aphids.

4.11.3.7 *Nanovirus* (Faba bean necrotic stunt virus-FBNSV) is an ssDNA, isometric, encapsidated virus with the monopartite genome. The particles are eighteen nanometres in diameter. It has been recorded from Ethiopia (Abraham et al., 2010) as a new virus. The name has been proposed and submitted for approval by the ICTV. It has a narrow host range, confined to members of family Fabaceae only. On faba bean plants, it causes yellowing of leaves and stunted growth. It is transmissible through aphid *Aphis craccivora* Koch in a persistent circulative manner.

4.11.4 Persistent propagative aphid-borne viruses

The virus genera *Nucleorhabdovirus* (Strawberry latent crinkle virus) and *Cytorhabdovirus* (Lettuce necrotic yellows virus, Strawberry crinkle virus, Coriander feathery red vein virus) and *Reovirus* (Raspberry latent virus) are known to be aphid-transmissible.

4.11.4.1 *Nucleorhabdovirus* (Raspberry crinkle latent virus-RpCLV) is always found in perinuclear spaces and the formation of the envelope is from the inner nuclear membrane. This genus includes the Raspberry crinkle latent virus which is a bacillus, rod-shaped, enveloped ssRNA virus with a particle size of seventy-four to eighty-eight nm in diameter and measures 68 x 190–380 nm. It was first described in 1950. Its genome contains five proteins including nucleoproteins (forty-five kDa), viral matrix proteins (seventy-seven kDa), non-structural phosphoprotein (fifty-five KDa), and glycoprotein (twenty-three kDa). It was first recorded in *Fragaria* spp. in Japan and North America in 1986. *Fragaria vesca* is a susceptible host of this virus. It can be recognized by the severe epinasty of newly emerged leaves in young plants. In addition, the diseased plants are dwarfed, and bear mottled and distorted leaves. It is transmissible by *Chaetosiphon fragaefolii* Cockerell aphid species in a persistent propagative manner. With an acquisition access of twenty-four hours and a latent period of ten to fifty-nine days in the vector the virus is known to persist for two weeks. Otherwise, the retention of virus in vector aphids is for life (Krezal, 1982). The symptoms of the disease are expressed forty-eight weeks after inoculation in plants. The disease is widespread, covering Asia, Europe, North America, Canada, etc. It is a different virus from strawberry crinkle virus.

4.11.4.2 *Cytorhabdovirus* (Lettuce necrotic yellows virus-LNYV) was first recorded in *Lactuca sativa* in Australia in 1963. This virus normally obtains its envelope from the outer membranes. The type species of this genus is Lettuce necrotic yellows virus. It is ssRNA (–), enveloped, bullet-shaped, bacilliform (277 nm x 68 nm long); a linear genome (130 kb in length) virus with monopartite segmentation that replicates in the cytoplasm. The disease can be identified by dull green and chlorotic leaves. The outer canopy leaves are flaccid and bronze in appearance. The virus is transmissible through aphid *Hyperomyzus lactucae* Linnaeus. The virus is carried through molts, multiplies in vectors and the vector-virus relationship is of propagative nature. It is mechanically transmitted but seed transmission is negative. It is prevalent in Australia and New Zealand and is known to attack members of families, Chenopodiaceae, Compositae, Leguminosae, Liliaceae, and Solanaceae.

4.11.4.3 *Reovirus/Rasalvirus* (Raspberry latent virus-RpLV)(Proposed) has ten dsRNA segments and is transmissible by the strawberry aphid, *Amphorophora agathonica* Hottes in a persistent propagative manner (Quito-Avila et al., 2012). Its genome has 26,128 nucleotides. It is prevalent in the Pacific North-West, USA and in BC, Canada. The serial passage was positive, and the virus has a latent period of six days. The virus particles were detected in aphid vector after twenty-four to forty-eight hours of acquisition access on virus source.

4.12 Aphids and Fungi

The aphids are efficient vectors of plant viruses but are also strongly associated with the spread of fungi. Rust fungi, *Puccinia punctiformis* has a mutual relationship with aphids, *Aphis fabae* sp. *cirsiiacanthoidis*, and *Uroleucon cirsii*. The development of these aphid species is faster on the diseased plant as compared to a healthy plant. The large, robust colonies are found on rust infected plants (Kluth et al., 2002). The woolly apple aphid is a gall former and feeds on bark. It produces warts on the submerged parts of an apple. These warts crack under the influence of freezing temperatures and the cracks provide suitable sites for another fungus to invade the plant. The fungi *Gloeosporium perennans* attacks such sites and the aphids subsequently spread the conidia by carrying them externally on the body as in *Cladosporium cladosporioides* and *Pullularia pullulans*.

4.12.1 Rust disease is a disease caused by a fungus, *Puccinia punctiformis* (Basidiomycota; Actinomycetes; Pucciniales). There are 168 genera and 7000 species of this fungus. They infect leaves, petioles, shoots, stems, fruits, etc., and colored powder can be seen on affected plant parts. It is known to cause deformities in plant parts that resemble stunted growth and witches broom. Additionally, hypertrophy, canker and gall formation are visible on plants. The scattered powder is nothing but a collection of aeciospores. These spores, in early summer or late spring, produce telia that grow on leaves or emerge through the bark of plants. These produce teliospores which grow aerially to give rise to basidiospores. The spread of these spores is through insects and other agents such as air or water. The fungus has two types of life cycles, i.e., macrocyclic (fungi producing five kinds of spores viz. pycnidiospores, aeciospores, urediniospores, teliospores, and basidiospores) and microcyclic (fungi producing three types of spores viz. urediniospores, teliospores and basidiospores).

4.13 Aphids and Bacteria

In nature, the pathogenic bacteria are associated with aphids, but the aphids are rarely vectors. However, the aphids act as primary hosts for plant pathogenic bacteria and are associated with their spread. The important species of the pea aphid, *Acyrthosiphon pisum* Harris, is a primary host of several bacteria known to inflict diseases in plants. These bacteria include *Erwinia aphidicola* (causal organism of leaf spot disease in the common bean) (Santos et al., 2009), *Pantoea stewartii* (causal organism of stewarts wilt) (Stavrinides et al., 2010), *Dickeya dadantii* (causal organism of soft rot of maize and potato) (Granier et al., 2006) and *Pseudomonas syringae* (causal organism of canker in several plants). The pathogenic bacteria like *Pseudomonas syringae* sp. *savastanoi* are known to cause canker in various plant species like olives, chestnuts, sweet cherry, etc. The bacteria live and multiply in the body of insects like *Acyrthosiphon pisum* Harris and the spread is brought about in this manner. All of the above-mentioned bacteria are plant pathogenic but they do cause damage to their host as the bacteria inhabit the gut. They either modify the behavior of the aphid, prevent molting or cause mortality of aphid hosts. Besides these pathogenic bacteria, there are bacteria which are in a symbiotic relationship with insect hosts. One such symbiotic relationship between aphids and bacterium *Buchnera aphidicola* (non-pathogenic) is an important association. The bacteria live in aphids and multiply inside their bodies. The bacterium is responsible for converting the diet, rich in essential amino acids, for aphids, whereas the aphids harbouring facultative bacteria (*Wolbachia, Erwinia,* and *Pantoea*) are protected against natural enemies. Heat tolerance, color change and reproductive alterations help the bacteria to reproduce and increase in numbers (Gauthier et al., 2015). These bacteria are lying on leaf surfaces and the aphids, while piercing leaf surface with stylets, ingest the bacteria along with sap. These bacteria survive in the body of aphids and multiply. These bacteria are secreted along with the honeydew. *Acyrthosiphon pisum* Harris acts as a host and vector of *Pseudomonas syringae PV syringae* B728a (Psy B728a) and *P. syringae* PV *tomato* DC3000 strains of bacteria and also *Pantoea agglomerans* bacteria (Danhorn and Fuqua, 2007).

4.13.1 Bacterial canker

This bacterium, *Pseudomonas syringae PV syringae* (PsyB728 a), is rod-shaped, gram-negative, obligate, aerobic bacterium known to infest wheat, barley, peas, etc. It causes cankers and galls. Evidence of contamination includes bark becoming brown and an absence of sprouts during the spring season. If there is an emergence of sprouts, the wilting of sprouts is also common. Sunken stem lesions appear gradually on the stems after one or two years. The pea aphid, *Acyrthosiphon pisum* Harris acquires the bacteria while feeding on colonizing bacteria of the infected plant. These bacteria colonize the digestive tract where they multiply and are later on excreted in the honeydew (Stavrinides et al., 2009). Thus, the spread of bacteria is brought about by aphids. In addition, the warty galls on roots of apple are caused by the feeding behaviour of the woolly apple aphid, *Eriosoma lanigerum*. The severe winter causes bursting of galls and the injured or cracked galls are exposed to fungi, *Gloeosporium perennans*, which are further responsible for inflicting fruit rot and canker. When the warm weather returns in summer, the host repairs the damaged tissues and covers it properly for the establishment of aphids, thus the mutual relationship becomes quite clear.

References

Abraham AD, Bencharki B, Torok V, Katul L, Varrelmann M and Vetten JH (2010). Two distinct *Nanovirus* species infecting faba bean in Morocco. Archives of Virology, 155: 37–46.

Ahlawat YS, Pant RP, Lockhart BEL, Srivastava M, Chakravorty NK and Varma A (1996). Association of *Badnavirus* with Citrus mosaic disease in India. Plant Disease, 80: 590–592.

Altenbach D and Bitterlin W (2011). Rapid immune-test combined with magnetic bead technology for on-site detection of Potato leafroll virus. Phytopathology, 101.6 p. S6.

Backus EA (1988). Sensory systems and behavior which mediate hemipteran plant feeding—a taxonomic overview. Journal of Insect Physiology, 34: 151–157.

Backus EA (1990). History, development, and applications of the AC monitoring system for insect feeding. In: Ellsbury EM, Backus EA and Ullman DL (eds). History, Development, and Application of AC electronic insect feeding monitors. Lanham, MD: Thomas Say Publications in Entomology, Entomological Society of America, 1990: 14–40.

Backus EA and Bennett WH (1992). New AC electronic insect feeding monitor for fine-structure analysis of waveforms. Annals of Entomological Society of America, 85: 437–444.

Baulcombe D, Lloyd J, Nannoussopa YN and Harrison BD (1993). Signal for *Potyvirus* – dependent aphid transmission of Potato aucuba mosaic virus and its effect on its transfer to potato virus X. Journal of General Virology, 74: 1245–1253.

Blanc S, Schmidt I, Kuhl G, Esperandieu P, Lebeurier G, Hull R, Cerutti M and Louis C (1993). Paracrystalline structure of Cauliflower mosaic virus aphid transmission factor produced both in plants and in a heterologous system and relationship with a solubilized active form. Virology, 197: 283–292.

Blanc S, Lopez Moya JJ, Wang RY, Garcia S, Lampasona S, Thornbury DW and Pirone TP (1997). A specific interaction between coat protein and helper component correlates with aphid transmission of a *Potyvirus*. Virology, 231: 141–147.

Blanc S, Ammar ED, Garcia-Lampasona S, Dolja VV, Llave C, Baker JM and Pirone TP (1998). Mutations in the *Potyvirus* helper component protein: effects on interactions with virions and aphid stylets. Journal of General Virology, 79: 3119–3122.

Blanc S, Hebrard E, Drucker M and Froissart R (2001). *Caulimovirus*. pp. 143–166. In: Harris KF, Smith OP and Duffus JE (eds). Virus–Insect-Plant Interactions. New York Academic Press.

Blanc S and Drucker M (2010). Functions of virus and host factors during vector-mediated transmission. pp. 103–120. In: Caranta C, Aranda MA, Tepfer M and Lopez-Moya JJ (eds). Recent Advances in Plant Virology, Caister: Academic Press.

Blanc S, Drucker M and Uzest M (2014). Localizing viruses in their insect vectors Annual. Review of Phytopathology, 52: 403–425.

Bonning BC, Pal N, Liu S, Wang Z S, Dixon PM, King GF and Miller WA (2014). Toxin delivery by the coat protein of a Sivakumar aphid-vectored plant virus provides plant resistance to aphids. Nature Biotechnology, 32: 102–105.

Bouvaine S, Bonham N and Doughlas AE (2011). Interactions between a *Luteovirus* and the GroEL chaperonin protein of the symbiotic bacterium *Buchnera aphidicola* of aphids. Journal of General Virology, 92: 1402–1425.

Brault V, van den Heuvel JM, Verbeek M, Ziegler-Graff V, Reutenauer A, Herrbach E, Garaud JC, Guilley H, Richards K and Jonard G (1995). Aphid transmission of Beet western yellows *Luteovirus* requires the minor capsid read-through protein P74. European Molecular Biology Organization Journal, 14: 650–659.

Brault V, Mutterer J, Scheidecker D, Simonis MT, Herrbach E, Richards K and Ziegler-Graff V (2000). Effects of point mutations in the read through the domain of the Beet western yellows virus minor capsid protein on virus accumulation in planta and on transmission by aphids. Journal of Virology, 74: 1140–1148.

Brault V, Ziegler-Graff V and Richards KE (2001). Viral determinants involved in *Luteovirus*-aphid interactions. pp. 207–232. In: Harris KF, Smith OP and Duffus, JE (eds). Virus-Insect-Plant Interactions. New York: Academic Press.

Brault V, Bergdoll M, Mutterer J, Prasad V and Pfeffer S (2003). Effects of plant mutations in the minor capsid protein of Beet Western yellows virus on capsid formation, virus accumulation, and aphid transmission. Journal of Virology, 77: 3247–3256.

Brault V, Perigon S, Reinhold C, Erdinger M and Scheidecker D (2005). The *Polerovirus* minor capsid protein determines vector specificity and intestinal tropism in the aphid. Journal of Virology, 79: 9685–9693.

Bressan A and Watanabe S (2011). Immunofluorescence localization of Banana bunchy top virus (family: Nanoviridae) within the aphid vector, *Pentalonia nigronervosa* suggests a virus tropism indistinct from aphid transmitted *Luteoviruses*. Virus Research, 155: 520–525.

Calabrese EJ and Edwards LJ (1976). Light and gravity in leaf side selection by the green peach aphid, *Myzus persicae*. Annals of Entomological Society of America, 69: 1145–1146.

Chavez JD, Cilia M, Weisbrod CR, Ju HJ, Eng JK, Gray SM and Bruce JE (2012). Cross-linking measurements of thepotato leafroll virus reveal protein interaction topologies required for virion stability, aphid transmission, and virus-plant interactions. Journal of Proteome Research, 11: 2968–2981.

Chen B and Francki RIB (1990). *Cucumovirus* transmission by the aphid *Myzus persicae* is determined solely by the viral coat protein. Journal of General Virology, 71: 939–944.

Danhorn T and Fuqua C (2007). Biofilm formation by plant-associated bacteria. Annual Review of Microbiology, 61: 401–422.

Day M and Irzykiewicz H (1954). On the mechanism of transmission of semipersistent phytopathogenic viruses by aphids. Australian Journal of Biological Sciences, 7: 251–273.

Day MF and Fenner F (1953). Mechanical transmission of virus disease by arthropods. VI Congress of International Microbiology Riass, Communication, 2: 586–587.

Day MF and Irzykiewicz H (1953). Feeding behavior of aphid, *Myzus persicae* and *Brevicoryne brassicae* studied with radiophosphorus. Australian. Journal of Biological Sciences, 6: 98–108.

Demler SA, de Zoeton GA, Adam G and Harris KF (1996). Pea enation mosaic *Enamovirus*: properties and aphid transmission. pp. 303–344. In: Harrison BD and Murant AF (eds). The Plant Viruses, Polyhedral Virions and Bipartite RNA Genomes (Vol 5). New York; Plenum Press.

Dietzgen RG, Callaghan B, Wetzel T and Dale JL (2006). Completion of the genome sequence of Lettuce necrotic yellows virus, type species of the genus *Cytorhabdovirus* disease of coconut plant in South Asia—an overview. Archives of Phytopathology and Plant Protection, 45: 2085–2093.

Dombrovsky A, Sobolev I, Chijanovsky N and Raccah B (2007). Characterization of RR-1, RR-2 cuticular protein from *Myzus persicae*. Comparative Biochemistry and Physiology Part B: Biochemistry and Molecular Biology, 146: 256–264.

Drucker M, Froissart R, Hebrard E, Uzest M, Ravallec M, Esperandieu P, Mani JC, Pugniere M, Roquet F, Fereres A and Blanc S (2002). Intracellular distribution of viral gene products regulate a complex mechanism of Cauliflower mosaic virus acquisition by its aphid vector. Proceedings of National Academy of Sciences, USA, 99: 2422–2427.

Duffus JE (1963). Possible multiplication in the aphid vector of Sowthistle yellow vein virus; a virus with an extremely long latent period. Virology, 21: 194–202.

Elnagar S and Murant AF (1976). The role of the helper virus, Anthriscus yellows in the transmission of Parsnip yellow fleck virus by the aphid, *Cavariella aegopodii*. Annals of Applied Biology, 84: 169–181.

Elnagar S and Murant AF (1978). Relations of Carrot red leaf and Carrot mottle viruses with their aphid vector, *Cavariella aegopodii*. Annals of Applied Biology, 89: 237–244.

Fereres A and Moreno A (2009). Behavioural aspects influencing plant virus transmission by homopteran insects. Virus Research, 141(2): 158–168.

Ferriol I, Rubio L, Perez-Panades J, Corbonell EA, Davino S and Belliure B (2012). Transmissibility of Broad bean wilt virus by aphids: Influence of virus accumulation in plants, virus genotypes, and aphid species. Annals of Applied Biology, 162: 71–79.

Filichkin SA, Brumfield S, Filichkin TP and Young MJ (1997). *In vitro* interactions of the aphid endosymbiotic symbionts with Barley yellow dwarf virus. Journal of Virology, 71: 569–577.

Franz AW, Van Der Wilk F, Verbeek M, Dullemans AM and Van den Heuvel JF (1999). Faba bean necrotic yellows virus (genus; *Nanovirus)* requires a helper factor for its aphid transmission. Virology, 262: 210–219.

Garran J and Gibbs AJ (1982). Studies on Alfalfa mosaic virus and alfalfa aphids. Australian Journal of Agricultural Research, 33: 657–664.

Garret A, Kerlan D and Thomas D (1996). Thomas ultrastructural study of acquisition and retention of Potato leafroll luteovirus in the alimentary canal of its aphid vector, *Myzus persicae* Sulz. Archives of Virology, 141: 1279–1292.

Gauthier JP, Outreman Y, Mieuzet L and Simon JC (2015). Bacterial communities associated with host-adapted populations of pea aphid revealed by deep sequencing of 16 Ribosomes' DNA. PloS One, 10(3): e0120 664; doi:org/10 1371/journal pone0120664.

Ghosh DK, Bhose S, Mukherjee K, Aglave B, Warghane AJ, Motghare M, Baranwel VK and Dhar AR (2014). Molecular characterization of Citrus yellow mosaic badnavirus (CYMB) isolates revealed the presence of two distinct strains infecting citrus in India. Phytoparasitica, 42: 681–689.

Gildow FE (1999). Luteovirus transmission and mechanism regulating vector specificity. In: Smith HG and Barker H (eds). The luteoviridae. Wallingford: CABI Publication, 297: 88–112.

Gildow FE, Reavy B, Mayo MA, Duncan GH and Woodford JAT (2000). Aphid acquisition and cellular transport of Potato leafroll virus-like particles lacking P5 read through the protein. Phytopathology, 90: 1153–1161.

Govier DA and Kassanis B (1974). A virus-induced component of plant sap needed when aphids acquire Potato virus Y from purified preparations. Virology, 61: 420–426.

Govier DA and Kassanis B (1974). Evidence that a component other than the virus particle is needed for aphid transmission of Potato virus Y. Virology, 57: 285–286.

Gray S and Gildow FE (2003). *Luteovirus*-aphid interactions. Annual Review of Phytopathology, 41: 539–66.

Gray S, Cilia M and Ghanim M (2014). Circulative, "non-propagative" virus transmission: an orchestra of virus-, insect-, and plant-derived instruments. Advances in Virus Research, 89: 141–49.

Gray SM, Power AG, Smith DM, Seaman AJ and Altman NS (1991). Aphid transmission of barley yellow dwarf virus acquisition access periods and virus concentration requirements. Phytopathology, 81: 539–545.

Granier F, Duport G, Pages S, Condemine G and Rahbe Y (2006). The phytopathogen, *Dickeya dadanti* (*Erwinia chrysanthemi*) 3937 is pathogen of the pea aphid. Application Environmental Microbiology, 72: 1956–1965.

Gronenborn B (2004). Nanoviruses: genome organization and protein function. Veterinary Microbiology, 98: 103–109.

Guo DY, Rajamaki ML, Saarma M and Valkonen JPT (2001). Towards in protein interaction map of *Potyviruses* protein interaction matrix is of two *Potyviruses* based on the yeast two-hybrid system. Journal of General Virology, 82: 935–939.

Hoefert LL and Steinkamp MP (1977). Annulate lamellae in phloem cells of virus infested Sonchus plants. Journal of Cell Biology, 74: 111–118.
Hogenhout SA, Van der Wilk F, Verbeek M and Goldbach RW (1997). Potato leafroll virus binds to the equatorial domain of the aphid endosymbiotic GroEL homolog. Journal of Virology, 72: 358–65.
Hogenhout SA, Wan der Wilk F, Verbeek M, Goldbach RW and Van den Heuvel JF (2000). Identifying the determinants in the equatorial domain of Buchnera GroEL implicated in binding Potato leafroll virus. Journal of Virology, 74: 4541–48.
Hoh F, Uzest M, Drucker M, Plisson-Chastang C, Bron P, Blanc S and Dumas Hohn T (2007). Plant virus transmission from the insect point of view. Proceedings of National Academy of Sciences. USA (PNAS), 104: 17905–17906.
Hoh F, Uzest M, Drucker M, Chastang CP, Bron P, Blanc S and Dumas C (2010). Structural insights into the molecular mechanisms of Cauliflower mosaic virus transmission by its insect vector. Journal of Virology, 84: 4706–4713.
Hull R (2002). Matthews Plant Virology. Academic Press, New York, NY, USA.
Hull R (2014). Plant Virology, 5TH Edition, Academic Press.
Irwin MF, Kampmeier GE and Wiesser WW (2007). Aphid movement; Process and Consequences. pp. 153–186. In: Van Emden HF and Harring R (eds). Aphids as Crop Pests CABI Publishing Wallington.
Jagadish KS and Jayaramaiah M (2004). Biology of VFC tobacco aphid (Homoptera; Aphididae). Journal of Economic Biology, 16: 93–97.
John O, Kiare N, Solomon S, Charles L, Nyongesa M, Muthoni J, Otino S, Mbiyu M and Oyoo L (2014). Potato virus Y (PVY) and Potato virus X (PVX) resistance breeding in Kenya: Applicability and control approaches. Agriculture and Biology Journal of North America, 4(4): 398–495.
Kaplan IB, Lee L, Ripoll DR and Gray SM (2007). Point mutations in the Potato leafroll virus major capsid protein alter virion stability and aphid transmission. Journal of General of Virology, 88: 1821–1830.
Kassanis B and Govier DD (1971). The role of helper virus in aphid transmission of Potato aucuba mosaic virus and Potato virus C. Journal of General Virology, 13: 221–228.
Katis NI, Tsitsipis JA, Stevens M and Powell G (2007). Transmission of plant viruses, pp. 353–390. In: VanEmden R and Harrington (eds). Aphids as Crop Pests. CAB International, 2007.
Kennedy JS, Day MF and Eastop VF (1962). A conspectus of aphids as vectors of plant viruses. Commonwealth Institute of Entomology, London.
Khelifa M, Journou S, Krishnan K, Gargani D, Esperandieu P, Blanc S and Drucker M (2007). Electron-lucent inclusion bodies are structures specialized for aphid transmission of Cauliflower mosaic virus. Journal of General Virology, 88: 2872–2880.
Kirchner SM, Doering TF and Saucke H (2005). Evidence for trichromacy in the green peach aphid, *Myzus persicae* (Sulz) (Hemiptera: Aphididae). Journal of Insect Physiology, 51: 1235–1260.
Kluth S, Kruess A and Tscharntke T (2002). Insect vectors of plant pathogens: Mutualistic and antagonistic interactions. Oecologia, 133: 193–199.
Krezal H (1982). Investigation on the biology of strawberry aphid (*Chaetosiphon fragaefolii*), the most important vector of strawberry viruses in Germany. Acta Horticulturae, 129: 63–68.
Lee L, Kaplan IB, Ripoll DR, Liang D, Polukaitis P and Gray SN (2005). A surface loop of the Potato leafroll virus coat protein is involved in virion assembly, systemic movement, and aphid transmission. Journal of Virology, 79: 1207–1214.
Leonard SH and Holbrook FR (1978). Minimum acquisition and transmission times for Potato leafroll virus by the green peach aphid. Annals of Entomological Society of America, 71: 493–495.
Li C, Lod D, Cox-Foster D Gray SM and Gildow F (2001). Vector specificity of Barley yellow dwarf virus (BYDV) transmission, identification of potential cellular secretions binding BYDV-MAF in the aphid, Sitobion avenue. Virology, 286: 125–133.

which can be seen on plant parts. The spittlebug nymphs are always found covered with white froth. The froth is visible on leaves and stems and can be spotted from a distance as well. The adults have a blunt head (narrower than the large pronotum) and short, bristle-like antennae. Tibiae are long and adapted for leaping like frogs and the hind tibia possesses one or two stout marginal spines. The life history is a simple one, like the other hemipterans. The adults are seven millimetres in size. The eggs are laid by adult females in autumn, as the eggs must overwinter before hatching can occur in spring (April–May). The nymphs, after hatching, develop into adults in about five to eight weeks. The adults feed on hosts during the whole of the summer season and again lay eggs. There is only one generation of this variety of insect in a year. They were earlier identified as a vector of Pierce disease virus of grapevine, but this is now categorized as a bacterial disease (Delong and Severin, 1950).

5.1.5 Psyllids

These are important insects with respect to the spread of plant pathogens. Two species, namely Asian psyllid (*Diaphorina citri*) and African psyllid (*Trioza erytreae*) are involved in the transmission of plant pathogens. They have piercing and sucking mouthparts. The Asian psylla is three to four millimetres in length with a mottled brown body, but a black coloured head is a unique feature of African psylla. Mottled forewings with a brown-white band at the apical end (Asian psylla) and broad forewings with a brown band in middle (African psylla) are prominent characteristic features of these two species of psyllids. Further, the antennae of Asian psylla are black at the tip with two brown spots in the middle, while the antennae of African psylla are totally black. These are again hemimetabolous insects with piercing and sucking mouthparts. Females lay about 800 tapered, almond-shaped eggs. The eggs hatch to produce nymphs which pass through five instars in order to become adults. Adults live for several months. The total life cycle is completed in about fifteen to forty-seven days depending on the environmental conditions. There are ten generations of the insect in a year.

5.1.6 Mealybugs

The females are soft-bodied, pinkish, oval and segmented insects measuring about 1/20th to 1/50th of an inch. A mealybug's body is always covered with waxy secretions. They belong to the family Pseudococcidae in the order Hemiptera. The females have functional legs. The males are winged individuals and possess a non-functional mouth. They don't feed and their sole function is to mate with females and fertilize them. On maturation, the waxy secretions look like appendages around the body margins. In some species, it forms a long tail which is sometimes used in the identification of different species of mealybugs. The females lay eggs on plant parts and are always covered with waxy material. Some of the species lay eggs in egg sacs found inside the body. The egg laying process is completed in about ten to twenty days. The females die after egg laying. The fecundity of a female is generally between 100–200 eggs. The hatching of eggs takes place in about five to ten days. After hatching, the crawlers mature and turn into adults within six to nine weeks. There are two to six generations in a year. These insects overwinter as eggs or first instar crawlers under the bark of trees. In spite of their toxic saliva, the insects are known to act as vectors of plant viruses in all four virus genera, namely *Badnavirus, Ampelovirus, Closterovirus,* and *Vitivirus*. These insects spread the viruses in semi-persistent foregut-borne manner. The species, *Pseudococcus longispinus* and *Planococcus ficus*, are identified as vectors of grapevine leafroll-associated viruses (GVLRaV-3 and GVLRaV-1), respectively. These insects transmit the plant viruses with their piercing and sucking mouthparts.

5.1.7 Scale insects

The scale insects are also from the order Hemiptera and belong to two families viz. Coccidae (soft scales) and Diaspididae (armored scales). The adults are soft-bodied, sessile, legless, unsegmented and oval/circular in shape and the body always remains covered with waxy material. The intermingling of wax and cast of skin on drying makes the crust over the body. The males are winged, mobile and with clear body segmentation and live for one to two days. The eggs are laid under the waxy crust. After hatching from eggs in three to four weeks' time, the legged crawlers move out and select feeding sites to feed upon. The sexes can be identified in the immature stage. The female scales are oval, circular and without dorsal ducts, while males are elongate and oval but lack dorsal ducts. There are two to three immature stages (nymphs) of female scale but the pseudo pupal stage is missing while the male scale has four immature stages (nymphal instars) plus two pupal stages. These insects possess piercing and sucking mouthparts and have recently been included as vectors of plant viruses. When reared at 30°C, the scaled insect produced 117 offsprings and, thus, identified 30°C as a favourable temperature for the efficient reproduction of scaled insects (Kruger et al., 2013).

5.2 Feeding Mechanism of Hemipterans

5.2.1 Leafhoppers

The leafhoppers are diverse in their feeding behavior as they feed on phloem, xylem and mesophyll tissues. Due to this diversity, they can transmit semi-persistent, persistent and propagative viruses or *Spiroplasma*.

...Table 5.1 contd.

Sl. No.	Pathogen	Characters	Taxon	Vector species	Mechanism	Source(s)
31	*Phytoplasma*/Pear decline (PD)	*Candidatus phytoplasma* pyri, prokaryote with genome size 530–1350 kb	Psylla	*Psylla pyricola; P pyri; P. pyrisuga*	Persistent propagative	Vereijssen and Scott, 2013
32	*Phytoplasma*/Citrus greening (CG)	*Candidatus liberibacter*	Psylla	*Diaphorina citri; Bactericera cockerelli*	Persistent propagative	Hansen et al., 2008
33	Peacan bacterial leaf spot (PLS)	Bacteria *Xylella fastidiosa*	Spittlebug and leafhopper	*Clastophera achatma; Leptoyronia quantification*	Persistent propagative	Sanderlin and Melanson, 2010
34	*Topocuvirus*/Tomato pseudo curly top virus (TPCTV)	Icosahedral, geminate, non-enveloped, ssDNA, elongated capsid	Treehopper	*Micrutalus malleifera*	Persistent circulative	Bridden et al., 1996
35	Xylem inhabiting Bacteria/ Pierce disease virus (PD)	*Xylella fastidiosa* Gram –ve bacteria	Spittlebug and Sharpshooters	*Philaenus spumarius now spread glassy-winged sharpshooter, Homalodisca vitripennis*	Persistent propagative	Delong and Severin, 1950
36	Rhabdovirus? unconfirmed/Colocasia bobone disease virus (CBDV)	Bacilliform large particles resemble lettuce necrotic yellow virus (Need confirmation)	Planthopper	*Tarophagus proserpina*	Yet to investigate	Palomar, 1987

leafhopper vector, *Nephotettix virescens*. The leafhopper, *N. virescens,* would transmit the RTBV only if leafhoppers are given acquisition access first on plants infected with RTSV followed by RTBV. Thus, the acquisition of RTBV is dependent on RTSV for the purpose of transmission by a leafhopper. RTSV, however, is not dependent on RTBV for transmission through the vector. Thus, the phenomenon of helper viruses is operating in Rice tungro viruses as one virus is assisted by another for the purpose of transmission by utilizing helper component. In addition, these tungro viruses are transmissible through *N. impicticeps* and *Recilia dorsalis* as well. Both the nymphs and adults of *N. impicticeps* can acquire and transmit the virus but the minimum acquisition access is thirty minutes. With the increase in access period, the transmission efficiency is also increased in the case of the tungro virus. The inoculation threshold is fifteen minutes as mentioned in the literature. Another virus from Waikavirus genus is Maize chlorotic dwarf virus (MCDV), transmissible by leafhoppers in a semi-persistent manner. Maize chlorotic dwarf virus is acquired and inoculated in two hours each by *Graminella nigrifrons*. It is also transmitted by another species of leafhopper, *Amblysellus grex*. On feeding the vectors the purified virions of these viruses, the transmission obtained was negative. This confirmed the involvement of helper component as well. These findings further showed that the helper component is also not vectored species-specifically. Thus, the helper component is required for Maize chlorotic dwarf virus (MCDV) transmissible by *Graminella nigrifrons* (Creamer et al., 1993; Hunt et al., 1988). The experiments were conducted using two strains (mild-ms and severe-ss) of Maize chlorotic dwarf virus in order to find out the requirement of helper component. The purified version of MCDV, when fed to leafhopper vector *G. nigrifrons* through paraffin membrane, failed to transmit. However, on feeding the leafhopper vector first on plants infected with ms-strain of MCDV followed by ss-strain, both the strains were transmitted successfully. On reversing the trend, first feeding the vector on plants infected with a severe strain of MCDV, the vector transmitted the mild strain, but not the severe strain of the virus. The results showed that helper component is essential for transmission and is not virus strain-specific. The helper component acts as a bridge between virion coat protein and sites of attachment in the vector. The virus particles are attached to the linings of the pharynx, cibarium and precibarium in the foregut of inoculating vector along with helper component. With the passage of time, the virus detaches (virus detachment is faster than helper component) and passes on to the midgut without transmission. However, a small amount of virus is brought back to the healthy plants while feeding. The presence of helper component has also been assessed to be present in the cuticle linings of the esophagus, ciborium, and precibarium and sometimes in the maxillary food canal of *G. nigrifrons* (efficient) and *Dalbulus maidis* (inefficient) vectors but not in

5.5.7 Psylla

5.5.7.1 Phytoplasma

Pear decline, a widespread disease in America and Europe, is caused by *Phytoplasma* and transmissible by three species of citrus psylla namely, *Psylla pyricola*, *P. pyri* and *P. pyrisuga*. It is identified from the small and light-green rough leaves and the trees die slowly.

5.5.7.2 Fungi

Besides the plant viruses, the leafhoppers are also instrumental in the spread of fungal pathogens of plants. The leafhoppers are associated with the spread of fungi either by carrying it internally or externally on/in the body. The bud blast of *Rhododendron* is a fungal disease (*Pycnostysanus azaleae*) spread by leafhopper, *Graphocephala coccinea*.

5.5.7.3 Bacteria

Bacterial leaf scorch is caused by a bacterium, *Xylella fastidiosa* in mulberry, maple, elm trees, etc. (Nachappa et al., 2012; Purcell, 1980). It is characterized by necrosis of leaf margins of the lower canopy leaves. Later on, the disease covers the entire upper canopy leaves. The spread of the disease is through leafhoppers, *Homalodisca insolita*, *H. vitripennis*, spittlebugs, *Graphocephala versuta*, *H. coccinea* and various treehoppers. The bacterium is carried internally, and it multiplies inside the body of insects. The xylem inhabiting disease also includes Leaf scorch diseases of pear, maple, elm mulberry, oak, oleander, coffee, almond and Phony peach and Plum scald. In addition, Bunchy top of papaya and Clover club leaf diseases (xylem-borne bacterial diseases) are transmitted by leafhoppers. The transovarial transmission is positive in leafhoppers.

5.6 Homopterous Borne Pathogens/Diseases

5.6.1 Leafhopper-borne diseases

These insects are known to transmit plant viruses belonging to categories of transmission viz. non-persistent through ingestion–egestion method (Semi-persistent/foregut-borne) (Harris et al., 1981), persistent-circulative and persistent-propagative. The diseases caused by these viruses are as under.

5.6.1.1 Semi-persistent/foregut-borne viruses

5.6.1.1.1 Rice tungro

It is a disease caused by two stylet-borne viruses, i.e., RTSV (*Waikavirus*) and RTBV collectively. The disease is spread through Leafhoppers (*Nephotettix impicticeps* and *N. apicalis*), this has already been discussed earlier. It is an isometric virus with a diameter of a particle between thirty to thirty-three nm. It occurs in two strains that are severe (S) and mild strain (M) It is a pathogen on plants belonging to the family Gramineae specially *Oryza* spp. It is prevalent in Philippines, Malaysia, Pakistan, Thailand, and India. Another virus from this category is Maize chlorotic dwarf disease has also been dealt with earlier.

5.6.1.2 Persistent circulative

5.6.1.2.1 Maize streak virus MSV *(Mastrevirus)*

The disease was first recognized in 1901 in Saharan Africa as Mealie variegation and is caused by *Mastrevirus*. It caused an epidemic in South Africa and is now prevalent in Nigeria and Kenya, as well. It is ssDNA, monopartite circular genome and has four proteins, encapsidated 22 x 38 nm geminate particles measuring 18 x 30 nm. It has a single protein, thirty-two kDa.The size of the genome is 2.7 kb and it has 2690 nucleotides. It is known to attack plants of family Gramineae including eighty species of grasses. The notable plant hosts of the virus are wheat, millets, rye, barley, sorghum, and sugarcane. The disease is identified by the pale spots or specks which are 5.2 mm in diameter and coalesce to form long stripes running along the veins. It is transmissible by leafhoppers, namely *Cicadulina mbila, C. storeyi, C. arachidis, C. bipunctata, C. latens, C. parazeae, C. similis, C. ghaurii* and *C. dabrowski* (Shepherd et al., 2010). Trans-stadial transmission is positive. The relationship is the circulative persistent type. The acquisition access is one hour but the virus can be acquired in fifteen seconds and the inoculation can be in five minutes. The latent period of the virus is six to twelve hours in the body of the vector.

5.6.1.2.2 Sugarbeet curly top virus-BCTV (*Curtovirus*)

It is the second most well-worked virus after Tobacco mosaic virus. It is an ssDNA encapsidated, quasi-isometric virus with a small, circular, monopartite genome and 6 ORF. First discovered in 1888 in the USA and now prevalent in Mexico, South America, The Old World, the Mediterranean basin, and the Middle East. The initiation of replication of the virus in host cells is carried out by C1 and replication is completed by C3 coded protein. The C2 coded protein is meant for pathogenic reaction while C4 is responsible for production of symptoms in host plants. The plants of both monocotyledonous and dicotyledonous groups are its hosts for the most part. The important plants serving as virus hosts are beans, celery, spinach, peppers, squash, cucumbers, and tomato. In these plants, it induces symptoms such as hyperplasia, yellowing, and curling of leaves with purple veins and severe distortion. In addition, the pathogen is responsible for the swelling of leaf veins and for the premature shedding of fruits. It is transmissible through leafhopper *Circulifer tenellus* (Baker) in a circulative non-propagative manner in nature and follows the route from the digestive tract to salivary glands (Stanley, 2008).

5.6.1.3 Persistent propagative

5.6.1.3.1 Maize rayado fino virus MRFV (*Marafivirus*)

A virus belonging to *Marafivirus* genus (Tymoviridae) is responsible for causing this disease in maize. This virus is a small, icosahedral (particle size 31–33 nm in diameter), ssRNA virus known to restrict its host range to family Gramineae only. It is prevalent in Central America, Mexico, Peru, Uruguay, and Columbia (Games and Leon, 1988). It initially causes chlorotic spots in young leaves and fine stipple-stripping of veins. Finally, with the advancement of the disease, the chlorotic spots coalesce to form stripes. There is a poor formation of grains in the spikes; as a result, the yield is greatly reduced. The spread of the virus is through leafhoppers, *Dalbulus maidis* and *Graminella nigrifrons* and the relationships with both are persistent propagative. The acquisition and inoculation thresholds are six and eight hours, respectively. The virus has a latent period of eight to thirty-seven days (Edwards et al., 2016; Edwards et al., 2015). Female leafhoppers are more efficient vectors than males.

5.6.1.3.2 Rice dwarf virus-RDV (*Phytoreovirus*)

The virus is prevalent in Japan, China, Philippines, and Korea. It has dsRNA, is icosahedral (seventy to seventy-fivenm in diameter) and contains thirty-two capsomeres and twelve segments. Each segment is of 4.4 kb to 0.83 kb pairs with a total genome size of 25.13 kb pairs'. G segment namely S1, S2, S3, S5, S7, and S8 are known to code functional protein. It has three strains viz. O, D84 and S; O being the most severe. The diseased plants are stunted in growth and show chlorotic flecks, shortening of internodes and poor growth of roots. It is vectored by leafhoppers *Nephotettix cincticeps* and *N. apicalis* in a persistent propagative manner. Young nymphs are more efficient vectors of this virus. Nymphs can acquire the virus with an acquisition access period of one minute on the virus source and can inoculate in an inoculation access period of thirty minutes. The virus has a latent period of ten to thirty-five days in vectors and eight to twenty-seven days in plants. Transovarial transmission in leafhopper vectors is common in this virus.

5.6.2 Planthoppers

5.6.2.1 Persistent circulative

5.6.2.1.1 Coconut foliar decay virus (CFDV) (*Nanovirus*)

This is a disease caused by a virus from the *Nanovirus* (F: Nanoviridae) genus. It is a small ssDNA, icosahedral virus with 1291 nucleotides; the particle is twenty nm in diameter (Randles and Hanold, 1989). It has segmented genome, circular genomic arrangement, ORP for six proteins and each member with four segments encoding coat protein of thiry-three kilo Dalton (kDa). The particles are located in leaves and meristems. It has *Cocos nucifera* coconut palm as its host. On palms, it causes yellowing of leaflets from a seven to eleven position of fronds. The fronds start dying and are found hanging in the branches of the tree. The defoliation occurs due to necrosis of petioles which causes shedding of leaves. With the passage of time, the infection reaches the central crown and causes its death. The tree trunk becomes narrow and such trees bear poor quality fruits and fewer fruits than healthy trees. In nature its spread is through planthopper *Haplaxius (Myndus) taffini* and the relationship is the persistent circulative type (Wefels et al., 2015).

5.6.2.1.2 Northern cereal mosaic virus-NCMV (*Cytorhabdovirus*)

An ssRNA virus belonging to *Cytorhabdovirus*, it is bacilliform, enveloped with a particle 350 x 60 nm in size and prevalent in Japan, Korea, and China. Its potential hosts are oat, wheat and barley species of plants belonging to the family, Gramineae. It was first identified in 1910 in Japan through observation of symptoms such as chlorotic spots which coalesce to form chlorotic stripes or mosaic. It is synonymous with Cereal northern mosaic virus or Wheat rosette virus. It is a planthopper-borne virus with a persistent circulative relationship. The major vector is *Laodelphax striatellus (Delphacodes striatella)* but it is also transmissible through *Unkanodes albifascia, U. sapporona*, and *Muellerianella fairmairei*. The leafhopper can inoculate the virus in a minimum period of two to ten minutes. The hoppers remain viruliferous up to sixty days and transovarial transmission of the virus through the vector does occur.

5.6.2.2 Persistent propagative

5.6.2.2.1 Rice ragged stunt virus-RRSV (*Oryzavirus*)

It is a dsRNA virus with eight segments belonging to genus *Oryzavirus*. The virion is sixty-three to sixty-five nm in diameter with five proteins in phloem and gall cells. The disease is prevalent in China, Japan, India,

Srilanka, Malaysia, Thailand, Taiwan, Indonesia, and the Philippines. The disease can be identified from dwarfed plants, dark green serrated margins of leaves and twisting of leaf blades at the tips. The ragged appearance of leaves is conspicuous and such leaves are yellowish in color. The swellings on leaves and sheaths are also visible in diseased plants. The flag leaves are twisted and malformed. The diseased plants bear delayed flowers and earheads have partial grain filling. The disease is called Kardil hampa in Thailand. The virus is transmissible through brown planthopper *Nilaparvata lugens* in a circulative propagative manner. The 5th instar nymphs are more efficient vectors and can acquire the virus in twenty-four hours and retain it for one day. The virus has host plants from the family Gramineae, including rice and grasses.

5.6.2.2.2 Barley yellow striate mosaic virus-BYSMV (*Cytorhabdovirus*)

This *Cytorhabdovirus* is a synonym of Cereal striate mosaic virus, Wheat chlorotic streak mosaic virus and Wheat chlorotic striate virus and is bacilliform RNA virus with virion size of 55 x 330 nm. The infected plants have narrow leaves with chlorotic stripes. The apical margins of leaves are yellowish. It has been recorded in France, Morocco, Australia, and China. Its host range is confined to the Gramineae family (wheat, rye, and oat). Its transmission is through planthoppers *Javesella pellucida* and *L. striatellus* Fallen and the mechanism is circulative propagative in which the vector can acquire it within one hour of feeding on virus source. There is a latent period of nine to twenty-nine days, after which, the planthopper can inoculate within ten to fifteen minutes of feeding on healthy plants. Once the planthoppers become viruliferous, the virus is retained in the body of vector until death. The virus has also been demonstrated to be transovarially transmitted.

5.6.2.2.3 Maize mosaic virus disease-MMV (*Nucleorhabdovirus*)

The Maize mosaic virus disease is also caused by *Rhabdovirus* which is a bullet shaped/bacilliform virus having a particle size of 225 x 90 nm in diameter. The virus is a synonym of Corn mosaic, Corn yellow stripe, Corn stripe, Corn virus 1, maize stripe virus and zea virus. The disease is characterized by the stunted growth of plants, exposed kernels with shortening of husk and leaf veins with white bold streaks. The disease is a devastating one, appearing first in Hawaii (USA) in 1961 and causing a 100% loss in corn yield. The farmers were unable to harvest a single cob during that year. It is transmissible by *Peregrinus maidis*. It has a persistent-type relationship with the leafhopper vector. The virus can be acquired in minimum acquisition access of two hours with an inoculation threshold of fifteen min. The incubation period has been reported as fourteen to thirty-one days (McEwen and Kawanishi, 1967).

5.6.2.2.4 Oat sterile dwarf disease-OSDV (*Fijivirus*)

This disease is caused by oat sterile dwarf virus (genus: *Fijivirus*). It is known by other names, such as Arrhenatherum blue dwarf virus disease, or Lolium enation virus disease and is now prevalent in Britain, Czechoslovakia, Germany, Sweden, Finland, Norway and Poland. It is a polyhedral virus with a particle size of sixty-five to seventy nm in diameter.The virion has double shell of proteinsand contains dsRNA. The particle is icosahedral with spikes, non-enveloped, and replicates in the cytoplasm. The genome is linear, with a total size of 29.91 kb, and has ten segments. It produces symptoms like dark green grass (bushy plants) in the members of Gramineae family. In addition, leaf malformation and leaf enations are present on the opposite side of leaves on swollen veins. The leaves are misshapen, ragged and notched. The overall plants are deep blue-green, devoid of panicles. Planthoppers *Javesella pellucida, J. obscurella, J. discolor, J. dubia* and *Dicranotropis hamata* are vectors. The efficient vector *J. pellucida* can acquire the virus with an acquisition access of thirty to sixty minutes and requires a latent period of three to four weeks for successful inoculation of virus. The virus multiplies in vector, planthoppers and the relationship is persistent propagative.

5.6.2.2.5 Rice stripe virus disease-RSV (*Tenuivirus*)

This viral disease is prevalent in China, Japan, and Korea. It infects plants of family Poaceae (corn and wheat). It is a *Tenuivirus,* with four negative ssRNA segments and 8970 nucleotides. The particle size is 8 nm x 500–2000 nm. The affected plants show chlorotic stripes of pale yellow color, or blotches or brown to gray stripes of necrotic tissues, sparse tillers and sterile panicles devoid of grains. It is transmissible through planthopper *Laodelphax striatellus* in a persistent propagative manner. The acquisition and inoculation thresholds are fifteen and three minutes, respectively. Both nymphs and adults act as vectors, where nymphs are much more efficient vectors. Likewise, females are more efficient than males in the transmission of the virus.

5.6.3 Treehopper-borne diseases

5.6.3.1 Tomato pseudo-curly top virus disease-TPCTV (Topocuvirus)

Tomato pseudo-curly top virus, a Geminiviridae virus, hailing from *Topocuvirus* genus, is responsible for the production of this disease. It has geminate virion (thirty X eighteen nm in diameter) and is a quasi-icosahedral, non-enveloped, single-stranded, circular DNA virus. The virion also has 110 copies of 26.9 kDa protein. In addition to natural hosts, the virus is responsible for the production of diseases in dicotyledonous plants like nightshade, ragweed, Datura, chickweed, etc. The disease in tomato is recognized from the vein clearing initially, followed by swollen veins with leathery, brittle, curled and rolled leaves and finally chlorosis of plants. Its host range is limited to dicotyledonous plants and the important hosts are tomato, beans, chickweed (*Stellaria media*), *Datura stramonium*, ragweed (*Ambrosia* sp.) and night shade (*Solanum nigrum)* In nature, the virus spread is through treehopper *Micrutalis malleifera* (Simons, 1962) and the relationship is of semi-persistent type. The vector can acquire and inoculate the virus within an access of thirty minutes each on virus source. It has a latent period of twenty-four to forty-eight hours in vector body. The virus retention in the vector is for life.

5.6.4 Froghoppers/spittlebugs borne diseases

5.6.4.1 Pecan bacterial leaf scorch disease (PBLS)

The etiology of this disease remained a mystery for a long time. It was only in the year 1998 when it was identified as a disease caused by a bacterium found in the xylem vessels. The pathogen was identified as xylem-inhabiting bacteria, *Xylella fastidiosa*, and is now widespread in pecan growing areas in the USA. The disease can be identified from the brown tanning of leaflets from the tips and margins and it extends towards the center covering the entire leaflet. The discoloration of leaflets causes their drop. The bacterium lives in hardwood trees and causes clogging of xylem vessels after multiplication. The spread of this disease is via spittlebugs and leafhoppers but the main vectors are the spittlebugs (Purcell, 1980). The population of leafhoppers and spittlebugs is low in the USA and their testing has not been done to a desirable level (Sanderlin and Melanson, 2010). The pecan spittlebug, *Clastoptera achatina*, diamond-backed spittlebug, *Lepyronia quadrangularis*, and leafhoppers are still under investigation.

5.6.5 Scale insect-borne diseases

5.6.5.1 Little cherry virus disease-LCV (Ampelovirus)

A disease caused by a Little cherry virus-1, -2 or Western X *Phytoplasma*. It plays havoc with the cultivation of cherries in North America. It caused extensive damage in British Columbia between 1940–1950 and destroyed about 30,000 trees.The little cherry virus-2 contains RNA with a genome sequence of 15042 nucleotides lengthwise and eleven ORF (Closteroviridae; *Ampelovirus*) (Mekuria et al., 2013). The little cherry virus-1 is placed under the genus *Velarivirus*. Both viruses are found in phloem and parenchyma cells of diseased plants. It is a disease characterized by small, dusty-coloured cherries of insipid flavor, accompanied with the early ripening of fruits. The spread of this disease is through grapevine mealybug (*Pseudococcus maritimus*) and apple mealybugs (*Planococcus aceris*). Since the disease is caused by Western x *Phytoplasma*, the spread of it under such situations is through leafhoppers (*Scaphytopius acutus* and *Fieberiella florii*) (Munyaneza, 2010). However, the disease transmission is controversial as both mealybugs and scale insects have been roped in as vectors (Rolt and Jelkmann, 2005). Additionally, grapevine leafroll-associated viruses (GLRaV-1,-3) have been reported to be scale insect-transmissible as well as mealybug-transmissible. The scale insects involved in the spread of grapevine leafroll-associated viruses one and three are grapevine scale (*Parthenolecanium persicae*), nigra scale (*Parasaissetia nigra*), black scale (*Saissetia* spp.) and European scales, namely wax scale (*Pulvinaria innumerabilis*) and mirid scale (*Pulvinaria vitis*) in Italy. In addition, the soft scale *Ceroplastes rusci*, has also been reported to be a vector of Grapevine leaf roll-associated virus-3 and Grapevine leafroll-associated virus-5 in grapevines (Mahfoudhi et al., 2009).

5.6.6 Mealybug-borne virus diseases

5.6.6.1 Grapevine leaf roll associated virus diseases-GLRaV (Closterovirus; Ampelovirus)

The leaf roll disease of grapevines is collectively caused by a group of ten viruses from the family Closteroviridae (*Closterovirus* and *Ampelovirus*). Of these viruses, GLRaV-2 is *Closterovirus*, while all others (GLRaV-1, -3, -5, -7, etc.), are ampeloviruses. The disease is identified by the discolored inter-venal areas, yellow chlorotic mottling and the downward-rolling of leaves in white cultivars of grapes. Besides, the vines bear small sized loose clusters. These viruses are positive ssRNA viruses and are spread through different species of mealybugs viz. grape mealybug (*Pseudococcus maritimus*), long-tailed mealybug (*Pseudococcus longispinus*), obscure mealybug (*Pseudococcus viburni*), citrophilous mealybug (*Pseudococcus calceolariae*) and vine mealybug (*Planococcus ficus*) (Tsai et al., 2010). The transmission mechanism between mealybugs and viruses is semi-persistent (Cid et al., 2007). Previously, the leaf roll disease was thought to be the outcome of viral infection spread through mealybugs alone. Recently, however, GLRaV-3 has been reported as being soft scale transmissible. More detailed investigations are needed in order to confirm the real vector/vectors of this malady.

5.6.6.2 Citrus yellows mosaic Badnavirus—CYMBV (Badnavirus)

This virus is bacilliform, non-enveloped dsDNA, a circular genome with six ORF. A member of the family Caulimiviridae, it infects citrus plants, producing mosaic symptoms (Ghosh et al., 2014). Amongst citrus, it has been found mainly on mandarin and sweet orange. It is transmissible by the mealybug, *Planococcus citri*. The mealybugs can acquire the virus within a feeding period of forty-eight hours on the virus source and inoculate in test plants within four days of inoculation access.

5.6.6.3 Grapevine virus A, -GV, -A (Vitivirus)

It is a flexuous filament +ve ssRNA virus with, 7.4 um length and five ORF falls under Betflexiviridae family. Grapevine virus D, *Heracleum* latent virus and Grapevine virus C are types of species of this virus. It has been mentioned as transmissible through both mealybugs and scale insects. This controversial point is yet to be resolved (Alabi et al., 2014).

5.6.7 Mealybugs/scale insects and fungi

The sooty mold fungi develop on the honeydew secreted by sucking pests like mealybugs/scale insects/whiteflies/aphids. These insects get smeared with mold and also ingest the molds. Besides, the spread of *Cephalosporium dieffenbachiae*, causing leaf spot of *Dieffenbachia*, is through the feeding punctures of mealybug *Pseudococcus citri*.

5.7 Mollicutes and Insect Relationships

The prokaryotic bacteria without cell walls and reduced genome, put under the order Acholeplasmatales (Class: Mollicutes; Division Tenericutes), are called Mollicutes. The class Mollicutes includes genera namely *Phytoplasma* (now known as *Candidatus Phytoplasma; Candidatus liberibacter*), *Spiroplasma* and *Mycoplasma*. The lethal yellowing disease of the coconut palm is a well-known disease caused by a *Phytoplasma*. The *Phytoplasma* are obligate bacterial parasites of phloem tissues of the plant that lack cell walls. Instead, they are bound by triple membrane of protein, have a pleomorphic/filamentous shape and are 200–800 nm in size. These organisms were discovered in 1967 as *Mycoplasma Like Organisms* and were named as *Phytoplasma* by a committee, International Organization for Mycoplasmology (ogenhout et al., 2008). The Sub-committee on Taxonomy of *Phytoplasma* in 1992 renamed *Mycoplasma Like Organisms as "Phytoplasma"* and placed them in the order, Acholeplasmatales. Like the other pathogens, insects are also efficient vectors of this category of pathogens. Insects belonging to families Cicadellidae (leafhoppers), Fulgoridae (planthoppers) and Psyllidae (Jumping lice), Cixiidae, Psyllidae, Delphacidae, and Derbidae are vectors of *Phytoplasma* and their diseases (Weintraub and Beanland, 2006; Bertaccini et al., 2014). The nymphs and adults have similar characteristic features including feeding on plant tissues and a pathogen-vector relationship of persistent propagative type; also, they both contain obligate symbionts.

Besides viruses, the planthoppers also act as vectors of *Phytoplasma* and *Spiroplasma.* The Lethal yellowing disease in coconut palms is caused by *Phytoplasma* and is prevalent in the USA, the Caribbean Islands and West Africa. In this disease, the leaves of the lower canopy turn yellow, followed by browning and shedding. The premature dropping of fruits is common. The spread of this pathogen is via planthopper *Myndus crudus;* the pathogen multiplies in the vector. The diseases caused by mollicutes are separated into two categories viz. caused by bacteria and caused by *Phytoplasma*. The bacteria are associated with canker (Citrus canker caused by *Xanthomonas axonopodis PV citri*), Bacterial leaf blight (leaf blight caused by *X. oryzae PV oryzae*), Bacterial wilt (wilt caused by *Ralstonia solanacearum*), Soft rot (rot of fruits/vegetables/ornamentals caused by *Erwinia caratovora*), Crown gall (caused in woody plants/herbaceous plants by *Agrobacterium tumefaciens*) and Greening disease (Caused by *Candidatus Liberibacter asiaticus*) and the other category is *Phytoplasma* etiology (lethal yellowing of coconuts and Citrus stubborn). The *Phytoplasma* are polymorphic bacteria lacking a cell wall and their size varies between 200–800 nm. This *Phytoplasma* can be recognized from broad symptoms which include virescence, malformation, phyllody, sterility of flowers, yellowing of leaves, proliferation of buds to produce witches broom symptoms and elongation and etiolation of internodes (Bertaccini and Duduk, 2009; Bertaccini, 2007). The diseases linked to *Phytoplasma* include Grapevine yellows (Flavescence doree and Bois Noir), Fruit trees yellows (Apple proliferation-AP, Pear decline-PD, European stone fruit yellows-ESFY), Citrus witches broom (Witches broom of lime-WBDL) and Palm lethal yellowing (Coconut palm lethal yellowing). In addition to 91 diseases already known, five new cases of *Phytoplasmal* etiology were identified (Weintraub, 2007). Those new findings include Potato purple top (*Circulifer tenellus*) (Munyaneza et al., 2007), Apple proliferation (*Fieberiella florii*) (Krezal et al., 1988), Lettuce phyllody (*Neoaliturus fenestratus)* (Salehi et al., 2006), Cotton phyllody (*Orosius cellulosus)*, Little leaf of beeline/sweet potato (*Orosius lotophagorum*), Beet witches broom (*Orosius orientatus*) (Mirzaie et al., 2007), Sugarcane white leaf (*Yamatotettix flavittatus*) (Hanboonsong et al., 2002) and Maize red leaf (*Reptalus panzeri*) (Jovic et al., 2007) and the spread of this *Phytoplasma* is through leafhoppers, planthoppers, and psyllids. In addition to viruses, the leafhoppers are also vectors of *Phytoplasma* and *Spiroplasma* in a persistent propagative manner. Mollicutes are known to cause diseases in crop plants and their spread is both through vegetative propagation and aerial insect vectors. *Mycoplasma* is another genus of class Mollicutes. These are cell wall-less but bounded by triple-layered membrane organisms of 1.2 um size with a genome length between 539–2220 kb and they replicate in insects (Leafhoppers and psyllids) (Bosco and Tedeschi, 2013; Weintraub and Beanland, 2006) and in sieve tubes of host plants (Gasparich, 2010). These are gram (+)ve bacteria with both small genome and low content of guanine and cysteine of genome DNA (Gamier et al., 2001). These two genera can be differentiated with respect to motility and culture on artificial media as the *Spiroplasma* are motile, helical, wall-less prokaryotes with genome size between 780–2220kb, cultivable on nutrient-rich artificial media; whereas *Phytoplasma* is wall-less non-helical, pleomorphic and uncultivable prokaryotes, amenable on cultured media and a cause of diseases in crop plants. The transmission of *Phytoplasma* is through leafhoppers and the important ones include Apple proliferation (*Cacopsylla picta; C. melanoneura; Fieberiella florii*), Lethal yellowing of palm/Coconut Cacopsylla palm lethal yellowing (American cixiid *Haplaxius = Myndus*) *crudus*), Bois Noir (planthopper *Hyalesthes obsoletus*), Flavescence doree (leafhopper *Scaphoideus titanus)*, Pear decline (*Cacopsylla picta; C. melanoneura; C. pyricola; C. pruni*). Aster yellows (*Scaphoideus titanus*), Mulberry dwarf (*Hishimonoides sellatiformis*), Sugarcane white leaf (*Matsumuratettix hiroglyphicus*), and Plum decline (*C. melanoneura*). The important genera of class mollicutes are discussed in the following paragraphs.

5.7.1 *Spiroplasma*

Spiroplasma was first discovered in 1973 as helical bacteria associated with Corn stunt disease in France and California, also known to attack plants and animals. *Spiroplasma* is associated with insects belonging to six orders (Coleoptera, Hemiptera, Hymenoptera, Lepidoptera, Odonata and Diptera) and fourteen families. Of these, leafhoppers and planthoppers from order Hemiptera are vectors of plant pathogenic organisms; the rest transmit arthropod-pathogenic organisms. With advancement of research, many more species of *Spiroplasma* were discovered such as *Spiroplasma citri* in citrus (Claran et al., 1989), causal organism of Citrus stubborn disease and Citrus brittle root disease in horse radish and Carrot purple disease in carrot, *S. phoeniceum* causing Periwinkle yellows in periwinkle (Saillard et al., 1987) and *S. kunkelii* which is a causal organism of Corn stunt (Saggio et al., 1973; Whitcomb et al., 1972). *Phytoplasma/Spiroplasma* belong to order Acholeplasmatales, family Acholeplasmataceae and genus *Candidatus Phytoplasma / Spiroplasma* and there are many species of these two genera (Regassa and Gasparich, 2006). The species *Spiroplasma kunkelii* and *S. floricola* were identified in 1975. Among these diseases, Corn stunt and Citrus stubborn are the important disorders of *Spiroplasma* etiology. Corn stunt caused by *Spiroplasma citri* was first discovered in 1973 and cultured in 1986 and the causal organism was identified as *S. kunkelii*. In the same year, another *Spiroplasma phoeniceum* was isolated from periwinkle plants (causing Periwinkle yellows) in Syria. Besides these three *Spiroplasma*, the others were found infecting animals in nature. The animal infecting *Spiroplasma* include, *S. melliferum* and *S. apis* from honey bees, *S. floricola* from cockchafer, *Melolontha melolontha* causing Lethargy disease and *S. poulsonii* from *Drosophila* and *S. mirum* from rabbit ticks (Bove, 1997). Of these, the Corn stunt is caused by *Spiroplasma kunkelii* and spread by leafhoppers namely *Dalbulus maidis, D. elimatus.* Similarly, Citrus stubborn is caused by *S. citri* and spread by leafhoppers *Circulifer tenellus, Scaphytopius nitrides* and *Neoaliturus haemoceps.* Corn stunt is recognized by yellow streaks in freshly emerged leaves in bushy plants. The latent period after acquisition of pathogen is two to three weeks. Similarly, Citrus stubborn caused by *Spiroplasma citri* is another disease of *Spiroplasma* etiology easy to recognize from the upright growth of plants and bitter taste of fruits unsuitable for consumption. It is vectored by many species of leafhoppers, such as *Circulifer tenellus, Scaphytopius nitrides* and *Neoaliturus haemoceps.*

5.7.1.1 Corn stunt disease

It was earlier known to be a disease of viral origin reported by Kunkel in 1946. It was the year 1975 when its etiology was discovered to be of *Spiroplasma kunkelii* (Williamson and Whitcomb, 1975). It can be recognized by the small chlorotic stripes which appear at the bases of young leaves and spread towards tips of leaves and finally reach the older leaves. In the meanwhile, there is a proliferation of secondary shoots and leaf axils and dwarfing of plant growth. The spread of disease is through leafhoppers *Dalbulus maidis* and *D. elimatus.* The latent period of *Spiroplasma* in vector is around twenty days and retention is for forty-five days. The plants show symptoms of the disease about three weeks after inoculation of organism. The pathogen is phloem-borne in nature. The disease is prevalent in USA and Mexico.

5.7.2 Candidatus Phytoplasma

The organisms previously known as *Mycoplasma Like Organisms* (MLO) were subsequently divided into *Candidatus Phytoplasma* and *Spiroplasma.* The application of modern technology led to the identification of *Candidatus Phytoplasma* after resolving the etiology of MLO's and rules for future identification of species were framed (IRPCM, 2004). As a result, the first classification of *Candidatus Phytoplasma,* based on restriction fragment length polymorphism (RELP) and analysis of polymerase chain reaction (PCR) (amplified 16Sr DNA), was framed (Lee et al., 2000; Lee et al., 1998a). Accordingly, fifteen groups and forty sub-groups of *Candidatus Phytoplasma* were established and this classification was widely acknowledged (Al Saady et al., 2008; Lee et al., 2006a; Arocha et al., 2005; Lee et al., 2004b). In due course of time, the detection methods were identified, and PCR is now used extensively. The electron microscope is commonly used for indexing of *Phytoplasma.* Like viruses, the *Candidatus Phytoplasma* also inflict diseases with virescens, phyllody, malformation of floral parts, witches broom-type symptoms along with stunted growth of plants (Bertaccini, 2007). These prokaryotes are transmitted through insects belonging to the families Cicadellidae, Cixiidae, Psyllidae, Delphacidae and Derbidae and are responsible for over 300 diseases (Hoshi et al., 2007). Aster yellows is an important disease caused by *Phytoplasma* in vegetable crops like potato, onion, carrot, tomato, etc. (Marcone et al., 2000). The diseased plants show acute chlorosis, excessive axillary branching, stunted growth, sterility of flowers and malformation of fruits. The spread in nature is via an important leafhopper *Macrosteles fascifrons*. The leafhopper retains the *Phytoplasma* throughout its life and it multiplies in the insect vector. Tomato big bud and Apple proliferation are of *Phytoplasma* etiology and their spread is via brown leafhopper, *Orosius argentatus*, and planthoppers species such as *Philaenus spumarius, Aphrophora alni, Lepyronia coleoptrata, Artianus insterstitalis,* and *Fieberiella florii.* The other economically important diseases include Coconut lethal yellowing, Peach X disease, and Grapevine yellows. Another disease caused by *Mycoplasma Like Organisms* is the Papaya bunchy top, recognized in 1931 for the first time in Puerto Rico. It is now prevalent in widespread areas covering central and South America. It is spread via two species of leafhoppers namely, *Empoasca papayae* and *E. stevensi*. Papaya is the only host of this organism that is MLO. Peach X disease of peaches is caused by *Mycoplasma Like Organism, Xanthomonas arboricola (= campestris) pv. pruni.* The major source of spread of this disease is choke cherries. The diseased peaches show yellowing and upward curling of leaves with serrated margins, presence of water soaked red lesions, quick defoliation and fewer fruits with pale and leathery skin. The disease is also known as cherry buck skin disease. It is transmissible by eight to ten species of leafhoppers notably the choke cherry leafhopper (*Colladonus montanus*), saddleback leafhopper (*C. clitellarius*), sharp nosed leafhopper (*Scaphytopius acutus*) and cherry leafhopper (*Fieberiella florii)*. After the lapse of latent period (twenty to twentyfive days), the leafhoppers remain infective throughout their life (thirty-five to forty days). Likewise, Flavescence doree Phytoplasma (FDP) (Martini et al., 2002) is a *phytoplasma*-transmissible organism from this category; transmissible via leafhopper, *Scaphoideus titanus*. Bois Noir *Phytoplasma* in Italy is also from this category of organisms (Botti and Bertaccini, 2007). Presently, many diseases are still being added to the list of *Phytoplasma* diseases. In the recent past, Citrus huanglongbing, linked to aster yellows in china (16SrI) (Teixeira et al., 2009), and Pigeon pea witches broom, linked to *Phytoplasma* (16SrIX) in Brazil (Chen et al., 2008), have been identified. Besides the earlier records, two more species, *Candidatus Phytoplasma australiense* (causing Lethal decline disease in *Coprosma* spp., Lethal yellows in strawberry, Sudden decline in *Cordyline* spp. and Yellow leaf in *Phormium* spp.) and *Candidatus Phytoplasma pruni* (causing Branch inducing *Phytoplasma* in *Euphorbia* spp.) were discovered and identified and added to the list (Winks et al., 2014; Beever et al., 2004; Liefting et al., 1998). Recently thirty-three groups with more than 100 sub-groups have been constituted (Bertaccini et al., 2014). These groups are Aster yellows witches broom group (Aster yellows witches broom, tomato big bud, Aster yellow mild strain, Aster yellows, Clover phyllody, Paulowania witches broom, Blueberry stunt, Strawberry witches brooms, Soybean purple and yellow stem, Cherry little leaf, Mexican potato purple top, Peach rosette-like disease and Tomato brote grande) (Lee et al., 2013; Arocha-rosete et al., 2011; Santos-cervantes et al., 2010; Arocha-rosete et al., 2007; Lee et al., 2006a; Bai et al., 2006; Valiunas et al., 2005; Oshima et al., 2004; Lee et al., 2004a; Seruga et al., 2003; Lee et al., 2002; Beanland et al., 1999; Jomantiene et al., 1998; Lee et al., 1992); Peanut witches broom group (Lime witches broom, Lime witches broom, Faba bean phyllody, Papaya mosaic, Pichris phyllody and Cotton phyllody) (Wei et al., 2007; Seemuller et al., 1998; White et al., 1998; Zreik et al., 1995; Gundersen et al., 1994), Peach X disease group (Peach X disease, Clover yellow edge, Peach bunch, Golden rod yellow, Spiraea stunt, Milkweed yellows, Walnut witches broom, Poinsettia branch inducing, Virginia grapevine yellows, Chayote witches broom, Strawberry leaf fruit, cassava frog skin disease, Potato purple top, *Dandelion virescence*, Black raspberry witches broom, Sweet and sour cherry phytoplasma, Cirsium white leaf and passion *Phytoplasma*) (Davis et al., 2013; Davis et al., 2012; Alvarej et al., 2009; Valiunas et al., 2009; Davis et al., 2003; Jomantiene et al., 2002; Montano et al., 2000; Jomantiene et al., 1998), Coconut lethal yellows group (Coconut lethal yellowing, Yukatan coconut lethal decline and Tanzanian coconut lethal decline) (Harrison et al., 2002; Harrison et al., 1994), Elm yellows group (Elm yellows, Jujube witches broom, Flavescence doree, and Blanite witches broom) (Win et al., 2013; Torres et al., 2005; Lee et al., 2004b; Martini et al., 2002; Jung et al., 2003; Daire et al., 1992), Clover proliferation group (Clover proliferation, Strawberry multiplier disease, Ililinois yellows, Periwinkle little leaf, Cantarurex virescence, Catharanthus phyllody, Partulaca little leaf and Passion fruit phytoplasma) (Valiunas et al., 2009; Hiruki and Wang, 2004; Faggioli et al., 2004; Martini et al., 2002; Siddique et al., 2001; Griffith et al., 1999; Jomantiene et al., 1998), Ash yellows group (Ash yellows, engeron witches broom and Argentina alfa witches broom) (Conci et al., 2005; Barros et al., 2002; Griffith et al., 1999), Loofah witches broom group (Loofah witches broom, pigeon pea witches broom, Noxos periwinkle virescence, Junipercis witches broom, Almond witches broom, Almond and stone fruit witches broom) (Molino et al., 2011; Duduk et al., 2008; Verdin et al., 2003; Ho et al., 2001; Gundersen et al., 1996), Apple proliferation group (Apple proliferation, Europe stonefruit yellows, Pear decline, Spartium witches broom and Black older witches broom) (Seemuller and Schneider, 2004;

Marcone et al., 2003a), Rice yellows dwarf group (Rice yellows dwarf, sugarcane white leaf and leafhopper-borne phytoplasma) (Jung et al., 2003; Seemuller et al., 1994), Stolbur group (Stolbur, Australian grapevine yellows, strawberry lethal yellows, yellow diseased strawberry, Bos Noir, Windweed yellows and Japanese hydrangea phyllody (Quaglino et al., 2013; Martini et al., 2012; Marcone et al., 2000; Sawayanagi et al., 1999; Davis et al., 1997; Padovan et al., 1995), Mexican periwinkle virescence group (Mexican periwinkle virescence and Strawberry green petal) (Jomontiene et al. 1998; Gundersen et al., 1994), Bermuda grass white leaf group (Bermuda grass white leaf and Bermuda grass white leaf Iran) (Salehi et al., 2009; Marcone et al., 2003b), Hibiscus witches broom group (Hibiscus witches broom and Gauzima witches broom) (Montano et al., 2001; Villalobes et al., 2011), Sugacane yellow leaf syndrome group (Sugarcane yellow leaf syndrome) (Arocha et al., 2005), Papaya bunchy tops group (Papaya bunchy tops) (Lee et al., 2006a), American potato purple top wilt group (American potato purple top wilt) (Lee et al., 2000; 2006b), Chest nut witches broom group (Chest nut witches broom) (Jung et al., 2002), Rhamus witches broom group (Rhamus witches broom) (Marcone et al., 2003a), Pinus white plama group (Pinus white plama) (Schneider et al., 2005), Lethal yellow disease group (Lethal yellow davigroup (Buckland velley yellows)/Sorghum bunchy shoot group (Sorghum bunchy shoot)/Weeping tea witches broom group (Tea witches broom),Sugarcane phytoplasma group (Sugarcane phytoplasma/Sugarcane white stem), Derbi phytoplasm group/(Derbi sugarcane white stem) (Wei et al., 2007), Cassia witches broom group (Cassia witches broom) (Al Saady et al., 2008), Salt cedar witches broom (Salt cedar witches broom) (Jhao et al., 2009), Soybean stunt group (Soybean stunt) (Lee et al., 2011), Malaysian periwinkle virescence and phyllody group (Malaysian periwinkle virescence, Malaysian yellow dwarf phytoplasma and Malaysian oil palm phytoplasma) (Nejat et al., 2012; Nejat et al., 2009) and Allocasuarina phytoplasma group (*Allocasuarina phytoplasma*) (Marcone et al., 2003a).

5.8 Mechanism of Transmission of Mollicutes

The spread of Mollicutes is via leafhoppers (Cicadellidae), planthoppers (Fulgoridae) and psyllids (Psyllidae) in a circulative and propagative manner. These prokaryotes contain antigenic protein on the cell surface that normally reacts with microflora inhabiting their intestines and plays a vital role in transmission and infection (Hoshi et al., 2007; Suzuki et al., 2006). Like plant viruses, these organisms remain confined to their intestines and possibly influence the fitness of the vector (Sugio et al., 2011;

Christensen et al., 2005). These organisms are ingested with sap, enter the intestine and get into the hemolymph through absorption, finally, they invade the salivary glands and initiate a latent period of several weeks. These pathogens are transovarially transmitted, such as Aster yellows in *Scaphoideus titanus* (Danielli et al., 1996), Mulberry dwarf in *Hiishimonoides sellatiformis* (Kawakita et al., 2000), Sugarcane white leaf in *Matsumuratettix* = *Matsumura hiroglyphicus* (Hanboonsong et al., 2002) and Apple proliferation in *Cacopsylla melanoneura* (Tedeschi et al., 2006), through dodder and seeds of coconut, lime and tomato infected with *Phytoplasma* (16Sr I, 16Sr XII, 16Sr II) (Botti and Bertaccini, 2006; Khan et al., 2002) and through vegetative propagation means (cuttings and micropropagation, etc.). While feeding on healthy plants, the vector injects the pathogens and these organisms, being small in size, easily pass through the sieve pores to reach the sieve tube elements. From there, the *Phytoplasma* reaches the sink tissues, i.e., immature leaves and roots. In these tissues, there is a change in the volume of oxygen and carbon that influence the *Phytoplasma*. Additionally, the change in the concentration of carbohydrates in roots, phloem and leaves and photosynthetic pigments, hormone levels and soluble proteins also takes place (Musetti et al., 2005; Maust et al., 2003; Bertaccini and Nedunchezhian, 2001). These alterations are linked to restriction in phloem transport and photosynthesis and cause symptoms of the disease (Maust et al., 2003; Lepka et al., 1999). These organisms are difficult to detect but there are methods to locate their presence in plants. Earlier, ELISA was in use but soon PCR and RELP were deployed to detect and identify the pathogen. In the nineties, after the cloning of DNA of *Phytoplasma*, nucleic acid probes were put to use to detect such plant pathogens. The primer-based conserved sequence (16Sr RNA gene ribosomal protein gene operon, tuf, and Sec Y genes) is now the quick method for detection, identification and classification of Mollicutes (Martini et al., 2007; Wei et al., 2004; Marcone et al., 2000; Schneider et al., 1997). The details regarding different groups and sub-groups are presented (based on Lee et al., 1998a). Furthermore, several species of *Candidatus phytoplasma* have been identified and these are *Ca p japonicum*/Hydrangea phyllody (Sawayanagi et al., 1999) (*Ca p castaneae*/Chestnut witches broom (Jung et al., 2002), *Ca p pini*/pinus decline (Schneider et al., 2005), *Ca Phytoplasma cynodontis* (Marcone et al., 2003b), *Ca p rhamni*/Rhamnus witches broom (Marcone et al., 2003a), *Ca p allocasuarinae*/Allocasuarina yellows (Marcone et al., 2003a), *Ca p fragariae*/Strawberry yellows (Valiunas et al., 2006), *Ca p lycopersici*/Brote grande of tomato (Arocha et al., 2007), *Ca p palmicola* (Dickinson, 2014) and *Ca p tamaricis*/Salt cedar witches broom (Zhao et al., 2009).

5.9 Psyllid-borne Phytoplasma Diseases

5.9.1 Candidatus liberibacter

Psyllid-borne diseases viz. Citrus greening (bacteria) and Pear decline (*Phytoplasma)* are a limiting factor in the successful cultivation of citrus and pear throughout the globe. *Candidatus Liberibacter* is an important genus that includes plant pathogenic species *Candidatus Liberibacter solanacearum* (*Solanum lycopersicum, S. tuberosum, S. nigrum, S. aviculare, S. betaceum, S. laciniatum, Capsicum annum, Bidens* sp., *Lycium ferocissimum* and *Physalis peruviana*) (Vereijssen and Scott, 2013; Liefting et al., 2009; Liefting et al., 2008a and b; Munyaneza et al., 2007; Pearson et al., 2006) and *Ca Liberibacter europaeus (Cytisus scoparius*) (Thompson et al., 2013) recently identified from New Zealand. Of these, *Ca L solanacearum* is transmissible through tomato/potato psyllid, *Bactericera cockerelli*, while *Ca L europaeus* is vectored by scotch psyllid, *Arytainilla spartiophila*. Psyllids are vectors of *Phytoplasma* and the important diseases spread by psyllids include Citrus greening, Pear decline and Zebra chip of potato. Psyllids (*Diaphorina citri* and *Trioza erytreae*) are insects belonging to the family Psyllidae under the order Hemiptera and have piercing and sucking mouthparts and are thus also considered to be efficient vectors of plant pathogens. Psyllids are vectors of Citrus greening (Huanglongbing). The disease was recently introduced into Brazil and Florida. The disease is identified by the stunted growth of plants, branches showing dieback and bearing thin and chlorotic foliage and green bands appearing along main veins of leaves initially, which eventually turn completely chlorotic with green spots. Fruits are small and misshapen. The causative agent of disease is phloem-inhabiting bacteria. The bacteria can be acquired within a feeding period of fifteen to thirty minutes on the infected source and inoculated in a healthy plant with an inoculation access of fifteen minutes. With the increase in inoculation access period to one hour, the transmission efficiency increases, until it reaches 100 per cent with the passage of time. The 4th and 5th instar nymphs (Xu et al., 1988) can also acquire the bacterium but the transmission is only through adults. The adults can retain the bacteria for life and transovarial transmission (Vanden-Berg et al., 1992) is positive. The bacteria are present in three forms, *Candidatus liberibacter africanus, Candidatus L asiaticus* and *Candidatus L americanus.* The Asian psyllid, *D citri* is thre to four mm long. The egg laying is confined to new growth in the folds of leaves. The female lays between 800–1000 pale, almond-shaped eggs during its entire lifespan. The eggs hatch into nymphs within a period of three days and the nymphs pass through five nymph instars. The nymphs become adults in about ten to forty days and adult longevity is 50–80 days. These insects are phloem feeders and the bacterium is also present in the phloem. While feeding in the phloem, the bacteria are also picked up along with sap.

5.9.1.1 Zebra chip of potato

Besides the Citrus greening, the Zebra chip disease of potato caused by *Bactericera cockerelli* is transmissible by African psylla, *T. erytreae*. The bacterium *Candidatus liberibacter*, associated with disease of carrot (Carrot purple leaf disease), is transmitted by *T. apicalis* and *Daucus carota fsp. sativus* in Northern Europe (Munyaneza et al., 2010). *Candidatus liberibacter solanacearum* also infects tomato, pepper, eggplant, tobacco, tomatillo (*Physalis peruviana*), and tomarillo (*Solanum betaceum*) in addition to tomato. The disease is prevalent in the USA, Sweden, Canara Island, Spain, France, Norway, Guatemala, Nicaragua, and New Zealand. It can be identified by general chlorosis, erectness, and cupping of leaves with thickened apical internodes. The fruits are small and misshapen. The disease is also known by different names such as psyllid yellows, Punta morada, papa rayada, etc., depending upon countries of prevalence. Its spread is via psyllid species viz. *Bactericera cockerelli, B. trigonica* and *Trioza apicalis* in various different countries.

5.9.1.2 Citrus greening

It is a devastating disease of citrus, known the world over as Huanglongbing disease (HLB) or yellow dragon disease. Initially, the symptoms remain confined to one side of the tree, covering one branch or stem. Such symptoms very much resemble the symptoms caused by a deficiency of zinc in citrus trees. It has been an economically important disease in Florida since its detection in 2005. The state would previously produce up to 242 million boxes of fruits, but this sunk down to 104 million boxes in 2014. Its development is very much dependent on temperature. It is caused by *Candidatus liberibacter* species *Ca. l. asiaticus* (temperature of up to 35°C) and *Ca. l. africanus/Ca. l. americanus* (both develop at a temperature range of 20–25°C). It exists in three forms, namely the heat tolerant Asian form, and the heat sensitive African and American forms. It is now widespread in distribution and prevails in Asia, Africa, South America, Brazil, Mexico, etc., though it was first identified in 1929 in China. It is spread by Asian psylla (*Diaphorina citri*) and African psylla (*Trioza erytreae*) in nature (Manjunath et al., 2008). Both nymphs and adults can acquire the bacterium in an acquisition access of fifteen to thirty minutes to 5 hours. The psyllids can retain the bacterium throughout their lives once acquired. Another species of psyllid, *Bactericera cockerelli*, has been identified as a carrier of bacterium *C. liberibacter*, as the pathogen has been found in its hemolymph and salivary glands (Hansen et al., 2008). Initially, the diseased plants show asymmetrical yellowing and mottling of foliage (blotchy mottling), followed by appearance of green islands and finally dieback of young twigs. The decay of rootlets and development of lateral roots are also conspicuous in such diseased trees. The trees are stunted in growth and bear fewer fruits. The fruits borne on diseased trees continue to be of green color and juice of such fruits is of bitter taste and is unsuitable for human consumption. In addition, the diseased trees also bear small and misshapen fruits.

5.9.1.3 Apple proliferation

A disease that produces fruit tree yellows such as witches broom, dented leaves, enlarged stipules and small, flattened fruits with elongated panicles. Early leaf-reddening is evident in diseased trees. The causal organism, *Phytoplasma*, is vectored by *Cacopsylla picta* and *C. melanoneura* in Italy. Also in this category, the European stone fruit yellows is another malady which is transmissible through psyllids, *Cacopsylla pruni* (Tedeschi et al., 2006).

5.9.2 Candidatus Phytoplasma diseases

5.9.2.1 Lethal yellowing of coconut palms (LY)

A pandemic disease caused by *Mycoplasma Like Organisms*. Extensive damage to coconut palms in areas of its prevalence is reported (Cuba, Jamaica, Bahamas, Florida, Mexico, Belize, Kenya, Mozambique, Tanzania, Nigeria, and Ghana). It was first recorded in 1891 in Jamaica. The diseased plantation initially shows yellowing of lower canopy leaves with streaks, followed by the upper canopy. Defoliation, blackening, and necrosis of inflorescence ensues, and the trunk becomes devoid of branches and leaves. The coconut palm dies within a short span of 3–7 months after the infection of *Phytoplasma*. The causative organism (proposed) is *Candidatus Phytoplasma Palmae* (member of group 16Sr DNA RFLP group 16Sr IV, sub-group-A) and its transmission is through the American palm cixiid planthopper, *Haplaxius crudus* (synonyms: *H. cocois/Myndus crudus*) and in Caribbean *H. taffini* in other regions of occurrence of the disease. The planthopper adults feed on coconut palm (*Cocos nucifera*) and other palms, while the immature stages feed on roots of turf grasses (*Stenotaphrum secundatum, Palspalum notatum, Cynodon dactylon*) grown in the vicinity of coconut palms. Under this category, Bois Noir is another disorder caused by *Phytoplasma* and is transmissible via planthopper *Hyalesthes obsoletus*. It belongs to 16Sr XII-A group of *Phytoplasma* and is transmissible through cixiid. In all, fifty-seven species of insects have been confirmed as vectors which include leafhoppers, planthoppers, true bugs, cicada, spittlebugs, lace bugs and stink bugs. All the insect species are phloem feeders except cicada and spittlebugs, which are xylem feeders. These two species of insects have strong frontoclypeus containing a powerful sucking pump used to suck large quantities of fluid. The key planthopper can acquires the pathogen in minutes to hours with a latent period up to eighty days (Gurr et al., 2015).

5.9.2.2 Pear decline

It is a serious disease of pear and is widely distributed in the world, particularly in Europe, North America and Canada. The disease is caused by the organism *Candidatus Phytoplasma pyri* (class Mollicute, genus *Phytoplasma*). The prokaryote has small genome 530–1350 kb with low content of G + C. In general, the plants show poor shoot growth and dieback. Premature leaf-reddening with upward-rolling of leaves is also a common feature of the disease. It has slow and quick decline forms; in quick decline, the trees suddenly scorch, wilt, and die. The leaves also become red on account of damage to root system. While in the slow form of disease, the trees become thinly foliated and show signs of arrested terminal growth. The upward-rolling of leaves with thickened veins is also visible in diseased trees. The characteristic feature of damage is the presence of dark phloem ring just below the graft union. Slow decline occurs on trees grafted on the same rootstock. The disease is transmissible through different species of pear psylla (*Cacopsylla agricola; C. quinli; C. pyrisuga; C. pyri; C. chinensis*). The organism is present in cephalic part of foregut of the vector psylla. The pathogen is acquired by psylla in a few hours of access to the source and the vectors can retain the pathogen in their bodies up to three weeks.

5.10 Rickettsia Like Organisms (RLO)

In addition to viruses and Mycoplasma, there is another category of organisms known to be transmissible through leafhoppers. These organisms are from the phylum Proteobacteria, class Alphaproteobacteria, sub-class Rickettsiae, order Rickettsiales, family Rickettsiaceae, genus Rickettsia and have many species. These organisms are non-motile, gram-negative, non spore-forming, highly pleomorphic bacteria and exist as cocci and threads. Rickettsia can be distinguished from *Mycoplasma* by the presence of rods, two tri-laminar membranes, the outer one being convoluted in contrast to the single smooth tri-laminar membrane that binds *Mycoplasma* cells. These organisms resemble bacteria but are smaller than bacteria and are obligate intracellular organisms. They cause diseases both in plants and other animals such as clover club leaf in the USA and unnamed disease of clover in England, rugose leaf curl in Australia and clover decline in France (unconfirmed). Besides, the other suspected cases of Rickettsia are Beet latent rosette, Bunchy tops of banana, Grapevine infectious necrosis, Grapevine pierce disease, Grapevine yellows, Witches broom of Larix sp., Phony peach, White clover disease, Ratoon stunt of sugarcane and Apple proliferation (Benhamou and Sinha, 1981). Another study carried out in Cuba indicated that bunchy tops of papaya are due to *Rickettsia Like Organism* (Acosta et al., 2013). Accordingly, the study showed that partial rickettsiae sequence was 100 per cent identical to that of rickettsiae associated with Papaya bunchy top in Puerto Rico. Likewise, Strawberry lethal yellows have also been reported as a suspected case of rickettsial etiology in Australia (Streten et al., 2005). A study carried out in East Bohemia demonstrated that carrot proliferation disease is due to a *Rickettsia Like Organism* (Franova et al., 2008). The blueish-white patches in sieve tube elements were noticed under florescence microscope, and further confirmed through transmission electron microscope.

5.10.1 Bunchy tops of papaya

This disease is caused by *Rickettsia Like Organism* (RLO) and is prevalent in the USA and the Caribbean Islands. Its spread is via species of leafhopper, *Empoasca papayae* and *E. stevensi* (Haque and Parasram, 1973). The incubation period of the disease is approximately one month. The pathogen is retained for life and it multiplies in the vector. The disease can be recognized by mottling, chlorosis, and necrosis of leaf margins. On account of further disease development, there is a shortening of internodes due to retardation in apical growth and plants become bushy in appearance as a result.

References

Acosta K, Zamora L, Pinol B, Fernandez A, Chavez A, Flores G, Mendez J, Santos ME, Leyva NE and Arocha Y (2013). Identification and molecular characterization *Phytoplasma* and *Rickettsia* pathogens associated with bunchy top symptoms (BTS) and Papaya bunchy top (BPT) of papaya in Cuba. Crop Protection, 45: 49–56.

Adam G and Hsu NT (1984). Comparison of structural proteins from two Potato yellow dwarf viruses. Journal of General Virology, 65: 991–994.

Agrios GN (2008). Transmission of plant diseases by insects. pp. 3853–3885. In: Capinera JL (ed). Encyclopedia of Entomology, 2nd Edition, Dordrecht: Kluwer Academic.

Alabi OJ, Ravahnih MI, Mekuria TA and Naidu RA (2014). Genetic diversity of grapevine virus an in Washington and California vineyards. Phytopathology, 104: 548–560.

Al-Saady NA, Khan AJ, Calari A, Al-subhi AM and Bertaccini A (2008). *Candidatus Phytoplasma omanense*, a *Phytoplasma* associated with witches broom of *Cassia indica* (Mil) Lam, Oman. International Journal of Systematic and Evolutionary Microbiology, 58: 461–466.

Alvarez E, Mejia JF, Llano GA, Loke JB, Calari A, Duduk B and Bertaccini A (2009). Detection and molecular characterization of *Phytoplasma* associated with Frogskin disease in cassava. Bulletin of Insectology, 60: 273–274.

Ammar D, Tsai DC, Whitfield AE, Redinbaugh MG and Hogenhout SA (2009). Cellular and molecular aspects of *Rhabdovirus* interactions with insect and plant hosts. Annual Review of Entomology, 54: 447–468.

Ammar ED and Nault LR (1985). Assembly and accumulation sites of Maize mosaic virus, in its planthopper. Intervirology, 24: 33–41.

Ammar ED and Nault LR (1991). Maize chlorotic dwarf virus-like particles associated with the foregut in vector and nonvector leafhopper species. Phytopathology, 81: 444–448.

Ammar ED (1994). Propagative transmission of plants and animal viruses by insects; factor affecting vector specificity and competence. Advances in Disease Vector Research, 10: 289–332.

Ammar ED and Nault LR (2002). Virus transmission by leafhoppers planthoppers and treehoppers (Auchenorrhyncha: Homoptera). Advances in Botanical Research, 36: 141–167.

Ammar ED, Khalifa EA, Mahmoud A, Abol-Ela SE and Peterschmitt M (2007). Evidence for multiplication of the leafhopper-borne Maize yellow stripe virus in its vector using ELISA and dot-blot hybridization. Archives of Virology, 152: 489–94.

Ammar ED and Hogenhout SA (2008). A neurotropic route of maize mosaic virus (Rhabdoviridae) in its planthopper vector, *Peregrinus maidis*. Virus Research, 131: 77–85.

Arocha–Rosete Y, Zunnoon-Khan S, Krukovets I, Crosby W, Scott J, Bertaccini and Michelutti R (2011). Identification and molecular characterization of the *Phytoplasma* associated with rosette-like a disease at the Canadian clonal gene bank based on the 16Sr RNA gene analysis. Canadian Journal Plant Pathology, 33: 127–134.

Arocha Y, Lopez M, Pinol B, Fernandez M, Picornell B, Ai-meida R, Palenzuela I, Wilson MR and Jones P (2005). *Candidatus Phytoplasma graminis* and *Candidatus Phytoplasma caricae;* two novel *Phytoplasma* associated with diseases of sugarcane, weeds, and papaya in Cuba. International Journal of Systematic and Evolutionary Microbiology, 55: 2451–2463.

Arocha YO, Antesana E, Montellano E, Franco P, Plata G and Jones P (2007). *Candidatus Phytoplasma lycopersici;* a *Phytoplasma* with Hoja de perejil disease in Bolivia. International Journal of Systematic and Evolutionary Microbiology, 57: 1704–1710.

Asanzei MC, Bosque-perez MA, Nault LR, Gordon DT and Thottapally G (1995). Biology of *Cicadulina* species (Homoptera: Cicadellidae) and transmission of Maize streak virus African. Entomologist, 3: 173–179.

Bai X, Zhang J, Ewing A, Miller SA, Radek AJ, Shevchenko DV, Tsukerman K, Walunas T, Lapidus A, Campbell JW and Hogenhout SA (2006). Living with genome instability: The adaptation of *Phytoplasmas* to diverse environment of their insect and plant hosts. Journal of Bacteriology, 188: 3682–3690.

Barros TSL, Davis RE, Resene RO and Dally EL (2002). Erigeron witches-broom *Phytoplasma* in Brazil represents new subgroup VII-B in16Sr RNA gene group VII, the Ash Yellows *Phytoplasma* group. Plant Disease, 86: 1642–1648.

Beanland L, Hoy CW, Miller SA and Nault LR (1999). Leafhopper (Homoptera: Cicadellidae) transmission of Aster yellow *Phytoplasma;* Does gender matter? Environmental Entomology, 28: 1101–1106

Beever RE, Wood GA, Andersen MT, Pennycook S, Sutherland PW and Forster RLS (2004). *Candidatus Phytoplasma australiense* in *Coprosma robusta* in New Zealand. New Zealand Journal of Botany, 42: 663–675.

Benhamou N and Sinha RC (1981). Association of a *Rickettsia-Like Organism* with a disease of white clover. Canadian Journal of Plant Pathology, 3: 191–196.

Bertaccini A (2007). *Phytoplasma;* diversity, taxonomy and epidemiology. Frontiers of Bioscience, 12: 673–689

Bertaccini A and Duduk B (2009). *Phytoplasma* and *Phytoplasma* diseases; a review of recent research. Phytopathologia Mediterranea, 48: 355–378.

Bertaccini A, Duduk B, Paltrinieri S and Contaldo N (2014). *Phytoplasmas* and *Phytoplasma* diseases; a severe threat to agriculture. American Journal of Plant Sciences, 5: 1763–1788.

Bertaccini M and Nedunchezhian N (2001). Effect of *Phytoplasma* stolbur–subgroup (Bois noir-BN) of photosynthetic pigments, saccharides, ribose-1,5 bisphosphate carboxylase, nitrate and nitrite reductases and photosynthetic activities in field grow grapevine (Vitis vinifera L. cv Chardonnay) leaves. Photosynthetica, 39: 119–122.

Bosco D and Tedeschi R (2013). Insect vectors transmission assays. Methods in Molecular Biology, 933: 73–85.

Botti S and Bertaccini A (2006). *Phytoplasma* infection through seed transmission; further observations. 16th International Congress of the IOM Cambridge, UK, 9–14 July 76, 113.

Botti S and Bertaccini A (2007). Grapevine yellows Northern Italy: Molecular gentrification of Flavescence doree *Phytoplasma* strains of Bois Noir *Phytoplasma*. Journal of Applied Microbiology, 103: 2325–2330.

Bove JM (1997). Spiroplasma: infection agents of plants arthropods and vertebrates. When Klin Wochenschr, 109: 604–612.

Braithwaite KS, Geijskes RJ and Smith GR (2004). A variable region of Sugarcane bacilliform virus (SCBV) genome can be used to generate promoters for transgene expression in sugarcane. Plant Cell Reports, 23: 319–326.

Briddon RW, Bedford ID, Tsai JH and Markham PG (1996). Analysis of nucleotide sequence of the treehopper transmitted *Geminivirus,* Tomato pseudo-curly top virus, suggests a recombinant origin. Virology, 219: 387–394.

Catherall PL (1970). Oat sterile dwarf virus. Plant Pathology, 9: 75–78.

Chang CJ, Garnier M, Zreik I, Rosette V and Bove JM (1993). Culture and serological detection of the xylem-limited bacterium, causing Citrus variegated chlorosis and its identification as the strain of *Xylella fastidiosa*. Current Microbiology, 27: 137–142.

Chatterjee S, Almeida RPP and Lindow S (2008). Living in two worlds: the plant and animal lifestyles of *Xylella fastidiosa*. Annual Review of Phytopathology, 46: 243–271.

Chen Q, Chen HY, Mao QZ, Liu QF, Shimzu T, Uehara-Ichiki T, Wu ZJ, Xie LH, Omura T and Wei TY (2012). Tubular structure induced by plant virus facilitates viral spread in its vector insect. PLoS Pathog, 2012, 8: e1003032.

Chen Q, Wang HT, Ren TY, Xie LH and Wei T (2015). Interaction between non-structural protein Pns ten of Rice dwarf virus and cytoplasmic actin of leafhoppers is correlated with insect vector specificity. Journal of General Virology, 96: 933–938.

Chen H, Chen Q, Omura T and Uehara-Ichiki T and Wei T (2011). Sequential infection of Rice dwarf virus in the internal organs of its insect vector after ingestion of virus. Virus Research, 160: 389–394.

Chen J, Pu X, Deng X, Liu S, Li H and Civerolo E (2008). A *Phytoplasma* closely related to the Pigeon pea witches broom *Phytoplasma* in the state of Sao Paulo, Brazil. Phytopathology, 98: 977–984.

Childress SA and Harris KF (1989). Localization of virus-like particles in the foreguts of viruliferous *Graminella nigrifrons* leafhoppers carrying the semi-persistent Maize chlorotic dwarf virus. Journal of General Virology, 70: 247–251.

Chiu RJ, Liu HY, Macleod R and Black LM (1970). Potato yellow dwarf virus in leafhopper cell culture. Virology, 40: 387–396.

Christensen NM, Axelsen KB, Nicolaisen M and Schulz A (2005). *Phytoplasmas* and their interactions with hosts. Trends in Plant Science, 10: 526–535.

Cid M, Pereira S, Cabaleiro C, Faora F and Segura A (2007). Presence of Grapevine leafroll-associated virus-3 in primary salivary glands of the mealybug vector (*Planococcus citri*) suggests a circulative transmission mechanism. European Journal of Plant Pathology, 118: 23–30.

Claran EC and Bove JM (1989). Biology of *Spiroplasma citri*. In: Whitcomb RF and Tully JG (eds). The *Mycoplasma.* Academic Press, San Diego, CA.

Conci L, Meneguzzi N, Galddeano E, Torres L, Nome C and Nome S (2005). Detection and molecular characterization of an Alfalfa *Phytoplasma* in Argentina that represents a new subgroup in the 16Sr DNA Ash yellows group (*Candidatus Phytoplasma fraxini*). European Journal of Plant Pathology, 113: 255–265.

Creamer R, Nault LR and Gingery RE (1993). Biological factors affecting leafhopper transmission of purified Maize chlorotic dwarf *Macluravirus*. Entomologia. Experimentalis et Applicate, 67: 65–71.

Daire X, Boudon–Padieu E, Berville A, Schneider B and Caudwell A (1992). Cloned DNA probes for detection of Grapevine Flavescence doree *Mycoplasma Like Organism* (MLO). Annals of Applied Biology, 121: 95–103.

Dale WT (1957). Insect transmission studies. Annual Report West Africa Cocoa Research Institute 1955–56.

Danielli A, Bertaccini A, Alma A, Bosco D, Vibio M and Arzone A (1996). May evidence of 16SrI-group-related *Phytoplasmas* in eggs, nymphs, and adults of *Scaphoideus titanus* Ball suggest their transovarial transmission? IOM Letters, 4: 190–191.

Davis RE, Dally EL, Gundersen DE, Lee IM and Habili N (1997). *Candidatus phytoplasma australiense*, a new taxon associated with Australian grapevine yellows. International Journal of Systematic Bacteriology, 47: 262–269.
Davis RE, Dally EL, Dally EL and Converse RH (2003). Molecular identification of *Phytoplasma* associated with Witches broom disease of blackberry in Oregon and its classification in group 16SrIII, New group Q. Plant Disease, 95: 1121.
Davis RE, Jhao Y, Dally EL, Jomantiene R, Lee IM, Wei W and Kitajima EW (2012). *Candidatus Phytoplasma sudamericanum*, a novel taxon and strain pass WB-Br4, a new subgroup 16Sr III-V *Phytoplasma,* from diseased passion fruit *Passiflora edulis f flavicarpa* Deg. International Journal of Systematic and Evolutionary Microbiology, 62: 984–989.
Davis RE, Jhao Y, Dally EL, Lee IM, Jomantiene R and Douglas SM (2013). *Candidatus Phytoplasma pruni*, a novel taxon associated with X-disease of stone fruits *Prunus* spp. *multilocus* characterization based on 16Sr RNA Sec Y and ribosomal protein genes. International Journal of Systematic and Evolutionary Microbiology, 63: 766–776.
Delong DM and Severin HHP (1950). Spittle insect vectors of Pierce disease virus: characters, distribution and food plants Hilgardia, 19: 339–356.
Deng JH, Li S, Hong J, Ji YH and Zhou YJ (2013). Investigation on subcellular localization of in its vector small brown planthopper by electron microscopy. Virology Journal, 10: 310.
Dickinson M (2014). *Candidatus phytoplasma palmicola;* a novel taxon associated with lethal yellowing type disease (LYD) of coconut (Cocos nucifera l.) in Mozambique. International Journal of Systematic and Evolutionary Microbiology, HTTP/dx.doi:org/10 10.1099/ ijso.0600130.
Duduk B, Mesia JF, Calari A and Bertacinni A (2008). Identification of 16SrIX Group *Phytoplasma* infecting Columbian periwinkles and molecular characterization of several genes. 17th Congress of International Organisation for Mycoplasmology (IOM), Tienjin, 6–11 July 2008,112,83.
Eden-Green SJ, Balfas J and Ramalius (1986). Transmission of xylem-limited bacteria causing Sumatra disease of cloves in Indonesia by tube building cercopids, *Hindola* sp. (Homoptera; Machaerotidae). Proc. 2nd International Workshop on leafhopper and planthopper of economic importance, Brigham Young University, Provo, Utah, USA, 28th July–1st August 1986, pp 101–107.
Edwards MC, Weiland JJ, Todd J and Sterwart LR (2015). Infectious maize rayado fino virus from cloned DNA. Phytopathology, 105: 833–739.
Edwards MC, Weiland JJ, Todd J, Stewart CR and Lu S (2016). ORF-43 of Maize rayado fino virus is dispensable for systemic infection of maize and transmission by leafhoppers. Virus Genes, pp 1–5, 2016.
Faggioli F, Pasquini G, Lumia V, Campobasso G, Widner TL and Quimby PC (2004). Molecular identification of an anew member of the Clover proliferation *Phytoplasma* group (16Sr VI) associated with Yellow sowthistle virescence in Italy. European Journal of Plant Pathology, 110: 353–360.
Falk BW and Weathers LG (1983). Comparison of Potato yellow dwarf virus serotypes. Phytopathology, 73: 81–85.
Franova J, Karesova R, Navratil M and Nebesarova J (2008). A Carrot proliferation disease associated with *Rickettsia-Like Organism* in the Czech republic. Journal of Phytopathology, 148: 53–55.
Freitag JH (1963). Cross-protection of the strains of Aster yellows virus in the leafhopper and in the plant. Netherland Journal of Plant Pathology, 69: 215.
Gal-On A and Shiboleth YM (2006). Cross-protection. pp. 261–288. In: Loebenstein G, Carr JP (eds). Natural Mechanism of Plant to Viruses, Springer Dordrecht, The Netherlands, Kluwer Academic Publishers.
Gamez R and Leon P (1988). Maize rayado fino and related viruses. pp. 213–233. In: Koening R (ed). Polyhedron Virion with Monopartite RNA Genomes. Plenum Press New York.

Gamier M, Foissac X, Gaurivaud P, Laiigret F, Renaudin J, Sailland C and Bove JM (2001). *Mycoplasma* plant insect vectors: a matrimonial triangle. CR Academy of Sciences iii, 324: 923–1008.

Gasparich GE (2010). *Spiroplasma* and *Phytoplasmas* microbes associated with plant hosts. Biologicals, 38: 193–203.

Ghosh DK, Bhose S, Mukherjee K, Aglave B, Warghane AJ, Motghare M, Barawal VK, and Dhar AK (2014). Molecular characterization of Citrus yellow mosaic Badnavirus (CYMBV) isolates revealed the presence of two distinct isolates infecting citrus in India. Phytoparasitica, 42: 681–689.

Griffiths HM, Sinclair WA, Smart CD and Davis RF (1999). The *Phytoplasma* associated with Ash yellows L Lilac witches broom, *Candidatus Phytoplasma fraxini*. International Journal of Systematic and Evolutionary Microbiology, 49: 1605–1614.

Gundersen DE, Lee IM, Rehner SA, Davis RE and Kingsbury DT (1994). Phylogeny of *Mycoplasma Like Organism (Phytoplasmas)*; A basis for their classification. Journal of Bacteriology, 176: 5244–5254.

Gundersen DE, Lee IM, Schaff DA, Harrrison NA, Chang CJ, Davis RF and Kinsbury DT (1996). Genomic diversity a and differentiation among *Phytoplasmas* strains in 16Sr RNA GroupI Aster yellows and related *Phytoplasmas*) and III (X-disease related *Phytoplasmas*). International Journal of Systematic Bacteriology, 46: 64–75.

Gurr GM, Bertaccini A, Gopurenko D, Kruger RR, Alhudaib KA, Liu J and Fletcher MJ (2015). *Phytoplasma* and their insect vectors; Implications for date palm. pp 287–314. In: Wakil W, Faleiiro JR TA and Miller FA (eds). Sustainable Pest Management in Date Palm: Current Status and Emerging Challenges. Springer Science Plus Business, Media Dordrecht, The Netherlands.

Hanboonsong Y, Choosal C, Panyim S and Damak S (2002). Transovarial transmission of Sugarcane white leaf *Phytoplasma* in insect vector, *Matsumuratettix hiroglyphicus* (Matsumura). Insect Molecular Biology, 11: 97–103.

Hansen AK, Trumble JT, Stouthamer R and Paine TD (2008). A new huanglongbing species of *Candidatus liberibacter pysillaurous* found to infect tomato and potato are vectored by the psyllid, *Bactericera cockerelli* (Sulc). Application of Environmental. Microbiology, 74: 5862–5865.

Harris KF, Treur B, Tsai I and Toler R (1981). Observations on leafhopper (Homoptera; Cicadellidae) ingestion–egestion behavior—its likely role in the transmission of nonpersistent viruses and other plant pathogens. Journal of Economic Entomology, 74: 446–453.

Harrison BD (1985). Advances in geminiviruses research. Annual Review of Phytopathology, 23: 55–82.

Harrison JN, Davis RE, Oropeza C, Helmick E, Narvaez M, Eden green S, Dollet M, Harrison NA, Womack M and Carpio ML (2002). Detection and characterization of a lethal yellowing (16Sr IV) Group *Phytoplasma* in Canary island date Palms affected by the Lethal decline in Texas Plant Disease, 86: 676–681.

Harrison NA, Richardson PA, Kramer JB and Tsai JH (1994). Detection of *Mycoplasma Like Organism* associated with Lethal yellowing disease of Palms in Florida by polymerase chain reaction. Plant Pathology, 43: 998–1008.

Haque SQ and Parasram S (1973). *Empoasca stevensi,* a new vector of bunchy top disease of papaya. Plant Disease Reporter, 57: 412–413.

Heikinheima O and Raatikainen J (1976). *Megadelphax sordidula* (Stal) (Homoptera) (Delphacidae) as a vector of Phleum green stripe virus. Annals of Agriculturae Fenniae, 1: 34–55.

Hibino H (1983). Transmission of two Rice hungry associated viruses and Rice *Waikavirus* from doubly or singly infected source plants by leafhopper vectors. Plant Disease, 67: 574–577.

Hiruki C and Wang KR (2004). Clover proliferation *Phytoplasma; Candidatus Phytoplasma trifolii.* International Journal of Systematic and Evolutionary Microbiology, 54: 1349–1353.

Ho K, Tsai C and Chung T (2001). Organisation of ribosomal RNA from a Loofah witches broom *Phytoplasma.* DNA and Cell Biology, 20: 115–122.
Hogenhout SA, Ammar ED, Whitfield AE and Redinbaugh MG (2008). Insect vector interaction with persistently transmitted viruses. Annual Review of Phytopathology, 46: 327–359.
Hogenhout SA, Oshima K, Ammar ED, Kakizawa S, Kingdom H et al. (2008a). *Phytoplasmas;* Bacteria that manipulate plants and insects. Molecular Plant Pathology, 9: 403–423.
Hoh F, Uzest M, Drucker M, Plisson-Chastang C, Bron, Blanc S and Dumas C (2010). Structural insights into the molecular mechanisms of Cauliflower mosaic virus transmission by its insect vector. Journal of Virology, 84: 4706–471.
Horn NM, Reddy SV and Ready DVR (1994). Virus vector relationship of Chickpea chlorotic dwarf *Geminivirus* and leafhopper, *Orosius orientalis* (Hemiptera; Cicadellidae). Annals of Applied Biology, 124: 441–450.
Hoshi A, Ishii Y, Kakizawa S, Oshima K and Namba S (2007). Host-parasite interaction of *Phytoplasmas* from a molecular biological perspective. Bulletin of Insectology, 60: 105–107.
Houston DR (1994). Major new disease epidemics: Beech bark disease. Annual Review of Phytopathology, 32: 75–87.
Huang HJ, Bao YY, Lao SH, Huang XH, Ye YZ, Wu TX, Xu HJ, Zhou XP and Zhang C (2015). Rice ragged stunt induced apoptosis affects virus transmission from its insect vector, the brown planthopper to rice plant. Scientific Reports, 5: 11413 (Groi:10.1438/Srep 14413).
Hull R (1994). Molecular biology of plant–virus-vector interactions. Advances in Disease Vector Research, 10: 361–386.
Hull R (2002). Matthews Plant Virology. Academic Press, New York, NY, USA.
Hull R (2014). Plant Virology, 5th Edition, Academic Press.
Hull WH (1996). Molecular biology of Rice tungro viruses. Annual Review of Phytopathology, 34: 275–297.
Hunt RE, Nault LR and Gingery RE (1988). Evidence for infectivity of Maize chlorotic dwarf virus and for helper component in its leafhopper transmission. Phytopathology, 78: 499–504.
Huo Y, Liu W, Zhang F, Chen X, Li L, Liu Q, Zhou Y, Wei T, Fang R and Wang X (2014). Transovarial transmission of a plant virus is mediated by vitellogenin of its insect vector. PLoS Pathogens, 10(4): e1004141. https/doi:org/10 1371/Journalppat 1004141/.
Hyung LS and Shikata E (1977). Occurrence of Northern mosaic virus in Korea. Korean Journal of Plant Protection, 16: 87–92.
IRPCM (2004). *Candidatus Phytoplasma;* a taxon for the wall-less non-helical prokaryotes that colonize plant phloem and insects. International Journal of Systematic and Evolutionary Microbiology, 54: 1243–1255.
Ishii Y, Matsuura YU, Kakizawa S, Nikoh N and Fakatsai T (2013). Diversity of bacterial endosymbionts associated with *Macrosteles,* leafhoppers vectoring phytopathogenic *Phytoplasmas.* Application of Environmental Microbiology, 79: 5013–5022.
Izadpanah K, Ahmadi AA, Parvin S and Jafri SA (1983). Transmission particle size and additional hosts of the *Rhabdovirus* causing maize mosaic in Shiraz, Iran. Phytopathologica Zeitscrift, 107: 283–288.
Jackson AO, Dietzgen RG, Goodin MM, Bragg JN and Deng M (2005). Biology of plant rhabdoviruses. Annual Review of Phytopathology, 43: 623–660.
Jhao Y, Sun Q, Wei W, Davis RE, Wu W and Liu Q (2009). *Candidatus Phytoplasma tamarcis,* a novel taxon discovered in witches broom diseased salt cedar (*Tamarix chinensis* Lour). International Journal Systematic and Evolutionary Microbiology, 59: 2496–2504.
Jia DS, Mao QZ, Chen HY, Wang AM, Liu YY, Wang HT, Xie LH and Wei TY (2014). Virus-induced tubule: a vehicle for the rapid spread of virions through basal lamina from midgut epithelium in the insect vector. Journal of Virology, 88: 10488–10500.
Jomantiene R, Davis RE, Maas J and Dally EL (1998). Classification of new *Phytoplasma* associated with disease of strawberry in Florida based on analysis of 16Sr RNA and ribosomal protein gene operon sequences. International Journal of Systematic and Evolutionary Microbiology, 48: 259–277.

Jomantiene R, Maas JE, Takeda F and Davis RF (2002). Molecular identification and classification of Strawberry phylloid fruit *Phytoplasma* in group 16Sr. 1 New sub-group. Plant Disease, 86: 920.

Jovic J, Cvrkovic T, Metrovic M, Krnjajic S, Redinbaugh MG, Pratt R, Gingery RE, Hogenhout SA and Tosevski I (2007). Roles of stolbur *Phytoplasma* and *Reptalus panzeri* (Cixiinae: Auchenorrhyncha) in the epidemiology of maize redness in Serbia. European Journal of Plant Pathology, 118: 85–89.

Jung HY, Sawayanagi T, Kakizawa S, Nishigawa H, Wei W, Oshima K, Miyata S, UgakI M, Hibi T and Namba S (2003). *Candidatus Phytoplasma ziziphi*, a novel *Phytoplasma* taxon associated with Jujube witches broom disease. International Journal of Systematic and Evolutionary Microbiology, 53: 1037–1041.

Jung HY, Sawayanagi T, Kakizawa S, Nishigawa H, Miyata S, Oshima K, Ugaki M, Joontak I and Namba S (2002). *Candidatus Phytoplasma castaneae;* a novel *Phytoplasma* taxon associated with Chestnut witches broom disease. International Journal of Systematic and Evolutionary Microbiology, 52: 1543–1549.

Jung HY, Sawayanagi T, Wongkaew P, Kakizawa S, Nishigava H, Wei W, Oshima K, Miyata S, Ugaki M, Hibi T and Nambe S (2003). *Candidatus Phytoplasma oryzae,* a novel *Phytoplasma* taxon associated with Rice yellow dwarf disease. International Journal of Systematic and Evolutionary Microbiology, 53: 1925–1929.

Kawakita H, Saiki T, Wei W, Mitsuhashi W, Watanabe K and Sato M (2000). Identification of Mulberry dwarf *Phytoplasmas* in the genital organs and eggs of leafhopper, *Hishimonoides sellatiformis.* Phytopathology, 90: 909–914.

Khan AJ, Botti S, Paltrinieri S, Al-Subhi AM and Bertaccini A (2002). *Phytoplasmas* in alfalfa seedlings; infected or contaminated seeds? 13 the Congress IOM, Vienna, July 07-12,2002,6.

Kokutse F (2008). Swollen shoot disease devastating cocoa trees. Inter Press Service News Agency, Retrieved 26 October, 2011.

Krezal G, Krezal H and Kunze L (1988). *Fieberiella florii* (Stal) a vector of apple proliferation agent. Acta Horticulturae (ISHS), 235: 99–106.

Kruger K and Smit ND (2013). Grapevine leafroll-associated virus-3 (GLRaV-3) transmission of three soft scale insects (Hemiptera: Coccidae) with notes on their biology. African Entomology, 21: 1–8.

Kunkel LO (1955). Cross-protection between strains of yellows type viruses. Advances in Virus Research, 3: 251–273.

Lee IM, Davis RE, Chen TA, Chiykowski LN, Fletcher J, Hiruki C and Schaff DA (1992). A genotype base system for identification of *Mycoplasma Like Organism* (MLO's) in the Aster yellows MLO strain cluster. Phytopathology, 82: 977–986.

Lee IM, Gundersen-Rindal DE, Davis RF and Bartoszyk IM (1998a). Revised classification scheme of Phytoplasmas based on RELP analysis of 16Sr RNA and ribosomal protein gene sequence. International Journal of Systematic Bacteriology, 48: 1153–1169.

Lee IM, Gundersen-Rindal DE and Bertaccini A (1998b). *Phytoplasma*: ecology and genome diversity. Phytopathology, 88: 1359–1366.

Lee IM, Davis RE and Gundersen-Rindal DE (2000). *Phytoplasmas;* Phytopathogenic Mollicutes. Annual Review of Microbiology, 56: 1593–1597.

Lee IM, Gundersen-Rindal D, Davis RF, Bottner KD, Marcone C and Seemuller E (2004a). *Candidatus Phytoplasma asteris;* a novel *Phytoplasma* taxon associated with Aster yellows and related diseases. International Journal of Systematic Bacteriology, 54: 1037–1048.

Lee IM, Martini M, Marcone C and Zhu SF (2004b). Classification of *Phytoplasma* strains in the Elm Yellows group (16Sr V) and proposal of *Candidatus Phytoplasma ulmi* for the *Phytoplasma* associated with Elm Yellows. International Journal of Systematic and Evolutionary Microbiology, 54: 337–347.

Lee IM, Zhu Y and Bottner KD (2006a). Sec Ygene sequence analysis for finer differentiation of diverse strains in the aster yellows *Phytoplasma* group. Molecular and Cellular Probes, 20: 87–91.

Lee IM, Bottner KD, Secor G and Rivera varas V (2006b). *Candidatus Phytoplasma americanum,* a *Phytoplasma* associated with Potato purple top wilt disease complex. International Journal of Systematic and Evolutionary Microbiology 56: 1593–1597.

Lee IM, Bottner-Parker KD, Zhao Y, Villalobos W and Moreira I (2011). *Candidatus Phytoplasma costaricanum;* a novel *Phytoplasma* associated with an emerging disease in soybean (Glycine max). International Journal of Systematic and Evolutionary Microbiology, 61: 2822–2826.

Lee IM, Martini M, Bottner KD, Dave RA, Black MC and Troclair N (2013). Ecological implications of a molecule analysis of *Phytoplasma* involved in an Aster yellows epidemic in various crops in Texas. Phytopathology, 93: 1368–1377.

Lee ME, Grau CR, Lukaesko LA and Lee LM (2002). Identification of Aster yellows *Phytoplasma* in soybean in Wisconsin based on RFLP analysis of PCR amplified products (16Sr DNA, S). Canadian Journal of Plant Pathology, 24: 125–130.

Leh V, Jacquot E, Geldreich A, Hermann T, Leclerc D, Cerutti M, Yot P, Keller M and Blanc S (1999). Aphid transmission of cauliflower mosaic virus requires the viral PIII protein. The European Molecular Biology Organization Journal, 18: 7077–7085.

LeMaguet J, Beuve M, Herbach E and Lenaire O (2012). Transmission of ampeloviruses and two vitiviruses to grapevine by *Planococcus aceris.* Phytopathology, 102: 717–723.

Lepka P, Stitt M, Moll E and Seemuller E (1999). Effect of phytoplasmal infection on concentration and translocation of carbohydrates and amino acids in periwinkle and tobacco. Physiological and Molecular Plant Pathology, 55: 59–68.

Liefting LW, Padovan AC, Gibb KS, Beever RE, Andersen MT, Newcomb RD, Beck DL and Forster RLS (1998). *Candidatus Phytoplasma australiense* in the *Phytoplasma* associated with Australian grapevine yellows, papaya dieback, and Phormium yellow leaf diseases. Australian Journal of Plant Pathology, 104: 619–623.

Liefting LW, Perez-Egusquiza ZC, Clover GRG and Anderson JAD (2008a). A new *Candidatus Liberibacter* species in *Solanum tuberosum* in New Zealand. Plant Disease, 92: 1474.

Liefting LW, Ward LI, Shiller JB and Clover GRG (2008b). A new *Candidatus Liberibacter* species in *Solanum betaceum* (tamarillo) and *Physalis peruviana* (cape gooseberry) in New Zealand. Plant Disease, 92: 1588.

Liefting LW, Sutherland PW, Ward LI, Paice KL, Weir BS and Clover GRG (2009). A new *Candidatus Liberibacter* species associated with diseases of Solanaceous crops. Plant Disease, 93: 208–2014.

Liu Y, Jia D, Chen H, Chen Q, Xie L, Wu Z and Wei T (2011). The P7-1 protein of Southern rice black-streaked dwarf virus, a *Fijivirus,* induces the formation of tubular structures in insect cells. Archives of Virology, 156: 1729–1736.

Mahfoudhi N, Digiaro M and Dhouibi MH (2009). Transmission of Grapevine leafroll-associated viruses by *Planococcus ficus* (Hemiptera: Pseudococcidae) and *Ceroplastes rusci* (Hemiptera: Coccidae). Plant Disease, 93: 959–1002.

Makkouk KM, Kumari SG, Ghulam W and Attar N (2004). First record of Barley yellow striate mosaic virus affecting wheat summer nurseries in Syria. Plant Disease, 88: 83.1.

Manjunath KC, Halbert SE, Ramadugu C, Webb S and Lee RF (2008). Detection of *Candidatus liberibacter asiaticus* in *Diaphorina citri* and its importance in the management of citrus huanglongbing in Florida.Phytopathology, 98: 387–396.

Maramorosch K (1958). Cross-protection between two strains of Corn stunt in an insect vector, Virology, 6: 448–459.

Marcone C, Gibb KG, Streten C and Schneider B (2003a). *Candidatus Phytoplasma spartic, Candidatus Phytoplasma rhamni* and *Candidatus Phytoplasma allocasuarinae;* respectively associated with Spartium witches broom, Buckthorn witches broom, and Allocasuarina yellows diseases. International Journal of Systematic and Evolutionary Microbiology, 54: 1025–1029.

Marcone C, Lee IM, Davis RE, Ragozzino A and Seemuller E (2000). Classification of aster yellows group *Phytoplasmas* based on combined analysis of r RNA and tuf gene sequences. International Journal of Systematic and Evolutionary Microbiology, 50: 1703–1713.

Marcone C, Schneider B and See Muller E (2003b). *Candidatus Phytoplasma cynodontis;* the *Phytoplasma* associated with Bermuda grass white leaf disease. International Journal of Systematic and Evolutionary Microbiology, 54: 1077–1082.

Martelli GP, Minafra A and Saldarelli P (1997). *Vitvirus;* a new genus of plant viruses. Archives of Virology, 142: 1929–1932.

Martelli GP (2000). Major graft transmissible disease of grapevines: Nature, diagnosis, and sanitation. Proc. 50th Ann Am Soc EnoViticulture Meeting, Seattle, WA. American Journal of. Enology Viticulture, 51: 42–48.

Martelli GP (2014). Grapevine infected viruses. Journal of Plant Pathology, 96: 7–9.

Martin DP and Shepherd D (2009). The epidemiology, economic impact, and control of Maize streak disease. Food Security, 1: 305–315.

Martini M, Botti S, Marcone C, Marzachi C, Casati P, Bianco PA, Benedetti R and Bertaccini A (2002). Genetic variability among Flavescence doree *Phytoplasma* from different origins in Italy and France. Molecular and Cellular Probes, 16: 197–203.

Martini M, Lee IM, Bottner KD, Zhao Y, Botti S, Bertaccini A, Harrison NA, Carraro L, Marcone C, Khan J and Osler R (2007). Ribosomal protein gene-based phylogeny for finer differentiation and classification of *Phytoplasmas*. International Journal of Systematic and Evolutionary Microbiology, 57: 2037–2057.

Martini M, Marcone C, Mitrovic J, Maixner M, Delic D, Myrta A, Ermacora P, Berthaccini A and Duduk B (2012). *Candidatus Phytoplasma convolvuli*, a new *Phytoplasma* taxon associated with bindweed yellows in four European countries. International Journal of Systematic and Evolutionary Microbiology, 62: 2910–2915.

Maust BE, Espadas F, Talavera C, Aguilar M, Santamaria JM and Oropeza C (2003). Changes in carbohydrate metabolism in coconut palms infected with the Lethal yellowing *Phytoplasma*. Phytopathology, 93: 976–981.

McEwen FL and Kawanishi CY (1967). Insect transmission of Corn mosaic: laboratory studies in Hawaii. Journal of Economic Entomology, 60: 1413–1417.

Mekuria TA, Smith TJ, Beers E, Watson GW and Eastwell KC (2013). First report of transmission of the Little cherry virus-2 to sweet cherry by *Pseudococcus maritimus* (Ehrhorn) (Hemiptera: Pseudococcidae). Plant Disease, 97(6): 851–851.

Mirzaie A, Hosseini AE, Nodoshan AJ and Rahimian H (2007). Molecular characterization and potential insect vector of a *Phytoplasma* associated with Garden witches broom in Yazd, Iran. Journal of Phytopathology, 155: 198–303.

Molino LM, Quaglino F, Abou-Jawdah Y, Choueiri E, Sobh H, Casati P, Tedeschi R, Alma RA and Bianco PA (2011). Identification of new 16Sr IX, subgroups F and G among *Candidatus Phytoplasma phoenicium* strains infecting almond, peach, and Nectarine in Lebanon. Phytopathologia Mediterranea, 50: 273–282.

Montano HG, Davis RF, Dally EL, Pimentel JP and Brioso PST (2000). Identification and phylogenetic analysis of a new *Phytoplasma* from diseased chayote in Brazil. Plant Disease, 84: 429–436.

Montano HG, Davis RF, Dally EL, Hogenhout S, Pimentel JP and Brioso PS (2001). *Candidatus Phytoplasma braziliense*, a new *Phytoplasma* taxon associated Witches broom disease. International Journal of Systematic and Evolutionary Microbiology, 51: 1109–1118.

Muller E (2016). Cacao swollen shoot virus (CSSV): History, biology and genome. pp. 337–358. In: Cacao diseases, BA Bailey and LW Meonharti (eds). Springer International Publishing, Switzerland.

Munyaneza JE, Crosslin JM and Upton JE (2007). Association of *Bactericera cockerelli* (Homoptera; Psyllidae) with Zebra chip, a new potato disease in Southern United states and Mexico. Journal of Economic Entomology, 100: 656–663.

Munyaneza JE (2010). Emerging leafhopper-transmitted *Phytoplasma* disease of potato. Southwestern Entomologist, 35: 451–456.

Munyaneza JE, Fisher TW, Sengoda VG, Garczynski SF, Nissinen A and Lemmetty A (2010). Association of *Candidatus liberibacter solanacearum* with the carrot psyllid, *Trioza apicalis* (Homoptera: Triozidae) in Europe. Journal of Economic Entomology, 103: 1060–1070.

Musetti R, Sanita L, DI T, Martini M, Ferrini F, Loschi F, Favali MA and Osler R (2005). Hydrogen peroxide localization and antioxidant status in the recovery of apricot plants from European stone fruit yellows. European Journal of Plant Pathology, 112: 53–61.
Muturi SM, Wachira FN, Karangia LS, Wambulua MC and Macharia E (2013). *Paracoccus burnerae* (Hemiptera; Planococcidae) as a vector of Banana streak virus. Journal of Experimental Biology and Agricultural Sciences, 1(5): 406–414.
Nachappa P, Shapiro AA and Tamborindeguy S (2012). Effect of *Candidatus Liberibacter solanacearum* on the fitness of its insect vector *Bactericera cockerelli* (Hemiptera: Triozidae) on tomato. Phytopathology, 102(1): 41–46.
Nault LR and Ammar ED (1989). Leafhopper and planthopper transmission of plant viruses. Annual Review of Entomology, 34: 503–529.
Nault LR, Gingery RE and Gordon DT (1980). Leafhopper transmission and host range of Maize rayado fino virus. Phytopathology, 70: 709–712.
Nejat N, Vademalai G, Davis RE, Harrison NA, Sijam K, Dickinson M, Abdullah SN and Jhao Y (2012). *Candidatus Phytoplasma malaysianum;* a novel taxon associated with virescence, phyllody of Madagascar periwinkle (*Catharanthus roseus*). International Journal of Systematic and Evolutionary Microbiology, 63: 540–548.
Nejat N, Sijam K, Abdullah S, Vadamalai G and Dickinson (2009). *Phytoplasma* associated with disease of coconut in Malaysia; phylogenetic groups and host plant species. Plant Pathology, 58: 1152–1160.
Nuss DL (2008). Molecular biology of Wound tumor virus. Advances in Virus Research, 29: 57–93.
Omura, T, Yan J, Zhong BX, Wada M, Zhu YF, Tomaru M, Maruyama W, Kikuchi A, Watanabe Y, Kimura I and Hibino H (1998). The P2 protein of rice dwarf *Phytoreovirus* is required for adsorption of the virus to cells of the insect vector. Journal of Virology, 72: 9370–9373.
Oshima K, Kakizawa S, Nishigawa H, Jung HY, Wei W, Suzuki S, Arashida R, Nakata D, Miyata S, Ugaki M and Namba S (2004). Reductive evolution suggested from the complete genome sequence of a plant pathogenic *Phytoplasma*. Nature Genetics, 36: 27–29.
Padovan AC, Gibb KS, Bertaccini A, Vibio M, Bontiglioli RE, Magarey PA and Sears BB (1995). Molecular detection of Australian grapevine yellows *Phytoplasma* and comparison with Grapevine yellows *Phytoplasma* from Italy. Australian Journal of Grape and vine Research, 1: 25–31.
Pearson MN, Clover GRG, Guy PL, Fletcher JD and Beever RF (2006). A review the plant virus, viroid and Mollicute records for New Zealand. Australasian Plant Pathology, 35: 217–252.
Palomar MK (1987). Relationship between Taro feathery mosaic disease and its vector, *Tarophagus proserpina*. Annals of Tropical Research, 9: 68–74.
Pennazio S, Ruggero R and Conti M (2001). A history of plant pathology; cross-protection. New Microbiology, 24: 99–114.
Posnette AF and Strickland AH (1948). Virus diseases of cocoa in West Africa III: Technique of insect transmission. Annals of Applied Biology, 35: 53–63.
Purcell AH (1980). Almond leaf scorch; leafhopper (Homoptera: Cicadellidae) and spittlebug (Homoptera: Cercopidae) vectors. Journal of Economic Entomology, 73: 834–838.
Quaglino F, Zhao Y, Bianco PA, Wei W, Casti P, Durante G and Davis RE (2009). New 16Sr subgroup and distinct single nucleotide polymorphism lineages among grapevine Bois Noir *Phytoplasma* population. Annals of Applied Biology, 154: 279–289.
Quaglino F, Zhao Y, Casati P, Bulgari D, Bianco PA, Wei W and Davis RF (2013). *Candidatus Phytoplasma solani;* a novel taxon associated with Stolbur, Bos Noir–related diseases of plants. International Journal of Systematic and Evolutionary Microbiology, 63: 2879–2894.
Quainoo AK, Wetten AC and Allainguillaume J (2009). Transmission of Cocoa swollen shoot virus by seeds. Journal of Virology Method. 150: 45–49.
Randles JW and Hanold D (1989). Coconut foliar decay virus particles are twenty nm icosahedra. Intervirology, 30: 177–180.
Regassa LB and Gasparich GE (2006). Spiroplasmas: evolutionary relationships and biodiversity. Frontiers in Bioscience, 11: 2983–3002.

Ren CM, Cheng ZB, Miao Q, FanYJ and Jhou YJ (2013). First report of Rice stripe virus in Vietnam. Plant Disease, 97: 1123.

Ridley AW, Dhileepan K, Walter GH, Johanson KN and Croft BJ (2008). Variation in the acquisition of Fiji disease virus by *Perkinsiella saccharicida* (Hemiptera; Delphacidae). Journal of Economic Entomology, 10: 17–22.

Roberts SJ, Eden-Green SJ, Jones P and Ambler DJ (1990). *Pseudomonas syzygii* sp. *nov.*, the cause of Sumatra disease of cloves. Systematic Applied Microbiology, 13: 34–43.

Rolt MF and Jelkmann W (2005). Little cherry virus-2: sequence and genome organization of an unusual member of Closteroviridae. Archives of Virology, 150: 107–123.

Ryan CC (1988). Epidemiology and control of Fiji disease virus of sugarcane. Advances in Disease Vector Research, 5: 163–176.

Saggio P, Hospital MC, Caifliche D, DuPont G, Bove JM, Tully JG and Freudt EA (1973). *Spiroplasma citri* gen and sp. *N*; A *Mycoplasma-like organism* associated with Stubborn disease of citrus. International Journal of Systematic Biology, 23: 181–244.

Saillard C, Vigneault JC, Bove JM, Raie A, Tully JG, Williamson DL, Foss A, Garnier M, Gadeau A, Carle P and Whitcomb RF (1987). *Spiroplasma phoeniceum* sp. *nov.*, a new plant pathogenic species from Syria. International Journal of Systematic Bacteriology, 37: 106–115.

Salehi M, Izadpanah K and Siampour M (2006). Characterization of a new almond witches broom *Phytoplasma* in Iran. Journal of Physiology, 154: 386–391.

Salehi M, Izadpanah K and Heydamejad J (2009). Molecular characterization and transmission of Bermuda grass white leaf *Phytoplasma* in Iran. Journal of Plant Pathology, 91: 615–661.

Sanderlin RS and Melanson RA (2010). Insect transmission of *Xylella fastidiosa* to Pecan. Plant Disease, 94: 465–470.

Santos-Cervantes ME, Chavez-Medina JA, Acostapardini J, Flores–Zamora GL, Mendez–Lozano J and Leyva–Lopez NE (2010). Genetic diversity and geographical distribution of *Phytoplasma* associated with Potato purple top disease in Mexico. Plant Disease, 94: 388–395.

Saponai M, Coconsole G, Cosmara D, Gokomi RK, Stradis AD, Boscia A, Martelli GP, Krueger RC and Purcelli F (2014). Infectivity and transmission of *Xylella fastidiosa* by *Philaenus spumarius* (Hemiptera; Aphrophoridae) in Apulia, Italy. Journal of Economic Entomology, 107: 1316–1319.

Sawayanagi T, Horikoshi N, Kanehara T, Shinohira M, Berthaccini A, Cousin MT, Hiruki C and Namba S (1999). *Candidatus Phytoplasma japonicum,* a new *Phytoplasma* taxon with Japanese Hydrangea phyllody. International Journal of Systematic and Evolutionary Microbiology, 49: 1275–1285.

Schneider B, Torres E, Martin M, P, Schroder M, Behnke HD and Seemuller E (2005). *Candidatus Phytoplasma pini;* a novel taxon from *Pinus sylvestris* and *Pinus halepensis*. International Journal of Systematic and Evolutionary Microbiology 55: 303–307.

Schneider B, Gibb KS and Seemuller E (1997). Sequence and RFLP analysis of the elongation factor Tu gene used in differentiation and classification of *Phytoplasmas*. Microbiology, 143: 3381–3389.

Seemuller E, Schneider B, Maurer R, Ahrens U, Daire X, Kison H, Lorenz K, Firrao G, Avinent L, Sears BB and Sfackebrandt E (1994). Phylogenetic, classification of phytopathogenic mollicutes by sequence analysis of 16Sr ribosomal DNA. International Journal of Systematic and Evolutionary Microbiology, 44: 440–446.

Seemuller E, Marcone C, Lauer U, Ragozzino A and Goschl M (1998). Current status of molecular classification of the *Phytoplasmas*. Journal of Plant Pathology, 80: 3–26.

Seruga M, Skoric D, Botti S, Paltrinieri S Juretic N and Bertaccini A (2003). Molecular characterization of *Phytoplasma* from Aster yellows (16Sr I) Group naturally infecting *Populus nigra* L. Italica Trees in Croatia. Forest Pathology, 33: 113–125.

Seemuller E and Schneider B (2004). *Candidatus Phytoplasma mali: Candidatus Phytoplasma pyri* and *Candidatus Phytoplasma pronotum;* the casual agents of Apple proliferation, Pear decline, and European stone fruit yellows, respectively. International Journal of Systematic and Evolutionary Microbiology, 54: 1217–1226.

Sether DM, Ullman DE and Hu JS (1998). Transmission of pineapple mealybug wilt-associated virus by species of mealybugs (*Dysmicoccus* spp.). Phytopathology, 88: 1224–1230.

Severin HHP (1950). Spittle insect vectors of Pierce disease of virus (ii) Life history and virus transmission. Hilgardia, 9: 357–382.

Shepherd DN, Martin DP, Van Derwalt E, Dent K, Varsani A and Rybicki EP (2010). Maize streak virus: an old and complex pathogen. Molecular Plant Pathology, 11: 1–12.

Sherwood JC (2007). Mechanism of cross-protection between plant virus strains CIBA Foundation Symposium 133-Plant resistance to viruses. David Evered, Sara, Harnett, pp 136–150, 2000.

Siddique A, Agrawal GK, Alam N and Krishnareddy M (2001). Electron microscopy and molecular characterization of *Phytoplasma* associated with Little leaf disease of brinjal (*Solanum melongena* L.) and periwinkle (*Catharanthus roseus*) in Bangladesh. Journal of Phytopathology, 149: 237–244.

Simons JN (1962). Life history and behavioral studies on *Micrutalis malleifera,* a vector, Pseudo curly top virus. Journal of Economic Entomology, 55: 363–365.

Stanley J (2008). Beet curly top virus. Encyclopedia of Virology (third edition, Academic Press, pp 301–307.

Streten C, Herrington ME, Hutton DG, Persley DM, Waite GK and Gibb KS (2005). Plant hosts of the *Phytoplasma* and *Rickettsia-like organism* associated with Strawberry lethal yellows and Green petal diseases. Australasian Plant Pathology, 34: 165–173.

Sugio A, Maclean AM, Kingdom HN, Grieve VM, Manikale R and Hogenhout SA (2011). Diverse targets of *Phytoplasma* effectors: from plant development to defense against insects. Annual Review of Phytopathology, 49: 172–195.

Suzuki S, Oshima K, Kakizawa S, Arashida R, Jung HY, Yamaji Y, Nishigawa H, Ugaki M and Namba S (2006). Interaction between the membrane protein of a pathogen and insect microfilament complex determines insect vector specificity. Proceedings of National Academy of Sciences, USA, 103: 4252–4257.

Ta HA, Nguyen DP, Causse S, Nguyen JD, Ngo VV and Hebrard E (2013). Molecular diversity of Rice grassy stunt virus in Vietnam. Virus Genes, 46: 384–386.

Takahashi M, Goto C, Ishikawa K, Matsuda I, Toryam S and Tsuchiya K (2003). Rice stripe virus 23.9 k protein aggregates and forms inclusion bodies in cultured insect cells and virus-infected plant cells. Archives of Virology, 148: 2167–2169.

Teakle DS, Hicks S, Karan M, Hacker JB, Greber RS and Donaldson JF (1991). Host range and geographic distribution of Pangola stunt virus and its planthopper vector in Australia. Australian Journal of Agricultural Research, 42: 819–826.

Tedeschi R, Ferrato V, Rossi J and Alma A (2006). Possible *Phytoplasma* transovarial transmission in the psyllids *Cacopsylla melanoneura* and *Cacopsylla pruni*. Plant Pathology, 55: 18–24.

Teixeira DC, Wulff NA, Martins EC, Kitajima EW, Bassanezi R, Ayres AJ, Eveillard S, Saillard C and Bove JM (2009). A *Phytoplasma* related to *Candidatus Phytoplasma asteris* detected in citrus showing Huanglongbing (yellow shoot disease) symptoms in Guangdong, PR, China. Phytopathology, 99: 236–242.

Thompson S, Fletcher JD, Zietall H, Beard S, Panda P, Jorgeneen N, Fowler SV, Liefting LW, Berry N and Pitman AR (2013). First report of *Candidatus Liberibacter* Europeans associated with psyllid infested scotch broom. New Disease Reports, 27: (http.///de.Doi. org/10.5197/j 2044 0588,2013.027.006).

Torres E, Botti S, Rahola J, Martini MP and Berthaccini A (2005). Grapevine yellows, Diseases in Spain; eight years; Survey of disease spread and molecule characterization of *Phytoplasma* involved. Anales del Jardin Botanico de Madrid, 62: 127–133.

Tsai CW, Chau J, Fernandez L, Bosco D, Daane KM and Almeida RPP (1998). Transmission of grapevine leafroll-associated virus 3 by grapevine mealybug (*Planococcus ficus*). Phytopathology, 98: 1093–1098.

Tsai CW, Rowhani A, Golino DA, Daane KM and Almeida RPP (2010). Mealybug transmission of Grapevine leafroll viruses: an analysis of virus vector specificity. Phytopathology, 100: 830–834.

Valiunas D, Jomantiene R and Davis RE (2005). A *Candidatus Phytoplasma asteris*–related *Phytoplasma* associated with Cherry little leaf disease represent a new subgroup (16SrI Q). Phytopathology, 95: S106.

Valiunas D, Staniulus J and Davis RE (2006). *Candidatus Phytoplasma fragariae,* a new *Phytoplasma* taxon discovered in Yellow diseased strawberry, Fragaria x ananassa. International Journal of Systematic and Evolutionary Microbiology, 56: 277–281.

Valiunas D, Jomantiene R, Ivanauskas A, Abraitiz R, Staniene G, Zhao Y and Davis RE (2009). First report of a new *Phytoplasma* subgroup 16Sr III-T associated: with decline disease affecting sweet and sour cherry trees in Lithuania. Plant Disease, 93: 554.

Vanden-berg MA, ran-Vuuren SP and Deacon VE (1992). Studies in Greening disease transmission by the citrus psylla, *Trioza erytreae* (Hemiptera; Triozidae). Israel Journal of Entomology, 25-26: 51–56.

Verdin E, Salar P, Danet JL, Choueiri F, El Zammar S, Gelie B, Bove J and Garnier M (2003). *Candidatus phytoplasma phoenicium*; a new *Phytoplasma* associated with an emerging lethal disease of Almond trees in Lebanon and Iran. International Journal of Systematic and Environmental Microbiology, 53: 833–858.

Vereijssen J and Scott I AW (2013). Psyllid can overwinter on non-crop host plants. New Zealand Grower 68: 14–15.

Villalobos W, Martini M, Garital L, Monozl M, Osler R and Moreira I (2011). *Guazuma ulmi* folia (Sterculiaceae) a new natural host of 16Sr xv *Phytoplasm* in Costa Rica. Tropical Plant Pathology, 36: HTTP/dx.doi org/10.1590/s 182-5676 201100007.

Wefels E, Morin JP and Randles JW (2015). Molecular evidence for a persistent circulative association between Coconut foliar decay virus and its vector *Myndus caffeine*. Australasian Plant Pathology, 44: 283–288.

Wei T, Hibino H and Omura T (2009). Release of rice dwarf virus from insect vector cells involve secretary exosomes derived from multivesicular bodies. Communication in Integrative Biology, 2: 324–326.

Wei W, Kakizawa S, Jung HY, Suzuki S, Tanaka M, Nishikawa H, Miyata S, Oshima K, Ugaki M, Hibi T and Namba S (2004). An antibody against the SecA membrane protein of one *Phytoplasma* reacts with those of phylogenetically different *Phytoplasmas*. Phytopathology, 94: 683–686.

Wei W, Davis RE, Lee IM and Zhao Y (2007). Computer-simulated RFLP analysis of 16Sr RNA genes Identification of ten new *Phytoplasma* groups. International Journal of Systematic and Evolutionary Microbiology, 57: 1855–1867.

Weintraub PG and Beanland L (2006). Insect vectors of *Phytoplasma*. Annual Review of Entomology, 51: 91–111.

Weintraub PG (2007). Insect vectors of *Phytoplasmas* and their control-an update. Bulletin of Insectology, 60: 169–173.

Whitcomb RF, Worley JF, Whitcomb RF, Ishijima T and Steere RL (1972). Helical filaments produced by *Mycoplasma-like organism* associated with Corn stunt disease, Science, 176: 521–523.

White DT, Blackall LL, Scott PT and Walsh KB (1998). Phylogenetic positions of *Phytoplasmas* associated with Dieback yellow crinkle and Mosaic diseases of Papaya and their proposed inclusion in *Candidatus australiense* and a new taxon *Candidatus Phytoplasma australasia*. International Journal of Systematic and Evolutionary Microbiology, 48: 941–951.

Williamson DI and Whitcomb RF (1975). Plant *Mycoplasma*, a cultivable *Phytoplasma* causes Corn stunt disease. Science, 188: 1018–1020.

Win NKK, Lee SY, Bertaccini A, Namba S and Jung HY (2013). *Candidatus Phytoplasma balanitae* associated with Witches broom disease of *Balanites triflora*. International Journal of Systematic and Evolutionary Microbiology, 63: 636–640.

Winks CJ, Andersen MT, Charles JG and Beever RE (2014). Identification of *Zeoliarus oppositus* (Hemiptera: Cixiidae) as a vector of *Candidatus Phytoplasma australiense.* Plant Disease, 98: 10–15.

Wu W, Zheng L, Chen H, Jia, D, Li F and Wei T (2014). Nonstructural protein NS4 of Rice stripe virus plays a critical role in viral spread in the body of vector insects. PLoS ONE 2014 (2)9: e88636 https/doi;10.371 journal pone oo8636.

Xu CF, Xia YH, Li KB and Ke C (1988). Further study of the transmission of citrus huanglongbing by a psyllid, *Diaphorina citri* Kuwayama. pp. 243–248. In: Timmer LW, Gamary SM and Navrao L (eds). Proceedings of the 10-th Conference of the International Organization of Citrus Virologists Valencia, Spain University of, Riverside. California.

Yao M, Liu X, Li S, Xu Y, Zhou Y, Zhou X and Tao X (2014). Rice stripe lentivirus NSvc2 glycoproteins targeted to the Golgi body by the N-terminal transmembrane domain and adjacent cytosolic twenty-four amino acids via the COP I- and COP II-dependent secretion pathway. Journal of Virology, 88: 3223–3234.

Zeigler RS and Morales FJ (1990). Genetic determination of replication of Rice Hoja blanca virus within its planthopper vector, *Sogatodes Oryza cola.* Phytopathology, 80: 559–566.

Zhao S, Zhang G, Dai X, Hou Y, Li M, Liang J and Liang C (2012). Processing and intracellular localization of Rice stripe virus Pc2 protein in insect cells. Virology, 429: 148–154.

Zhao Y, Sun Q, Wei W, Davis RE, Wu W and Liu Q (2009). *Candidatus Phytoplasma tamaricis;* a novel taxon discovered in witches broom of diseased salt cedar (*Tamarix chinensis,* Lour). International Journal of Systematic and Evolutionary Microbiology, 59: 2498–250 .

Zheng L, Mao Q, Xie L and Wei T (2014). Infection route of rice grassy stunt virus, a *Tenuivirus,* in the body of its brown planthopper vector, *Nilaparvata lugens* (Hemiptera: Delphacidae) after ingestion of virus. Virus Research, 188: 170–173.

Zreik L, Carle P, Bove JM and Garnier M (1995). Characterization of *Mycoplasma Like Organism* associated with Witches broom diseases of lime, and proposition of a *Candidatus* taxon for the organism. *Candidatus Phytoplasma aurantifolia.* International Journal of Systematic and Evolutionary Microbiology, 449–450.

5.12 QUESTIONS (EXERCISE)

Q 1. Describe how to differentiate leafhoppers from planthoppers and treehoppers from spittlebugs.

Q 2. Describe the feeding mechanism of Leafhoppers in detail and pinpoint how it is different from aphids.

Q 3. What are the categories of plant viruses transmitted through leafhoppers? Describe the semi-persistent mechanism in detail, with suitable examples.

Q 4. Write an essay on mealybugs as vectors of plant viruses.

Q 5. What is the difference between mycoplasma and *Phytoplasma*? List three *Phytoplasma* diseases transmissible through psyllids and discuss one of them.

Q6. Write a short note on rickettsia-like organisms, giving suitable examples.

Q 7. Elaborate on the role of transmission determinants in the spread of foregut-borne, persistent circulative and persistent propagative viruses through leafhoppers and planthoppers.

Q 8. Treehoppers and scale insects are vectors of plant viruses. Discuss.

CHAPTER 6

Whiteflies

6.1 Identification and Brief Biology

The whitefly as a vector of Cotton leaf crumple virus (CLCrV) was identified for the first time as late as 1930, by Kirkpatrick. The whiteflies are insects belonging to the family Aleyrodidae in the order Hemiptera. In all, 1550 species of whitefly are known, of these, a few species are identified as vectors of plant pathogens. The whiteflies have piercing and sucking mouthparts, hence, they are included in the category of efficient vectors along with various other hemipterans. They have incomplete metamorphosis with three development stages, the egg stage, the nymph stage and the adult stage. In reality, the whitefly has an additional stage in its life history, designated as a pseudo-pupal stage. The female lays about 100 elliptical eggs, tapering to a point from one side. The eggs, being stalked (pedicel), are found hanging on the lower surface of leaves. Hatching of eggs takes place within a week to produce crawlers. The first instar nymph is a crawler and has three pairs of legs, three-segmented antennae, and two small eyes. After hatching the crawlers continue to move on the leaf surface for a day or so in order to locate a suitable site to feed and shelter. After roaming for a day on the leaf surface, the crawlers settle down and initiate feeding. The subsequent nymphal instars viz. second, third and fourth remain sessile and fix themselves on the lower surface of the leaf and thrust their mouthparts into the sieve elements of phloem. The insect continues to feed on the plant sap during the second and third nymph stages. The fourth instar nymph does not feed at all and is regarded as a pseudo-pupa. The body of nymphs is opaque, through which the mycetomes cells and inhabiting bacterial symbionts become visible. The fourth instar nymph (pupa) is oval and scale-like with the two prominent red eyes of a developing adult. After passing through four instars, the adults emerge from the pseudo-pupae. The life cycle is completed in about fifteen to twenty-four days. There are eleven to fifteen generations in a year. The adults are about 1.25 mm long with whitish wings and yellowish body, dusted with white, waxy, mealy powder. All four wings are of equal size and transparent with poor wing venation. The adults are also weak fliers; they make trivial flights and rapidly settle down on the same plant or on an adjoining one.

6.2 Feeding Mechanism and Pathogen Transmission

Generally, these insects act as vectors of plant pathogens and possess piercing and sucking mouthparts. They have two types of feeding mechanisms: the sheath-feeders and the non sheath-feeders (some leafhoppers and thrips). Like the aphids, the whitefly is an insect that falls into the category of sheath-feeders. For this purpose, the stylet bundle, containing two maxillary and two mandible stylets responsible for the formation of food and salivary canals, is provided. Each mandible on its outer side has two dendrites. The tip of the mandible is curved and marked with ridges and furrows on the lateral sides. The interlocking of maxillary stylets forms two canals. At one point, these two canals join at the distal end to form a single canal. The material content of both these canals is mixed up at this juncture. The whiteflies are poor fliers and are carried through air currents. While flying, whitefly adults are attracted towards green color. Color cues are used during both takeoff and flight, and the adult flight is governed by green color. The adults whitefly perceive signals of 550 nm of green light (Isaac et al., 1999). Additionally, the presence of photoreceptors on the dorsal or ventral side of the compound eye, are also known to react to UV radiation, facilitating the recognition of a host (Doukar and Panyne, 2007). It is a pest of sucking category; therefore, the feeding is mainly confined to vascular tissues, the phloem or the xylem and their feeding is either intercellular or intracellular. The whiteflies feed on the sieve tissues of phloem vessels and generally follow the intercellular path. Unlike aphids, the whitefly adults do not taste the sap for preference. They make far fewer shallow, intracellular probes in the epidermis or in the mesophyll, each lasting ten to twenty three seconds. This results in the inoculation of viruses, such as Ipomoviruses, in these tissues. The adult feeding at this juncture is at a very low frequency. They make around six punctures, each less than one-minute duration. With this kind of probing, the stylets reach the sieve cells in the phloem tissues. After feeding in the phloem for five min or less, the insects salivate and start ingesting sap from the phloem, the whole process lasting for a few minutes to hours. The stylets require a minimum of sixteen minutes to reach the phloem, but most species require more than an hour to reach the target cells. It is an established fact that, on landing, 80–100% of the adult population settles down on the abaxial side of the plant leaf rather than on the adaxial side (Summers et al., 2004; 1996). These insects, like the other hemipterans, feed on a diet that is poor in nutrients, and their intake of water as fluid is enormous. In such situations, to get rid of the excess fluid in their diet, special adaptations and sites to inhabit the diet-enhancing bacterial symbionts are provided. The pleomorphic bacterium symbiont present in bacteriocytes (cells meant to harbor endosymbionts) is *Candidatus Portiera aleyrodidarum* (Thao and Baumann, 2004). These symbiotic bacteria are devoid of outer membranes of cell walls, as in gram –ve bacteria, but instead possess one membrane. These symbionts are of different categories viz. primary and secondary symbionts. The primary symbiont present in all species of whiteflies is Protiera. To protect the host whitefly from oxidative stresses, the bacteria have carotenoids synthesizing genes (Samos-Garcia et al., 2012; Sloan and Moran, 2012). These whiteflies have the highest diversity of secondary symbionts such as *Alfa Proteobacteria Rickettsia* (Rickettsiales), *Orentia-like Organism* (Rickettsiales), *Wolbachia* (Rickettsiales), *Gamma-Proteobacteria, Arsenophonus* (Enterobacteriales), *Hamiltonella* (Enterobacteriales), *Cardinium* (Bactereoidetes) and *Fritsche* (Chlamydiales). Of these, *Cardinium* and *Fritsche* are known to infect all species of whiteflies throughout the globe, excluding Israel's population (Bing et al., 2012; Gottlieb et al., 2008). There are also different biotypes of whiteflies; *Hamiltonella* is present in B type (33% population), while *Wolbachia* and *Arsenophonus* are inhabitants of Q type (87% population) whiteflies (Chiel et al., 2007). The primary symbionts are always found together in the same cell, except the Rickettsia which has been recorded in the body cavity. Generally, the secondary symbionts are present in special cells (Gottlieb et al., 2008). The symbionts present in the whitefly vectors synthesize GroEL protein (chaperone) homologue to aphids that ensures the safe transport of virion in the body of the vector. One such protein, known as Heat Shock Protein seventy (HSP70), has been recorded in *Bemisia tabaci*, a vector of Tomato yellow leaf curl virus (TYLCV). The interaction between TYLCV and protein synthesized by symbionts is responsible for the movement of virus inside the body of vector (Czosneck and Ghanim, 2012; Gotz et al., 2012). Similarly, another protein sixty-three-kDa synthesized by symbionts (*Hamiltonella*) in whitefly vector *B. tabaci*, interact with coat protein of TYLCV and bring about cell-to-cell movement of virion in the body of vector (Morin et al., 1999). The whiteflies are vectors of viruses that circulate in the vector. The whitefly-borne viruses are picked up from the phloem and require a latent period before they can be transmitted to new healthy plants. The latent period elapses during translocation of the virus from digestive tract to hemolymph and from hemolymph to salivary glands. The virions of virus pass along the food canal in the stylet to the esophagus, through which virions enter the hemolymph via filter chamber, a main site (Skaljac and Ghanim, 2010). The virus particles move through the cytoplasm of epithelial cells in the filter chamber and fuse with basal plasma membrane, then release in the membrane and basal lamina (Cicero and Brown, 2011b). The latent period is eight hours in whitefly (Ghanim et al., 2001). The other virus, Tomato yellow leaf curl Sardinia virus (TYLCSV) is found in midgut epithelial cells and cytoplasm of primary salivary gland cells. The same is true of *Nanovirus* (Squash leaf curl virus-SLCV), present in principal salivary glands but not in accessory salivary glands. The same cycle, involving virus ingestion along with sap, is followed; this allows the virus to enter the alimentary canal and access the hemolymph. The blood transports it to other tissues and finally it is picked up by accessory salivary glands, through which the virus is ejected into new cells of plant tissues. It follows the endocytic passage for the movement of virus from one cell to another cell in the body of vector. The whiteflies are known to transmit semi-persistent viruses, carried either on linings of foregut lumen or tip of maxillary stylet in the common duct, as the viruses are inoculated with egestion of ingested sap or salivation. These viruses are lost within few hours of acquisition. In all, 114 species of viruses have been identified as transmissible by whiteflies. These whitefly-borne viruses are from genera *Begomovirus* (90%), *Crinivirus* (6%) and *Closterovirus/Carlavirus / Torradovirus/Ipomovirus* (4%). Of these, 111 species are transmitted by *B. tabaci* alone. The remaining whitefly-borne viruses are transmitted by three species of whitefly, namely *Trialeurodes abutilonea, T. vaporariorum* and *T. ricini*) (Jones, 2003). With respect to species and strains of whiteflies, a lot of confusion has been experienced in the past. Now the method of classification and naming of begomoviruses based on pairwise genome sequence and considering biological characters has been suggested (Brown et al., 2015). The mechanism of transmission of begomoviruses (Geminiviradae) through Potato whitefly (*Bemisia tabaci* Gennadius) and Castor bean whitefly (*T. ricini* Misra) is circulative in nature, whereas the semi-persistent mechanism has subsequently been reported in criniviruses (Closteroviridae) transmissible through *B. tabaci,* Sweet potato whitefly, *B. afer*, Greenhouse whitefly, *Trialeurodes vaporariorum* Westwood, The banded whitefly and *T. abutilonea* Haldeman, ipomoviruses (Potyviridae) vectored by *B. tabaci*, carlaviruses (Betflexiviridae) transmissible by *B. tabaci* and torradoviruses (Secoviridae) transmissible through *B. tabaci* and *T. vaporariorum.* Among these species, *T. ricini* and *B. afer* are awaiting confirmation (Navas-Castillo et al., 2011)). The International Committee on Taxonomy of Viruses (ICTV) have identified 2000 virus species, six orders, sixty-seven families, 349 genera of viruses so far (Brown and Czosnek, 2002). Of these, now there are 800 viruses, in ninety genera of twenty families which are known to infect plants. It is important to note that of the emerging diseases, 47% dare of viral etiology (Anderson et al., 2004). The interaction between virus and vector via transmission determinants has opened up a new arena in the mode of transmission. The determinants (capsid protein and helper component) are known to play a vital role in the transmission of circulative non-propagative viruses. In *Begomoviruses*, CP is the only protein that interacts with the receptors in the body of the insect. The exchange of CP gene of whitefly-borne African cassava mosaic virus (ACMV) and leafhopper-borne Beet curly top virus (BCTV) and Sida golden mosaic (transmissible) and Abutilon mosaic virus (non-transmissible) was done and it resulted in the successful transmission of Abutilon mosaic and Beet curly top viruses via whiteflies. These molecular components play a role mainly in the acquisition, retention, and inoculation of plant viruses. The viruses could be enveloped or non-enveloped. The virion of non-enveloped viruses attaches directly to the specific domain of the coat protein in whiteflies. In this direction, Lettuce infectious yellows virus (LIYV) belongs to *Crinivirus* (Closteroviridae) genus and is transmitted by whitefly, *B. tabaci*, by utilizing the capsid strategy instead of helper component (Ng and Zhou, 2015; Stewart et al., 2010). In another study, the vector whiteflies were fed in a sequence, first on purified virus titer (LIYV antibodies culture), then they are allowed access to secondary antibodies containing virus concentration. The target virus was retained in the anterior foregut or in cibarium through capsid protein. It is therefore concluded that capsid protein is mandatory for whitefly-borne viruses (Chen et al., 2011). It is further predicted in geminiviruses that these viruses (e.g., TYLCV) touch the filter chamber at the juncture of the midgut, hindgut, and Malpighian tubules and are excreted to the outside environment (Ammar et al., 2009). The virus TYLCV has been detected in the cells of filter chamber (Medina et al., 2006). These viruses are known to cause typical symptoms like a mosaic, yellows, leaf curl, vein yellowing, etc., and have been categorized either based on symptoms, the transmission mechanism in whiteflies or considering the chemical nature/shape of viruses (Duffus, 1987; 1963). The acquisition access and inoculation access periods for non-persistent, semi-persistent and persistent categories of viruses are generally one and two to twenty-four hours, six to twenty-four and eight to twenty-four hours, and forty-eight to seventy two hours and twenty-eight to seventy-two hours, respectively (Polston and Capobianco, 2013).

6.3 Virus Vector Relationship

The viruses causing mosaic-type symptoms are said to be located in parenchyma tissues, while those associated with leaf curl symptoms are known to be in the phloem tissues. Most viruses transmitted by whiteflies are acquired in a comparatively longer period than those transmitted by aphids as these viruses are located in deeper tissues viz. phloem (as in leaf curl viruses). The transmission of mosaic viruses is almost the same in both categories of insects. The acquisition and inoculation thresholds are fifteen minutes (Infectious chlorosis of *Sida carpinifolia*) and ten minutes, respectively (Varma, 1963), but there could be a reduction in access period with fasting of vector. It has been demonstrated in the transmission of

Tomato leaf curl virus in tomatoes in India. The whiteflies were likely able to reduce the time taken for stylet to reach the target site due to fasting. The variable threshold periods in whitefly-borne viruses could be attributed to suitability of host, uneven virus titer in the target tissues and the occurrence of virus inhibitors in host (Silber Schmidt et al., 1957). Unlike the aphids, the longer the acquisition access on virus source, the more the transmission efficiency of virus increases. Since the whitefly-borne viruses circulate in the body of vector, they possess definite retention period (twenty days) in *B. tabaci*. Furthermore, the phenomenon of periodic acquisition also occurs in whiteflies, in which the whiteflies cannot acquire the additional virus titer (TYLCV) unless these insects exhaust the previously acquired virus titer (Cohen and Nitzany, 1966; Cohen and Harpaz, 1964). The retention of virus is for life of insect and it is roughly twenty days, but there are reports where the adults of whiteflies could transmit the leaf curl virus in tomatoes up to fifty-three days (Butter and Rataul, 1977). Besides, the females (86%) were identified as much more efficient vectors than males (56%) in the transmission of Tomato leaf curl virus (ToLCuV) (Butter and Rataul, 1977). It has been demonstrated in many more plant viruses vectored by whiteflies. The female whiteflies are known to pick up more sap and, therefore, more virus titre due to more egg production activity as compared to males. The immature stages of whiteflies can acquire the virus if they have had access to infected plants. They then transmit acquired viruses during adult stages. The immature stages, being sessile (except the first instar nymphs), do not play role in the spread of plant viruses. The insect can acquire the virus in nymphal stage but cannot transmit the virus. There is no evidence of transovarial and seed transmission except for one report (Keur, 1934) in which seed transmission of Abutilon variegation virus in the hybrid of two species of Abutilon was demonstrated. Through the use of modern techniques like PCR, Tomato yellow leaf curl virus (TYLCV-IL) has been demonstrated to be seed-transmissible for the first time (Kil et al., 2016). The virus was detected both in the seed and in the seedlings raised from infected seed.

6.4 Begomoviruses

More than 200 species of begomoviuses have been identified in four genera, namely *Topocuvirus, Curtovirus, Mastrevirus* and *Begomovirus*, based on the arrangement of the genome, vector species and identities (Brown et al., 2012). The important whitefly-transmissible virus genera are *Begomovirus* (Geminiviridae) (Table 6.1), *Crinivirus* (Closteroviridae), *Carlavirus* (Betaflexiviridae), *Ipomovirus* (Potyviridae), and *Torradovirus* (Sequiviridae) (Table 6.2; Fig. 6.1) (EFSA, 2013; Navas-Castillo et al., 2011; Jones, 2003). Among these genera, the viruses belonging to genus *Begomovirus* are whitefly-borne, circular, ssDNA and have an encapsidated genome

...*Table 6.2 contd.*

Sl. No.	Virus	Virus genera	Vector	Biotype	Source
13	Cucurbit yellow stunting disorder virus (CYSDV)	*Crinivirus*	*Bemisia tabaci*	A, B, Q	Wisler et al., 1998
14	Cucurbit yellow vein virus	Ipomovirus	*Bemisia tabaci*	Not known	Cohen and Nitzany, 1960
15	Diodia vein chlorosis virus (DVCV)	*Crinivirus*	*Trialeurodes abutilonea & Trialeurodes vaporariorum*	Not known	Tzanetakis et al., 2011
16	Acalypha yellow vein mosaic virus (EYVMV)	*Luteovirus* (P)	*Bemisia tabaci*	Not known	Bird and Maramorosch, 1978
17	Jasmine chlorotic ringspot virus (JCRSV)	*Carlavirus*	*Bemisia tabaci*	Not known	Wilson, 1972
18	Lettuce chlorosis virus (LClV)	*Crinivirus*	*Bemisia tabaci*	A, B	Tzanetakis et al., 2013a
19	Lettuce chlorosis virus (LClV)	*Crinivirus*	*Bemisia tabaci*	Biotype–B	Kubota and Ng, 2016
20	Lettuce infectious Yellows virus (LIYV)	*Crinivirus/Closterovirus* (non-persistent foregut-borne)	*Bemisia tabaci*	A, B	Cohen et al., 1992
21	Potato yellow vein virus (PYVV)	*Crinivirus*	*Trialeurodes vaporariorum*	Not known	Salazar et al., 2000
23	Squash vein yellowing virus (SVYV)	*Ipomovirus*	*Bemisia tabaci*	Not known	Webb et al., 2012; Baker et al., 2008
24	Strawberry pallidosis associated virus (SPaV)	*Crinivirus*	*Trialeurodes vaporariorum*	Not known	Tzanetakis et al., 2006
25	Sweet potato chlorotic fleck virus (SPCFV)	*Carlavirus*	*Bemisia tabaci*	Not known	Opiyo et al., 2010
26	Sweet potato chlorotic stunt virus (SPCSV)	*Crinivirus*	*Bemisia tabaci & Trialeurodes abutilonea*	B afer	Sim et al., 2000; Gamarra et al., 2010

27	Sweet potato disease (sweet potato chlorotic stunt (SPCSV) and sweet potato feathery mottle virus) (SpFMV)	*Crinivirus*	*Bemisia tabaci; Trialeuodes abutilonea*	Not known	Opiyo et al., 2010
28	Sweet potato mild mottle virus (SPMMV)	*Ipomovirus*	*Bemisia tabaci*	Not known	Valverde et al., 2004
29	Sweet potato mild speckling virus (SPMSV)	*Ipomovirus*	*Bemisia tabaci*	Not known	Opiyo et al., 2010
30	Sweet potato sunken vein virus (SPSVV)	*Closterovirus*	*Bemisia tabaci*	Biotype B	Cohen et al., 2001
31	Tomato chlorosis virus (ToCV)	*Crinivirus*	*Trialeurodes abutilonea, Trialeurodes vaporariorum* & *Bemisia tabaci*	A, B, Q	Navas-Castillo et al., 2000; 2011; Wintermental and Wisler, 2006
32	Tomato chocolate virus (ToChV)	*Torradovirus*	*Trialeurodes abutilonea*	Not known	Verbeek et al., 2014
33	Tomato infectious Chlorosis virus (TICV)	*Crinivirus*	*Trialeurodes vaporariorum*	Not known	Duffus et al., 1996
34	Tomato marchitez virus (ToMarV)	*Torradovrus*	*Trialeurodes abutilonea*	Not known	Verbeek et al., 2014
35	Tomato necrotic dwarf virus (TNDV)	*Nepovirus* (non-persistent)	*Bemisia tabaci*	Not known	Larsen et al., 1984
36	Tomato pale chlorosis virus (TPCV)/ strain of cowpea mild mottle virus	*Carlavirus* (non-persistent)	*Bemisia tabaci*	Not known	Antignus and Cohen, 1987
37	Tomato torrado virus (ToTV)	*Torradovirus*	*Ttialeurodes vaporariorum Bemisia tabaci*	Not known	Verbeek et al., 2014
38	Uganda cassava brown streak virus (UCBSV)	*Ipomovirus*	*Bemisia tabaci*	*B. afer*	Mbanzibwa et al., 2011
39	Watermelon vine decline virus (WmVDV) Squash vein yellowing virus (SVYV)	*Ipomovirus*	*Bemisia tabaci*	Biotype B	Adkins et al., 2010
40	Zinnia yellow net virus (ZYNV)	*Luteovirus* (Persistent)	*Bemisia tabaci*	Not known	Srivastava et al., 1977

Sweet potato leaf curl (Lotrakul and Valverde, 1999), Sweet potato leaf curl Georgia virus and Ipomoea yellow vein virus (Lotrakul et al., 2003) have also been included in the list of begomoviruses (Valverde et al., 2004). The whiteflies are generally instrumental in spreading viruses belonging to families Geminiviridae and Closteroviridae. The important genera are *Begomovirus* (Tomato yellow leaf curl virus), *Crinivirus* (Cucumber yellow stunting disorder virus), *Carlavirus* (Tomato pale chlorosis/Cowpea mild mottle virus), *Torradovirus* (Tomato torrado virus), *Ipomovirus* (Sweet potato mild mottle virus) and *Closterovirus* (Tomato infectious chlorosis virus). The mechanism of transmission is either foregut-borne non-persistent (Tomato torrado virus) or persistent-circulative (Sweet potato mild mottle virus and Tomato yellow leaf curl virus). The whiteflies acquire the viruses from the phloem tissues along with cell sap and the virus is passed on through the food canal with the sucking apparatus. The stylets then pave the way to the filter chamber through the cuticle linings of the esophagus. The filter chamber is located at the juncture of midgut and hindgut; therefore, the majority of virus titer is absorbed in this part of alimentary canal and some amount is absorbed by midgut epithelial cells, and the virus moves on to salivary glands in a circulative manner (Rosen et al., 2015; Czosnek et al., 2002). These viruses accumulate in the salivary glands, while the cuticula viruses are confined to a point of fusion of two canals. Like the other aphid-borne viruses, the survival of the virus in hemolymph is ensured by chaperone produced by endosymbionts (producing GroEL homologue protein of aphids) present in the specialized cells (bacteriosomes) (Czosnek and Ghanim, 2012). The virus's particles are transferred from the cytoplasm of filter chamber and are released between the basal plasma membrane and the basal lamina (Cicero and Brown, 2011a). However, the Tomato yellow leaf curl virus is located in midgut instead of hindgut, from where the virus comes into contact with salivary glands via hemocoel (Ghanim et al., 2009). To circulate in the body, the virus requires some time (90 minutes) from stylets to hemolymph after acquisition access on virus source; this is sometimes mistakenly identified as a latent period (Ghanim et al., 2001). The transovarial transmission studies indicated different views; however, the horizontal transmission involving the transmission of the virus from males to females of B biotype of *B. tabaci* is now confirmed (Ghanim and Czosnek, 2000). In contrast, there is no transfer of virus between opposite sexes of B and Q biotypes as these biotypes do not mate, which is a mandatory factor in this case (Pascual and Callejas, 2004). The begomoviruses are required to break the barriers of hindgut epithelial cells. The apical plasmalemma of salivary glands actually determine the transmission of the virus. The virus passes from the salivary gland through the ducts and is ejected outside the body into new plants. The cuticula viruses are mainly confined to the point where the fusion of two canals takes place. The differential role of biotypes in the transmission of *Geminiviruses* has been explored. The isolate of TYLCV from Spain is not differentially transmissible by B and Q biotypes of *B. tabaci* (Jiang et al., 2012) but on making comparison, the B biotype (80%) was found to be more efficient than the Q type (5%) with regard to the extent of transmission (Gottlieb et al., 2010). The differential transmission efficiency could be attributed to the presence of endosymbionts secrete chaperone responsible for preventing inactivation of virus in the hemolymph.

6.5 Crinivirus

These viruses belong to Closteroviridae family of viruses and are transmitted by whitefly (*B. tabaci* and *B. afer*) in a semi-persistent manner. The family contains several genera which are flexuous filaments rods (650–1000 nm long) with bipartite or tripartite segmented largest genome of encapsidated +ve ssRNA (15.3–17.7 kb) (Martelli et al., 2011; Liu et al., 2000). RNA proteins are present to take care of encapsidation, cell-to-cell movement of virus, and transmission by whitefly vectors as these are species specific (*Bemisia tabaci; Trialeurodes abutilonea; Trialeurodes vaporariorum*). These diseases are identified by the interveinal yellowing/chlorotic mottling, brittleness of leaves and senescence of plants. The incubation period of the virus in plants is between three to four weeks after inoculation of virus. The viruses belonging to the genus *Crinivirus* are Abutilon yellows virus (AYV), Beet pseudo yellows virus (BPYV), Blackberry yellow vein associated virus (BYVSV), Diodia vein chlorosis virus (DVCV), Potato yellow vein virus (PYVV), Strawberry pallidosis associated virus (SPSV), Bean yellow disorder virus (BYDV), Cucurbit chlorotic yellows virus (CCYV), Cucurbit yellow stunting virus (CYSV), Lettuce chlorosis virus (LCV), Sweet potato chlorotic stunt virus (SPCSV), Tomato chlorosis virus (ToCV), Lettuce infectious yellows virus (LIYV) and Tomato infectious chlorosis virus (TICV). Abutilon yellows virus particles are flexuous, filamentous, twelve nm x 810–900 nm in length and are transmissible by whitefly (*Trialeurodes abutilonea*) not by mechanical means (Wisler and Duffus, 2001; Liu et al., 1997). It is characterized by vein yellowing which appears within two to three weeks of inoculation. The virus-vector relationship studies showed transmission efficiency ranging from 4% to 81%. The whiteflies were allowed acquisition of twenty-four hour and inoculation of forty-eight hours. Using sets of one and fifty insects, transmission efficiencies of 19% and 77%, respectively, were recorded. The viruliferous whitefly retained the virus in its body for three days after acquisition access on virus source. Beet pseudo yellow virus has genome size between 15.5 to 15.9 kb with two to three and seven to eight open reading frames in RNA1 and RNA2, respectively, in strawberry isolate. The cucumber isolate, however, lacks open reading frames. The number of frames on RNA-2 are seven and eight in cucurbit and strawberry isolates, respectively. The diseased plants show

is known to attack sixty plant species belonging to ten families, especially Leguminosae.

6.7.7 *Citrus chlorotic dwarf virus–CCDV (Geminiviridae)*

CCDV is an ssDNA virus belonging to family Geminiviridae. It is characterized by puckered leaves in lemon, tangelo and sweet orange and is transmitted by bayberry whitefly *Parabemisia myrice* in Turkey (Loconsole et al., 2012). Details regarding virus-vector relationship are not readily available since it has only recently been discovered; it has, however been declared as an emerging serious disorder.

6.7.8 *Squash vein yellowing virus-SVYV (Ipomovirus)*

This is a dsRNA, rod-shaped (740–800 nm particle length) *Ipomovirus* in the family Potyviridae. It thrives on cucurbits, especially watermelon, squash, muskmelon, cantaloupes, cucumbers, pumpkins some ornamentals and weeds like *Momordica charantia* and *Melothria pendula*. The virus is widespread in the Northern part of the USA, covering Florida, Georgia, Indiana and South Carolina. It also appears in Peurto Rico, Israel, Jordon, Spain, Turkey and Europe. The disease can be slightly difficult to identify, compared to other viral diseases. The pathogen initially produces vein clearing symptoms followed by vein yellowing in affected cucurbits. The foliage turns brown in advanced stages of the disease and the flesh of fruit becomes deep red and has a bitter taste. Bisected fruits show peculiar blotches on the inner side of the rind. The whitefly *Bemisia tabaci* (B-biotype) is a vector of this virus. The adults can acquire the virus within an acquisition access of one to two days and require another two days to transmit the virus. The whiteflies will be viruliferous only for four to six hours, after which, the adults lose the virus. The relationship between virus and vector is of the semi-persistent type.

References

Adkins S, Webb SE, Roberts PD, Kousik CS, Stansly PA, Bruton BD, Achor D, Muchovj RM and Baker CA (2010). A review of ipomoviruses and Watermelon vine decline in Florida. Pest bionomics and management. A Global Pest, pp. 333–337.

Ahmad M (1978). Whitefly (*Bemisia tabaci*) transmission of Yellows mosaic virus disease of cowpea *Vigna unguiculata*. Plant Disease Reporter, 62: 224–226.

Allcai T, Omongo CA, Maruthi MN, Hillocks RJ, Bequma Y, Kawuki R, Bua B and Otim-Nape GW and Lolvin G (2007). Re emergence of Cassava streak disease in Uganda. Plant Disease, 91: 24–29.

Ammar D, Tsai DC, Whitfield AE, Redinbaugh MG and Hogenhout SA (2009). Cellular and molecular aspects of *Rhabdovirus* interactions with insect and plant hosts. Annual Review of Entomology, 54: 447–468.

Anderson PK, Cunningham AA, Patel NG, Morales FJ, Epstein PR and Daszak P (2004). Emerging infectious diseases of plants pathogen, pollution, climate change and agrotechnology drivers. Trends in Ecology and Evolution, 19: 535–544.

Antignus Y and Cohen S (1987). Purification and some properties of a new strain of Cowpea mild mottle virus in Israel. Annals of Applied Biology, 110: 563–569.

Baker C, Webb S and Adkins S (2008). Squash vein yellowing virus causal agent of watermelon vine decline in Florida. Plant Pathology circular no 407, July 2008.

Bing XL, Yang J, Zchori–Fein E, Wang XW and Liu SS (2012). Characterization of a newly discovered symbiont of the whitefly, *Bemisia tabaci* (Hemiptera: Aleyrodidae). Applied and Environmental Microbiology, 79: 569–575.

Bird J (1962). A whitefly-transmitted mosaic of *Rhychosia minima* and its relation to Tobacco leaf curl and other viral diseases of plants in Puerto Rico. Phytopathology, 52: 285–288.

Bird J, Perez JE, Alconero R, Vakili NC and Melendez PL (1972). A whitefly-transmitted Golden yellow mosaic virus of *Phaseolus lunatus* in Puerto Rico. Journal Agriculture University Puerto Rico, 56: 64–74.

Bird J, Sanchez J, Rodiguez R Land Julce FJ (1975). Rogaceous (whitefly transmitted) viruses in Puerto Rico. In: Bird J and Maramorosch K (eds). Tropical Diseases of Legumes Academic Press New York, pp 3–25.

Bird J and Maramorosch K (1978). Viruses and diseases associated with whiteflies. Advances in Virus Research, 22: 55–109.

Bisht RA and Mathur RS (1964). Occurrence of two strains of Jute mosaic virus in UP. Current Science, 33: 434–435.

Briddon RW, Pinner MS, Stanley J and Markham PG (1990). *Geminivirus* coat protein replacement alters insect specificity. Virology, 177: 85–94.

Brown JK, Zerbini FM, Navas-Castillo J, Moriones E, Ramos-Sobrinho R, Silva JCF, Fiallo-Olive E, Briddon RW, Hemandez-Zegede C, Idris A, Malathi VG, Martin DP, Rivera-Bustamate R, Veda S and Varsani A (2015). Revision of *Begomovirus* taxonomy based on pairwise sequence comparison. Archives of Virology, 160: 1593–1619.

Brown JK, Fauquet CM, Briddon RW, Zerbini FM, Moriones E and Navas-Castillo J (2012). Family Geminiviridae. pp 351–373. In: King AMQ, Adam MJ, Carsten EB and Lefkowitz EJ (eds). Virus Taxonomy, 9th Report of International Committee on Taxonomy of Viruses. Elsevier, Academic Press London.

Brown JK and Czosnek H (2002). Whitefly-transmitted viruses. In Advances in Botanical Research, 36: 65–100. New York, Academic Press.

Brown L, Brown JK and Tsai JH (1990). Lettuce infectious yellows virus. Plant Pathology, circular no 335, Sept 1990. Fl Dept of Agriculture Consumer Service, Division of Plant Pathology.

Butter NS and Rataul HS (1977). Virus vector relationship of the Tomato leaf curl virus (TLCV) and its vector, *Bemisia tabaci* Gennadius (Hemiptera: Aleyrodidae). Phytoparasitica, 5: 173–186.

Navas-Castillo J, Fiallo-Olive E and Sanchez-Campos S (2011). Emerging virus diseases transmitted by whiteflies. Annual Review of Phytopathology, 49: 219–248.

Chen AYS, Walker GP, Carter D and Ng JCK (2011). A virus capsid component mediates virion retention and transmission by its insect vectors. Proceedings of. National Academy of Sciences, USA, 108: 16777–16782.

Chen JC, Jiang CZ, Gookin T, Hunter D, Clark D and Reid M (2004). Chalone synthesis as a receptor in virus-induced gene silencing studies of flower senescence. Plant Molecular Biology, 55: 525–530.

Chenulu VV and Pathak HC (1965). Yellow mosaic of *Acalypha indica* L., a new whitefly-transmitted disease in India. Current Science, 34: 321–322.

Chiel E, Gottlieb Y, Schori–Fein E, Mozes-Daube N, Katzir N, Inbar M and Ghanim M (2007). Biotype dependent secondary symbiont communities in a sympatric population of *Bemisia tabaci*. Bulletin of Entomological Research, 97: 407–413.

Chu D, Zhang YJ, Brown JK, Cong B, Xu BY, Wu QJ and Zhu GC (2006). The introduction on the exotic Q biotype of *Bemisia tabaci* from Mediterranean region into China on ornamental crops. Florida Entomologist, 89: 168–174.

Cicero JM and Brown JK (2011a). Anatomy of accessory glands of the whitefly, *Bemisia tabaci* (Hemiptera: Aleyrodidae) and correlation to *Begomovirus* transmission. Annals of Entomological Society of America, 104: 280–286.
Cicero JM and Brown JK (2011b). Functional anatomy of whitefly organs associated with Squash leaf curl virus (Geminiviridae; *Begomovirus*) by the B type of *Bemisia tabaci* (Hemiptera; Aleyrodidae). Annals of the Entomological Society of America 104: 261–279.
Cohen J, Lapidot M, Loebenstein G and Gera A (2001). First report of sweet potato sunken vein virus occurring in Lisianthus. Plant Pathology, 85: 679.1–679.1.
Cohen S and Nitzany FE (1960). A whitefly-transmitted virus of cucurbits in Israel. Phytopathologia. Mediterranea, 1: 44–46.
Cohen S and Harpaz I (1964). Periodic, rather than the continual acquisition of a new tomato virus by its vector, the tobacco whitefly (*Bemisia tabaci* Genn.). Entomologia Experimentalis et Applicata, 7: 155–156.
Cohen S and Nitzany FE (1966). Transmission and host range of Tomato yellow leaf curl virus. Phytopathology, 56: 1127–1131.
Cohen S, Duffus JE, Larsen RC, Liu HY and Flock RA (1983). Purification serology and vector relationship of squash leaf curl whitefly transmitted *Geminivirus*. Phytopathology, 73: 1669–1873.
Cohen S, Duffus JE and Liu HY (1992). A new *Bemisia tabaci* (Gennadius) biotype in Southern California, United States and its role in Silver leaf squash and Transmission of Infectious yellows virus. Phytopathology, 82: 86–90.
Costa AS and Bennett CW (1950). Whitefly transmitted mosaic of *Euphorbia prunifolia* Phytopathology, 40: 259–272.
Costa AS and Carvalho AMB (2008). Mechanical transmission and properties of the abutilon mosaic. Journal of Phytopathology, 37: 259–273.
Czosnek H, Ghanim M and Ghanim M (2002). The circulative pathway of begomoviruses in whitefly vector *B tabaci*–insight from a study in Tomato yellow leaf curl virus. Annals of. Applied Biology, 140: 215–23.
Czosnek H and Ghanim M (2012). Back to basics: Are begomoviruses whitefly pathogens? Journal of Integrative Agriculture, 11(2): 225–234.
Dickson RC, Johnson MM and Laird EF (1954). Leaf crumple, a virus disease of cotton. Phytopathology, 44: 479–480.
Dombrovsky A, Sapkota R, Lachman O, Pearlsman M and Antignus Y (2012). A new aubergine disease caused by a whitefly-borne strain of Tomato mild mottle virus (TomMMoV). Plant Pathology, 62: 450–459.
Dombrovsky A, Reingold V and Antignus Y (2014). *Ipomovirus*-a typical genus in the family Potyviridae transmitted by whiteflies. Plant Management Science, 70(10): 1553–1567.
Doukar D and Panyne CC (2007). Greenhouse whitefly (Homoptera; Aleyrodidae) dispersal under different UV light environments. Journal of Economic Entomology, 100: 389–397.
Duffus JE (1963). Possible multiplication in the aphid vector of Sowthistle yellow vein virus, a virus with an extremely long insect latent period. Virology, 21(2): 194–202.
Duffus JE (1965). Beet pseudo yellows virus, transmitted by greenhouse whitefly (*Trialeurodes vaporariorum*). Phytopathology, 55: 450–453.
Duffus JE, Cid HY and Johns MR (1985). Melon leaf curl virus—a new *Geminivirus* with the host, serological variations from Squash leaf curl virus. Phytopathology, 75: 1312.
Duffus JE (1987). Whitefly transmission of plant viruses. pp. 73–91. In: KF Harris (ed.). Current Topics in Vector Research, Springer-Verlag, New York.
Duffus JE, Liu HY and Wisler GC (1996). Tomato infectious chlorosis virus–a new clostero like virus transmitted by *Trialeurodes vaporariorum*. European Journal of Plant Pathology, 102: 219–226.
EFSA (2013). Scientific opinion the risks to plant health posed by *Bemisia tabaci* species complex and viruses transmitted for the EU territory. European Food Safety Authority Journal 11(4): 3162.

Gamarra HA, Fuentes S, Morales FJ, Glover R, Malumphy C and Baker L (2010). *Bemisia afer sensu lato,* a vector of Sweet potato chlorotic stunt virus. Plant Disease, 94: 510–514.

Ghanim M, Morin S, Zeidan M and Czosnek H (1998). Evidence for transovarial transmission of Tomato yellow leaf curl virus by its vector whitefly, *Bemisia tabaci*. Virology, 240: 295–303.

Ghanim M and Czosnek H (2000). Tomato yellow leaf curl virus (TYLCV-Is) is transmitted among whiteflies, *Bemisia tabaci* in a sex-related number. Journal of Virology, 74: 4738–4745.

Ghanim M, Morin S and Czosnek H (2001). Rate of Tomato yellow leaf curl virus translocation in the circulative transmission pathways of its vector, the whitefly, *Bemisia tabaci.* Phytopathology, 91: 188–196.

Ghanim M, Bromin M and Popovsky S (2009). A simple, rapid and inexpensive method of localization of Tomato yellow leaf curl virus and Potato leafroll virus in plant-insect vectors. Journal of Virological Methods, 159: 311–314.

Ghanim M (2014). A review of the mechanisms and components that determine the transmission efficiency of Tomato yellow leaf curl virus (Geminiviridae, *Begomovirus*) by its whitefly vector. Virus Research, 186: 47–54.

Gil-Salas FM, Peters J, Boonham PN, Cuadrado I M and Janssen D (2012). Co-infection with Cucumber vein yellowing virus and Cucurbit yellow stunting disorder virus leading to synergism in cucumbers. Plant Pathology, 61: 468–472.

Gottlieb Y, Ghanim M, Gueguen G, Kontscdalov S, Vavre F and Fleury F (2008). Inherited intracellular ecosystem; Symbiotic bacteria share bacteriocytes in whiteflies. FASEB Journal, 22: 2591–2599.

Gottlieb Y, Zchori–Fein F, Mozes–Daube N, Konstedalov S, Skaljae M and Brumin M (2010). The transmission efficiency of Tomato yellow leaf curl virus by the whitefly, *Bemisia tabaci* is correlated with the presence of a specific symbiotic bacterium species. Journal of Virology, 84: 9310–9317.

Gotz M, Popovski K, Kollenberg M, Gorovitz R, Brown JK, Cicero JM, Czosnek H, Winter S and Ghanim M (2012). Implication of *Bemisia tabaci* heat shock protein 70 in *Begomovirus* –whitefly interactions. Journal of Virology, 86 13241–13252.

Guang LH, Tseng HH, Li JT and Chen TC (2010). First report of Cucurbit chlorotic yellows virus infecting cucurbits in Taiwan. Plant Disease, 94: 11682–11682.

Gyoutoku Y, Okazaki S, Furuta A, Etoh T, Mizobe M and Kuno K (2009). Chlorotic yellow disease of melon caused by Cucurbit chlorotic yellows virus, a new *Crinivirus*. Japan Journal of Phytopathology, 75: 109–111

Hoh F, Uzest M, Drucker M, Plisson-Chastang C, Bron P, Blanc S and Dumas Hohn T (2007). Plant virus transmission from the insect point of view. Proceedings of National Academy of Sciences, USA, 104: 17905–17906.

Issacs R, Wills MA and Byrne DN (1999). Modulation of whitefly take off and flight orientation by wind speed and visual cues. Physiology Entomology, 24: 31–318.

Jefferson O, Frazer J, De LaRosa D, Beever JS, Ahlquist P and Maxwell DP (1994). Whitefly transmission and efficient ssDNA accumulation of Bean golden mosaic *Geminivirus* require functional coat protein. Virology, 204: 289–296.

Jeyanandarajah P and Brunt AA (1993). The natural occurrence, transmission properties and possible affinities of Cowpea mild mottle virus. Journal of Phytopathology, 137: 148–158.

Jiang ZF, Xia F, Johnson KW, Bartom E, Tuteja JH, Stevens R, Grossman RL, Brumin M, White KP and Ghanim M (2012). Genome sequence of primary endosymbiont *Candidatus portiera aleyrodidarum* in whitefly, *Bemisia tabaci* B and Q biotypes. Journal of Bacteriology, 194: 6678–6679.

Jones DR (2003). Plant viruses transmitted by whiteflies. European Journal of Plant Pathology, 109: 195–219.

Keur JY (1934). Studies on the occurrence and transmission of virus diseases in the genus Abutilon. Bulletin of Torrey Botanical Club, 61: 53–76.

Kil EJ, Kim S, Lee YJ, Byun HS, Park J, Seo H, Kim CS, Shim JK, Lee JH, Kim JK, Choi HS and Lee S (2016). Tomato yellow leaf curl virus (TYLCV-IL) A seed transmissible *Geminivirus* in tomato. Science Reporter, 6: 19013; doi:10 1038/Srep.19013(2016).

Kubota K and Ng JCK (2016). Lettuce chlorosis virus P23, suppresses RNA silencing and induce local necrosis with increase severity and raised temperatures. Phytopathology, 106: 653–662.

Larsen RC, Duffus JE and Liu HY (1984). Tomato necrotic dwarf, a new type of whitefly-transmitted virus. Phytopathology, 74: 795.

Li J, Liang X, Wang X, Shi Y, Gu Q, Kuo YM, Falk BW and Yan F (2016). Direct evidence for the semi-persistent transmission of Cucurbit chlorotic yellows virus by a whitefly vector. Science Reporter, 4(6): 36604. doi:10. 10381/srep 36604.

Liu HY, Wisler GC and Duffus JE (2000). Particle length of whitefly-transmitted criniviruses. Plant Disease, 84: 803–805.

Liu S, Bedford ID, Briddon RW and Markham PG (1997). Efficient whitefly transmission of African cassava mosaic *Geminivirus* requires sequences from both genomic components. Journal of General of Virology, 78: 1791–1794.

Loconsole G, Saldarelli P, Doddapinen H, Savino V, Martelli GP and Seponari M (2012). Identification of India sheden DNA virus associated with Citrus chlorotic dwarf virus, a new member in family Geminiviridae. Virology, 432: 162–172.

Lotrakul P and Valverde RA (1999). Cloning of DNA–a-like genomic component of Sweet potato leaf curl virus: nucleotide sequence in phylogenetic relationships. Molecular Plant Pathology, Online htt//www bess UK/1999/0206 lotrakul.

Lotrakul P, Valverde RA, Clark CA and Fauquet CM (2003). Properties of begomoviruses isolated from sweet potato (*Ipomoea batatas* L.) infected with Sweet potato leaf curl virus. Revista Mexicana de Fitopatologia, 21: 128–138.

Lu S, Li J, Wang X, Song D, Bai R, Shi Y, Gu Q, Kuo YW, Falk BW and Yan F (2017). Asemi-persistent plant virus differentially manipulate feeding behavior of different sexes and biotypes of its whitefly vector. Viruses, 2017, 9: PCI E4; doi;10.3390/v 90 10004.

Martelli GP, Agranovsky AA, Bar-Joseph M, Boscia D, Candresse T and Coutts RNA (2011). Family Closteroviridae. pp 987–1001. In: King A, Adam MJ, Carsten EB, Lefkowitz EJ (eds). Virus Taxonomy, 9th Report of International Committee on Taxonomy of Viruses (Amsterdam Elsevier, Academic Press).

Martin G, Cuadrado TM and Janssen D (2001). Bean yellow disorder virus: Parameters of transmission by Bemisia tabaci and host range. Insect Science, 28: 50–56.

Martin G, Velasco L, Segunda E, Caudrado CM and Jenssen D (2008). The complete nucleotide and sequence of genome organization of Bean yellow disorder, a new member of the genus *Crinivirus*. Archives of Virology, 153: 999–1001.

Martin RR, Tzanetakis IE, Gergerich RC, Fernamdez G and Pesic Z (2004). Blackberry yellows virus associated virus; a new *Crinivirus* found in blackberries. Acta Horticulturae, 656: 137–142.

Martin RR, MacFarlane S, Sabanadzoic S, Quito DE, Poudel B and Tzanetakis I (2013). Viruses and virus diseases of Rubus. Plant Disease, 97: 168–182.

Maruthi MN, Hillocks RJ, Mtunda K, Raya MD, Muhanna M, Kiozia H, Rekha AR, Colvin J and Thresh JM (2005). Transmission of Cassava brown streak virus by *Bemisia* doi .org/10 1155/2012/795697.

Mathur RN (1973). Leafcurl in Zinnia elegans at Dehradun. Indian Journal of Agricultural Sciences, 3: 89–96.

Mathur RN (1964). A whitefly-transmitted Mosaic virus of *Wissadula amplissima*. Plant Disease Reporter, 48: 902–905.

Mbanzibwa DR, Tran YP, Tugume AK, Patil BL, Yadav JS, Bagewadi B, Abarshi MM, Alical T, Changadeya J, Muli MB, Mukasa SB, Tairo F, Baguma Y, Kyamanydia S, Kullaya A, Maruthi MN, Fauquet CM and Valkomen JPT (2011). Evaluation of Cassava brown streak disease-associated viruses. Journal of General Virology, 92: 974–987.

Medina V, Pinner MS, Bedford ID, Achon MA, Gemeno C and Markham PG (2006). Immunolocalization of Tomato yellow leaf curl virus in natural host plants and its vector, *Bemisia tabaci*. Journal of Plant Pathology, 88: 290–308.

Menzel W, Abang MM and Winter S (2011). Characterization of Cucumber vein clearing virus, a whitefly (*Bemisia tabaci* G.) transmitted *Carlavirus*. Archives of Virology, 156: 2309–23011.

Mohammed IU, Abarshi MM, Muli B, Hillocks RJ and Maruthi MN (2012). The symptoms and genetic diversity of Cassava brown streak virus infecting cassava in East Africa. Advances in Virology, 2012, article ID 79569710 pages http /dx/.

Morin S, Ghanim M, Zeidan M, Czosnek H, Verbeek M, Johannes FJ and Van den Heuvel JF (1999). A GroEL homologue from endosymbiotic bacteria of whitefly, *Bemisia tabaci* is implicated in the circulative transmission of Tomato yellow leaf curl virus. Virology, 256: 75–85.

Mukasa SB, Rubaihayo PR and Valkonen JPT (2003). Sequence variability within the 3 proximal part of the Sweet potato mild mottle virus genome. Archives of Virology 148: 487–496.

Muniyappa V, Gowda MMN, Babitha CR, Colvin J, Briddon RW and Rangaswamy KT (2003). Characterization of Pumpkin yellow vein mosaic virus from India. Annals of Applied Biology, 142: 323–331.

Muniyappa V, Reddy HR and Shivashanker G (1976). Studies on the Yellow mosaic disease of horse gram (*Dolichos biflorus* L.) II. Host range studies. Mysore Journal of Agricultural Sciences, 10: 611–614.

Naidu RA, Gowda S, Satyanaryana T, Boyko V, Reddy AS, Dawson WO and Reddy V (1998). Evidence of whitefly-transmitted Cowpea mild mottle virus belongs to *Carlavirus*. Archives of Virology, 143: 769–780.

Nair RR and Wilson KI (1970). Leafcurl of *Jatropha curcas* L. in Kerala. Science and Culture, 36: 569.

Nariani TK and Pathanian PS (1953). *Physalis peruviana* L. A new host of Tobacco leaf curl virus. Indian Phytopathology, 6: 143–145.

Nariani TK (1956) Leaf curl of papaya. Indian Phytopathology, 9: 151–155.

Nariani TK (1960). Yellow mosaic of mung (*Phaseolus aureus* L.). Indian Phytopathology, 13: 24–29.

Navas-Castillo J, Camero R, Bueno M and Morioness E (2000). Severe yellowing outbreaks in tomato in Spain associated with infections of Tomato chlorosis virus. Plant Disease, 84: 835–837.

Navas-Castillo J, Fiallo-Olive E and Sanchez-Campos S (2011). Emerging virus diseases transmitted by whiteflies. Annual Review of Phytopathology, 49: 219–248.

Ng JCK and Falk BW (2006). Virus–vector interactions mediating nonpersistent and semi-persistent transmission of plant viruses. Annual Reviews of Phytopathology, 44: 183–212.

Ng JCK and Zhou JS (2015). Insect vector-plant virus interactions associated with non-circulative, semi-persistent transmission: Current perspectives and future challenges. Current Opinion in Virology, 15: 48–55.

Nigam K, Srivastava S, Kumar D, Verma Y and Gupta S (2014). Detection and analysis of leaf curl virus from Jatropha. International Scientific Research Publication Vol 4, May 2014.

Noris E, Vaira AM, Caciagli P, Masenga V, Gronenborn B and Accotto GP (1998). Amino acids in the capsid protein of Tomato yellow leaf curl virus that are crucial for systemic infection, particle formation, and insect transmission. Journal of Virology, 72: 10050–10057.

Okuda M, Okazaki S, Yamasaki S, Okuda S and Sugiyama M (2010). Host range and complete genome sequence of Cucurbit chlorotic yellows virus, a member of genus *Crinivirus*. Phytopathology, 100: 560–566.

Opiyo SA, Ateka EM, Owuor PO, Manguro LOA and Miano DW (2010). Development of multiplex PCR technique for simultaneous detection of SPFMV and SPCV. Journal of Plant Pathology, 92: 363–367.

Pascual S and Callejas C (2004). Intra and interspecific competition between biotypes B and Q of *Bemisia tabaci* (Hemiptera: Aleyrodidae) from Spain Bulletin of Entomological Research, 94: 369–375.

Pirone TP and Blanc S (1996). Helper dependent vector transmission of plant viruses. Annual Review of Phytopathology, 14: 227–247.

Polston JE and Capobianco H (2013). Transmitting plant viruses using whiteflies. Journal of Visualized Experiments, 81; 4372 e4332: doi;10:3791/4332(2013).

Qiao Q, Jhang ZC, Qin YH, Jhang DS, Tian YT and Wang YJ (2011). First report of sweet potato chlorotic stunt virus infecting sweet potato in China. Plant Disease, 95: 356.

Qin L, Zhang Z, Qiao Z, Qiao Q, Zhang D, TianY and Wang S (2013). First report of Sweet potato leaf curl Georgia virus on sweet potato in China. Plant Disease, 97: 1388.3–1388.3.

Ramakrishnan K, Kandaswamy JK, Subramanian KS, Janarthanan R, Mariappan V, Samuel GS and Navaneathan G (1991). Investigations on virus diseases of pulse crops in Tamil Nadu. Final report Tamil Nadu Agricultural University, Coimbatore, India.

Rosen R, Kanakala S, Kliot A, Pakkianathan BC, AbuFarich B, Santana-Magal N, Elimelech M, Kontcdalov S, Lebedev G, Cila M and Ghanim M (2015). Persistent, circulative transmission of begomoviruses by whitefly vectors. Current Opinion in Virology, 15: 1–8.

Sahambhi HS (1958). Viral disease of sesamum and their control. In: Mycology Research Workers Conference, Simla, India.

Salazar LE, Muller G, Querei M, Zapote JL and Owens RA (2000). Potato yellow vein virus, its host range, distribution in South America and identification as a *Crinivirus* transmitted by *Trialeurodes vaporariorum*. Annals of Applied Biology, 137: 7–19.

Samos–Garcia D, Farnie PA, Beiti F, Zchori–Fein E, Vavr VIC, Re F and Mouton L (2012). Complete genome sequence of *Candidatus potiera aleyrodidarum* BT-Q, an obligate symbiont that supplies amino acids and carotenoids to *Bemisia tabaci*. Journal of Bacteriology, 194: 6654–6655.

Seal SE, Van den Bosch FC and Jeger MJ (2006). Factors influencing *Begomovirus* evolution and their increasing global significance implications for sustainable control. Critical Reviews in Plant Science, 25: 23–46.

Sela I, Assouline I, Tanne, Cohen S and Marco S (1980). Isolation and characterization of rod-shaped whitefly transmissible DNA containing plant virus. Phytopathology, 70: 226–228.

Silberschmidt K, Flores E and Thommasi CR (1957). Further studies on the experimental transmission of infectious chlorosis of Malvaceae. Phytopathologica. Zeitscrift, 30: 387–414.

Sim J, Valverde RA and Clark CA (2000). Whitefly transmission of Sweet potato chlorotic stunt virus. Plant Disease, 84: 1250.

Singh BP and Mishra AK (1971). Occurrence of Hollyhock yellow mosaic virus in India. Indian Phytopathology, 24: 213–214.

Skaljac M and Ghanim M (2010). Tomato yellow leaf curl virus and plant–virus–vector interactions. Israel Journal of Plant Science, 58: 103–111.

Sloan DB and Moran NA (2012). Endosymbiotic bacteria as a source of carotenoids in whiteflies. Biology Letters, 8: 986–889.

Srivastava KM, Swadesh Shren VC, Srivastava BN and Singh BP (1977). Zinnia yellow net disease–transmission, host range and agent–vector relationship. Plant Disease Reporter, 61: 550–554.

Stein VE, Coutts RHA and Buck KW (1983). Serological studies on Tomato golden mosaic virus, a *Geminivirus*. Journal of General Virology, 64: 2493–2498.

Stewart LR, Medina V, Tian T, Turina M, Falk BW and Ng JCK (2010). A mutation in the Lettuce infectious yellows virus minor coat protein disrupts whitefly transmission but not in planta systemic movement. Journal of Virology, 84: 12165–12173.

Storey HH (1931). A new virus disease of the tobacco plant. Nature, 128: 187–188.

Storey HH and Nicholas RFW (1938). Studies on the mosaic of cassava. Annals of Applied Biology, 25: 790–806.

Summers CG, Newton Jr AS and Estrada D (1996). Intraplant and interplant movement of *Bemisia argentifolii* (Hemiptera: Aleyrodidae) crawlers. Environmental Entomology, 25: 1360–1364.

Summers CG, Mitchell JP and Stapleton JJ (2004). Management of aphid-borne viruses and *Bemisia argentifolii* (Homoptera: Aleyrodidae) in Zucchini squash by using UV reflective plastic and wheat straw mulches. Environmental Entomology, 33: 1447–1457.

Suteri BD and Srivastava SB (1975). Protein content in healthy and Yellow mosaic infected soybean seeds. Current Science, 44: 205–206.

Tarr SAJ (1951). Leaf curl disease of cotton. Commonwealth Mycological Institute Kew Surrey England, 55 pp.
Thao ML and Baumann P (2004). Evolutionary relationships of primary prokaryotic endosymbionts of whiteflies and their hosts. Applied Environmental Microbiology, 70: 3401–3406.
Tzanetakis IE and Martin RR (2004). Complete nucleotide sequence of a strawberry isolate of Beet pseudo yellows virus. Virus Genes, 28: 239–246.
Tzanetakis IE, Wintermantel WM, Cortez AA, Barmes JE, Barrett SM and Bolda MP (2006). Epidemiology of Strawberry pallidosis associated virus and occurrence of pallidosis disease in North America. Plant Disease, 90: 1343–1346.
Tzanetakis IE, Wintermantel WM, Poudel B and Zhou J (2011). Diodia vein chlorosis virus as a group-1 *Crinivirus*. Archives of Virology, 156: 2033–2037.
Tzanetakis IE, Martin RR and Wintermantel WM (2013a). Epidemiology of criniviruses: an emerging problem in world agriculture. Frontiers in Microbiology, 4: 115.
Tzanetakis IE, Martin RR and Wintermantel WM (2013b). Expanding field of strawberry viruses which are important in North America. International Journal of Fruit Science, 13: 184–195.
Uzest M, Gargani D, Dombrovsky A, Cazevieille C, Cot D and Blanc S (2010). The "acrostyle": a newly described anatomical structure in aphid stylets. Arthropod Structure and Development, 39: 221–229.
Valverde RA, Sim J and Latrakul P (2004). Whitefly transmission of sweet potato viruses. Virus Research, 100: 123–128.
Van Der Laan PA (1940). Motschilduis en eupatorium als Oorzaken van Pseudo mosaic (whitefly and eupatorium as causes of pseudo mosaic) Vlugsckr Deli-Proefst, Medan, 67: 4.
Varma PM (1963). Transmission of plant viruses by whiteflies. Bulletin of National Institute of Sciences, India Bulletin, 24: 11–33.
Verbeek M, Petra J, Van Bekkum V, Dullemans A, Rene AA and Vlugt VD (2014). Torradoviruses transmitted in a semi-persistent and stylet-borne manner by three whitefly vectors. Virus Research, 186: 55–60.
Verma VS (1974a). Lupen leaf curl virus. Gartenbauwissenschaft, 39: 53.
Verma VS (1974b). Soapwort leaf curl virus. Gartenbauwissenschaft, 39: 567–568.
Verma HN, Srivastava KM and Mathur A (1975). A whitefly-transmitted Yellow mosaic virus disease of tomato from India. Plant Disease Reporter, 59: 494–498.
Wang L-L, Wei X-M, Ye X-D, Xu H-X, Zhou X-S, Wang X-W and Liu X (2014). Wang expression and functional characterization of a soluble form of Tomato yellow leaf curl virus coat protein. Pest Management Science, 70: 1624–1631.
Wei J, Zhao JJ, Zhang T, Li FF, Ghanim XP, Zhou XP, Ye GY, Liu SS and Wng XW (2014). Specific cells in the primary cells in primary salivary glands of whitefly, *Bemisia tabaci* control, retention and transmission of begomoviruses. Journal of Virology, 88: 13469–13468.
Webb SE, Adkins S and Reitz SR (2012). Semi-persistent whitefly transmission of Squash vein yellowing virus causal agent of watermelon vine decline. Plant Disease, 96: 839–844.
Wilson KY (1972). Chlorotic ring spot of jasmine. Indian Phytopathology, 25: 157–158.
Wintermantel WM and Wisler GC (2006). Vector specificity, host range and genetic diversity of Tomato chlorosis virus. Plant Disease, 90: 814–819.
Wisler GC and Duffus JE (2001). Transmission properties of whitefly-borne criniviruses and their impact on virus epidemiology. pp. 293–308. In: Harris KF, Smith OP and Duffus JE (eds). Virus-Insect-Plant Interactions (SanDiego: Acad Press).
Wisler GC, Duffus JE, Liu HY and Li RH (1998). Ecology and epidemiology of whitefly-transmitted closteroviruses. Plant Disease, 82: 270–280.
Wisler GC, Duffus JE, Liu HY, Li R and Falk BW (1997). New whitefly transmitted *Closterovirus* identified in tomatoes. California Agriculture, 51: 24–26.
Wu ZC, Hu JS, Polston JE, Ullman DE and Hebert E (1996). Complete nucleotide sequence of nonvector transmissible strain of Abutilon mosaic, *Crinivirus* in Hawaii. Phytopathology, 86: 508–513.

Yamashita S, Doi Y, Yora K and Yoshino M (1979). Cucumber yellows virus, its transmission by greenhouse whitefly, *Trialeurodes vaporariorum* (Westwood) and the yellowing disease of cucumber and muskmelon caused by the virus. Annals of Phytopathological Society of Japan, 45: 484–496.

QUESTIONS (EXERCISE)

Q 1. Narrate the life cycle of *Bemisia tabaci.*

Q 2 Describe the transmission mechanism of Lettuce infectious yellow virus utilizing transmission determinants.

Q 3. List the vector species of whiteflies and discuss the virus-vector relationship of one virus with *Trialeurodes vaporariorum.*

Q 4. Describe the seed-borne virus disease transmitted through whiteflies.

Q 5. Write brief notes on:

a) Periodic acquisition
b) Metamorphosis in whiteflies
c) Torradoviruses

CHAPTER 7

Heteropterous Bugs and Thrips

7.1 Heteropterous Bugs (True Bugs)

The Heteropterans are the true bugs and fall under the sub-order Heteroptera in the order Hemiptera. The families Miridae, Orsillidae, Piesmatidae and Pentatomidae contain vectors of plant viruses. The members of both sub-orders possess very similar mouthparts, but bugs are regarded as more inefficient vectors of plant viruses vis a vis the other members of order, Hemiptera. It is an established fact that the saliva of bugs is toxic and causes necrosis of leaf tissues or sometimes inactivates the virus. In all, seven families (Anthocoridae, Berytidae, Miridae, Oxycarenidae, Orsillidae, Pentatomidae and Tingidae) are known to spread *Phytoplasma*. Additionally, these bugs are responsible for spreading fastidious phloem-inhabiting bacteria (Coreidae and Piesmatidae), non-fastidious bacteria (Coreidae, Miridae, Pentatomidae and Phyrrhocoridae), fungi (Alytidae, Coreidae, Lygaeidae, Miridae, Nabidae, Orsillidae, Oxycarenidae and Pentatomidae, Phyrrhocoridae, Reduviidae, Rhopalidae, Scutelleridae and Tingidae) and Trypanosomatids (Coreidae, Largidae, Lygaeidae, Orsillidae, Pentatomidae, Reduviidae, Rhopalidae, Rhyparochromidae, Phyrrhocoridae and Stenocephalidae). Let us examine these two assumptions regarding their ability to act as vectors of plant pathogens.

7.1.1 Identification and brief biology

Of the two sub-orders of Hemiptera, Homoptera and Heteroptera, only the latter contains true bugs. These can be identified by the triangular scutellum that covers half of their bodies, and by half of the forewings being of hardened and thickened consistency. The legs of true bugs are thin and without spines. The tarsus of bugs is three-segmented. Like many other insects, bugs possess antennae; these antennae are generally five-segmented. They undergo incomplete metamorphosis. The freshly laid eggs are smeared with bacteria and the nymphs, after hatching, feed on the bacteria adhered to the shell of eggs. After feeding on the bacteria in the first instar, they pass through another four instars and become adults. The adult longevity is thirteen to forty-six days and twenty two to thirty-eight days for females and males, respectively. They have piercing and sucking mouthparts. On account of injection of toxic saliva while feeding, necrosis of tissues is caused that inhibits replication of virus in the dead tissues. The large injury caused by bugs is therefore responsible for reduction in transmission efficiency of plant pathogens.

7.1.2 Feeding apparatus and mechanism

The mouthparts of bugs consist of two maxillary and two mandible stylets that form a thickened tube-like structure capable of intracellular penetration of plant tissues. Like the other Hemipteran members, bugs feed by making use of two food and salivary canals. The salivary canal is lobed (Miles, 1999) the lobes of salivary canal are meant for releasing two kinds of saliva, one gel-like and the other is watery. The gel type sticky saliva is consumed to make the salivary sheath while watery saliva is meant to enable free movement of stylets. However, not all species of bugs construct salivary sheaths. The members that do not construct salivary sheath are in the families, Miridae and Tingidae. The bugs cause extensive damage to plant tissues with their thicker and larger stylets which penetrate intracellularly. As a result of damage, the chances of virus infection are reduced due to the extensive necrotic areas (Mitchell, 2004). In addition to thick, stout stylets, the enzyme-rich secretion of the salivary glands is also instrumental in the destruction of the host. As a result of bug feeding, larger necrotic spots or dead areas, in which viruses do not multiply, become conspicuous. The plant viruses require living tissues to multiply as they are obligate parasites. The bugs of both the categories, with salivary sheath (pentatomomorpha) and without salivary sheath (cimicomorpha), differ in their feeding habits. The destruction of cells in the former is caused by movement of stylets in watery saliva (macerate and flush method) (Miles, 1968) while in the latter, it is due to the collective action of stylet penetration and action of enzymes (lacerate and flush method) (Miles and Taylor, 1994). The production of a sheath is a characteristic feature of vascular feeding, while the absence of sheath formation is a distinct feature of intercellular feeding. The action of stylets, rate of ingestion of food and chemical analysis depends on the food. It has been seen in the case of *Dysdercus koenigii* that when bugs feed on liquid food, the stylets do not protrude beyond the labium and, as a result, the liquid is without any salivary secretions (Sexena, 1963). However, on exposing the stylets to solid food, the labium is with held outside and only the stylets enter into the food. When bugs feed on tough and non-porous food, they release an acidic salivary sheath and provide strength to the labium during insertion of stylets. These bugs ingest liquid food eight to twelve times faster than solid food. The solubility, thus, determines the rate of ingestion in such bug species. Like the other insects of order Hemiptera, bugs also assess the chemical composition of sap. To initiate feeding, the bugs are known to make probes using chemosensilla present on precibarium, rostrum tip and labial palpi in order to test the exuded sap. Bugs lacking the labial sensory structure make use of precibarial sensilla to assess the substratum. The chemosensory organs are always present on the rostrum tip of all bugs. After test probes, egestion is common in the members of Pentatomidae family. Without specialized equipment, it is very difficult to differentiate the damage caused by bugs from viral symptoms. On account of the similarity between damage caused by bugs and viral symptoms, the identification of many plant viruses could not be carried out until much more recent times (Carter, 1973). Once the phytotoxic symptoms are identified through experimentation, the task of diagnosis of plant viruses becomes easier.

7.1.3 Mechanism of pathogen transmission

Bugs are known to spread a large number of pathogens such as viruses, *Phytoplasma*, bacteria, fungi and Trypanosomatids.

7.1.3.1 Bugs and viruses

Of these plant pathogens, plant parasitic viruses belonging to the genera *Sobemovirus* (Velvet tobacco mottle virus and Sowbane mosaic virus), *Carlavirus* (Potato mosaic virus), *Luteovirus/polerovirus* (Potato leaf roll virus), *Potyvirus* (Centrosema mosaic virus) and *Rhabdovirus* (Beet leaf curl virus) are bug-borne (Table 7.1). Beet leaf curl disease inflicted by *Rhabdovirus* is transmissible via *Piesma quadratum* (Central Europe) and *P. cinereum* (United States of America) species of bugs. The virus is known to be propagative in bugs, based on the evidence of widespread presence of virus particles scattered in salivary glands, midgut, hemolymph and feces of vector species, in addition to a long latent period and lifelong retention of the virus in the vector (Proeseler, 1980; Proeseler, 1978). Both nymphs and adults can acquire the virus, but only adults can transmit the virus. Nymphs cannot transmit the virus because the latent period is longer than the nymphal period, so the nymphs become adults before the virus becomes transmissible. *P. cinereum* has also been identified as vector of another pathogen (Beet savoy disease), but subsequent studies showed a low efficiency rate. In another study, the possibility of same species being vector of another virus has been ruled out,

Table 7.1 Transmission of important plant pathogens through bugs.

S. No	Pathogen	Disease	Causal organism	Vector	Source(s)
1	Virus	Velvet mottle	Velvet Tobacco mottle Virus (*Sobemovirus*)	Mirid bug, *Engytatus nicotianae*	Gibb and Randles, 1991
2	Virus	Potato mosaic	Potato mosaic virus-M (*Carlavirus*)	*Lygus rugulipennis*; *Lygus pratensis*	Proeseler, 1980; Proeseler, 1978
3	Virus	Potato leaf roll	Potato leaf roll virus (*Polerovirus*)	*Lygus rugulipennis*; *Lygus pratensis*	Proeseler, 1980
4	Virus	Centrosema mosaic	Centrosema mosaic virus (*Potyvirus*)	*Nysius* spp.	VanVelsen and Crowley, 1961
5	Virus	Beet leaf Curl	Beet leaf curl virus (*Rhabdovirus*)	*Piesma quadratum*	Proeseler, 1980
6	Virus	Longan witches Broom	Longan witches broom virus (*Phytoplasma*?) unconfirmed	*Tessaratoma papillosa*	Chen et al., 2001
7	Virus	Beet savoy disease	Beet savoy virus (?) unconfimed	*Piesma cinereum*	Schneider, 1964
8	Virus	Sowbane mosaic	Sowbane mosaic virus (*Sobemovirus*)	*Halticus bractatus*; *Halticus citri*	Bennett and Costa, 1961
9	*Phytoplasma*	Paulownia witches broom	Paulownia witches broom *Phytoplasma*	*Halyomorpha halys*	Hiruki, 1999
10	*Phytoplasma*	Lethal wilt of coconut palms	*Phytoplasma*	*Stephanitis typica*	Mathen et al., 1990

11	*Phytoplasma*	Protea Witches broom	*Phytoplasma*	*Oxycarenus maculatus*	Wieczorek and Wright, 2003
12	*Rickettsia Like Organism*	Beet latent rosette disease	Beet rosette Rickettsia	*Piesma cinereum*	Nienhaus and Schmutterer, 1976
13	Bacteria	Cotton boll rot	*Xanthomonas campestris malvacearum*	*Lygus lineolaris*	Mendis, 1956
14	Bacteria	Bacterial blight	*Xanthomonas campestris malvacearum*	*Pseudatomoscelis seriatus*	Martin et al., 1987
15	Bacteria	Boll rot	*Xanthomonas campestris malvacearum*	*Lygus lineolaris; Hypselonotus fulvus*	Mendis, 1956
16	Bacteria	Black rot of rice	*Erwinia herbicola*	*Pentatomid bug, Nezara viridula*	Tanii et al., 1974
17	Bacteria	Fire blight of apple and pear	*Erwinia amylovora*	*Lygus elisus', Lygus lineolaris*	Wheeler, 2001
18	Bacteria	Soft rot of celery	*Erwinia carotovora*	*Lygus lineolaris*	Richardson, 1938
19	Bacteria	Ring rot of potato	*Clavibacter michiganensis sepedonicus*	*Lygus lineolaris*	Wheeler, 2001
20	Trypanosomatid (Protozoa)	Coffee phloem necrosis	*Phytomonas leptovasorum*	*Dicranocephalus agilis; Nezara viridula; Phthia picta*	Parthasarathy et al., 1976; Alves-Silva et al., 2013
21	Trypanosoma	Hartrot of coconut	*Phytomonas staheli*	*Dicranocephalus agilis; Nezara viridula; Phthia picta*	Alves-Silva et al., 2013; Camargo And Wallace, 1994
22	Trypanosoma	Sudden wilt or Marchitez of oil palm	*Marchitez sorpresiva*	*Dicranocephalus agilis; Nezara viridula; Phthia picta*	Alves-Silva et al., 2013; Camargo and Wallace, 1994

Table 7.1 contd. ...

...Table 7.1 contd.

S. No	Pathogen	Disease	Causal organism	Vector	Source(s)
23	Fungus	Boll rot of cotton	*Asbhya gossypii; Alternaria tenuis; Fusarium moniliforme*	*Edessa meditabunda; Lygus lineolaris; Lygus borealis; Dysdercus howardi*	Mendis, 1956; Frazer, 1944; Bagga and Laster, 1968
24	Fungus	Stigmatomycosis	*Nematospora coryli*	*Dysdercus intermedius; Leptoglossus gonagra*	Daugherty, 1967
25	Fungus	Pistachio stigmatomycosis	*Aureobasidium pullulans*	Leaffooted bugs (*Leptoglossus* sp.)	Michailides and Morgan, 1991
26	Fungus	Soybean leaf spot	*Nematospora coryli*	*Acrosternum hilare*	Daugherty, 1967
27	Fungus	Cotton lint rot	*Aspergillus flavus*	*Lygus hesperus*	Stephenson and Russell, 1974

taking into account the involvement of more taxa and the issue of specificity (Power, 2000). An earlier study concerning the transmission of plant viruses indicated definite association of bugs; among them, the involvement of both species of mirid bugs, *Lygus rugulipennis* and *L. pratensis*, in the spread of *Carlavirus* (Potato mosaic virus M) and *Luteovirus* (Potato leaf roll virus), has been confirmed (Turka, 1978). However, the details of the virus-vector relationship are yet to be explored. Logan witches broom virus is known to be circulative in its vector, *Tessaratoma papillosa* (Pentatomidae) (Chen et al., 2001). The presence of virus particles in salivary glands confirmed the use of circulative mechanism of spread through bugs. Two *sobemoviruses*, namely Tobacco velvet mottle virus and Sowbane mosaic virus, are transmissible via mirid bug (*Halticus bractatus*). Velvet tobacco mottle virus is unique, with regard to its mechanism of transmission, as it is neither propagative, nor circulative, in spite of the fact that the virus has a long latent period (nine days) and passes through the eggs. Its absence from salivary glands, however, raises questions as to its nature (Gibb and Randles, 1991). To draw a definite conclusion, it requires re-examination. Earlier, a non-persistent type of mechanism in Centrosema mosaic virus (*Potyvirus*) and its vector, *Nysius* sp. (Orsillidae) had also been observed (Van Velsen and Crowley, 1961).

7.1.3.2 Diseases

7.1.3.2.1 Sowbane mosaic virus disease-SoMV (*Sobemovirus*)

It is a virus disease caused by Sowbane mosaic virus and is known by different names, such as Apple latent virus-2, *Chenopodium* mosaic virus, *Chenopodium* seed-borne mosaic virus and *Chenopodium* star mottle virus. This virus infects plants of family Chenopodiaceae. It is an ssRN virus with spherical particles (twenty five to twenty-eight nm). It has two variants of 4.5 and 4 kb size. The disease is prevalent in Australia and produces symptoms like yellow mottling, leaf deformation, and necrosis in leaves of sowbane. Besides, yellow vein banding is also common in *Chenopodium* species and spinach. It is transmissible through mechanical means, seed/pollen, mirid bug (*Halticus bractatus; H. citri*) (Bennett and costa , 1961) and thrips; *Thrips tabaci* transmits it in a non-persistent manner. Besides bugs and thrips, other insects such as leaf-miner, *Liriomyza langei*, fleahoppers, *Halticus citri*, sugarbeet leafhopper, *Circulifer tenellus* and green peach aphid, *Myzus persicae*, can also act as a vector of this virus. Velvet tobacco mottle virus is also included in these genera.

7.1.3.2.2 Centrosema mosaic of passion fruits-CenMV (*Potyvirus*)

It is a virus belonging to *Potyvirus* genus of plant viruses. The virus is capable of causing mosaic diseases in legumes, *Crotalaria* spp., French beans and passion fruits. It is known to infect 44 species of dicotyledonous plants hailing from 5 families, including *Centrosema pubescent, Crotalaria anagyroides, C. retusa, C. goreensis, C. mucronata, Calopogonium mucunoides* and *Desmodium distortum* (Morales, 1994). It is an RNA virus (flexuous rods) with a particle size of 750 x 12 nm. In passion fruit, it produces mosaic, ring spots, rugosity and distortion of leaves in affected trees. It is prevalent in Australia, Surinam, Papua New Guinea, Colombia, Caribbean areas and Papua (Indonesia). It spreads via aphids (*Aphis gossypii, Myzus persicae, Brachycaudus helichrysi*), bugs (*Nysius* spp.), and Dodder. Of these, two species of aphids (*A. gossypii* and *M. persicae*) are the most efficient vectors of this virus.

7.1.3.2.3 Beet leaf curl virus disease-BLCV (*Rhabdovirus*)

It is primarily a disease of *Beta vulgaris*, caused by a Beet leaf curl virus (BLCV), containing negative ssRNA and being bacilliform in shape. It has been recorded in *Atriplex* spp., *Nicotiana* spp. and *Chenopodium* spp., as well as in sugar beet. *Tetragonia teragonoides* is its main host. The virion size is 75 nm x 180 nm and the genome is around 11–15 kb. It contains five types of proteins. Initially, the diseased plants show vein clearing, followed by crinkling of leaves and petioles. In the last stages of the disease, the veins are swollen and the young leaves show acute curling. The disease has been recorded in the Czech Republic, Slovakia, Germany, Poland, Turkey, Slovenia and Russia. It is transmissible through lacewing bug, *Piesma quadratum*. Both the nymphs and the adults can acquire and inoculate the virus in thirty and forty minutes of access, respectively. Once the virus is acquired, it is retained by the vector for life. There is a latent period of seven to thirty-five days in the body of vector and the relationship is persistent propagative. The virus has been found in a vector in salivary glands and the intestinal wall of hemolymph.

7.1.3.2.4 Potato mosaic virus-PMV (*Carlavirus*)

Potato crops suffer from a large number of mosaic viruses in the *Carlavirus* genus, including Potato virus S, Potato virus M, Potato virus H (Li et al., 2013), Potato latent mosaic virus and Potato virus E. These are identified by the mottling, mosaic, crinkling, rolling of leaves and stunted growth of plants. The mosaic diseases are prevalent throughout the potato growing areas of the world. The transmission of the viruses mentioned is through aphids (*M. persicae, Aphis frangulae, A. nasturtii, Macrosiphum euphorbiae*) in a non-persistent manner. Potato virus M is an ssRNA virus with curved filamentous particles (650 x 12 nm). The newly discovered virus of potato, Potato virus H, also has curved filamentous particles (Li et al., 2013). These viruses have also been reported as being bug-transmissible.

7.1.3.2.5 Potato leaf roll virus-PLRV (*Polerovirus*)

This virus is a positive ssRNA virus belonging to genus *Polerovirus*. It is characterized by downward-rolling with necrosis of leaf margins and stunted plants. The necrosis of tubers is also common in affected plants. It is known to infect members of family Solanaceae, including potato. It is transmissible by many aphid species but *Myzus persicae* is an efficient vector of this virus. After acquisition from the source, the virus passes through the gut and gets into the hemolymph and finally reaches the salivary glands. Thus, the mechanism involved is the persistent circulative type.

7.1.3.2.6 Longan witches broom virus disease-LonWBV (Virus/ Phytoplasma?) To cause a disease, three agencies are involved, namely Elm yellows (16Sr-V), Stolbur (16Sr-VII) *Phytoplasma* and a Longan virus, but Koch's postulates are yet to be proved; until such a time, it is considered as a suspected case. The Longan witches broom virus (LonWBV) is a filamentous virus with a particle measuring 700–1300 nm in length and prevalent in China, Brazil, Taiwan, and Thailand. It is characterized by small, light green leaves with curved margins. Subsequently, the leaves roll down and then become deformed in the advanced stage of the disease. As more time passes, the veins turn brownish in color. The flowers become detached from the petioles and the inflorescence looks like a broom; this can easily be identified. The disease infects Longan (*Dimocarpus longan* in Vietnam and *Euphoria longan* in Hongkong) and litchi in nature. In areas of its prevalence, 50–100% plants are found infected due to this disease. The spread of disease is through litchi stink bug *Tessaratoma papillosa*, longan psylla, *Cornegenapsylla sinica*, gall mite *Eriophyes dimocarpi*, seeds, bud wood and dodder.

7.1.3.3 Bugs and Phytoplasma

Mycoplasma Like Organisms (MLO) were discovered in 1967 by refuting the etiology of some disorders of viral origin which were wrongly reported. Some of the diseases identified as being of Mycoplasmic origin were wrongly identified and later reclassified as being linked to Phytoplasmic or Spiroplasmic agents. The diseases inflicted by such pathogens are known to produce symptoms such as yellowing, dwarfing, witches boom or phyllody in plants and vectored through leafhoppers. The involvement of bugs in the spread of *Spiroplasma/Phytoplasma* (Mollicute) in addition to viruses is now irrefutable. Plant bugs of seven families, namely Anthocoridae (*Orius* sp. and *Halticus minutus* for Protea witches broom), Oxycarenidae (*Oxycarenus maculatus* for Protea witches broom), Orsillidae (*Nysius vinitor* for unknown papaya disease), Berytidae (*Gampsocoris* sp. for Paulownia witches broom), Miridae (*Nesidiocoris tenuis* for Paulownia witches broom; *Lygus rugulipennis* for Tomato Stolbur and Sweet potato little leaf), Pentatomidae (*Halyomorpha* of vector. It is detected in mid gut, salivary glands, malpighian tubules, tracheae, and hemolymph of the vector. The disease is caused by *Xylella fastidiosa* phloem-colonizing bacteria and is widely prevalent in Germany.

7.1.3.5 Bugs and fungi

Fourteen families of bugs have been confirmed as being associated with the spread of fungal pathogens that cause diseases in soybean (Root rot/Yeast spot), rice (Sheath rot/discoloration), beans (Yeast spot), cotton (Stigmatomycosis/Boll rot/Lint rot), Opuntia (Joint lesions), pistachio (Panicle shoot blight/Stigmatomycosis), citrus (Fruit stains/ Stigmatomycosis), tomato (Stigmatomycosis), cassava (Candle stick disease), cacao (Canker/Dieback/Pod rot), cotton (Boll rot/Lint rot/ Stigmatomycosis), mustard (Pod/Seed lesions), alfalfa (*Verticillium* Wilt), carrot (Black rot), coffee (Bean rot) and oil palm (Leaf spot). Of these, the transmission via bugs of fungal pathogens involved in production of Boll rot in cotton, Stigmatomycosis in Pistachio, pod and Seed lesions in mustard, *Verticillium* Wilt in alfalfa, Stigmatomycosis in citrus, and Yeast spot in soybean/beans has been confirmed (Michailides and Morgan, 1991; Michailides and Morgan, 1990; Dammer and Grillo, 1990; Burgess et al., 1983; Grillo and Alvarez, 1983; Daugherty, 1967; Frazer, 1944). Boll rot in cotton is caused by *Fusarium moniliforme* and *Alternaria*, for which the predisposing factors are boll weevil (*Anthonomus grandis*) and tarnished plant bug (*Lygus lineolaris*). For *Aspergillus flavus*, the Lygus bug (*Lygus hesperus*) creates feeding punctures which allow the entry of fungus. In addition, the red and dusky cotton bugs create feeding punctures for the entry of lint staining fungus, *Nematospora gossypii,* into the developing boll of cotton. Cacao dieback is caused by *Fusarium rigiduscula* or *Botryodiplodia theobromae*, which normally enter through the wounds caused by capsid bugs (*Sahibergella singularis* and *Distantiella theobromae*), thus, the bugs act as predisposing factor for the entry of fungus by inflicting feeding punctures.

7.1.3.5.1 Stigmatomycosis

Stigmatomycosis is disease caused by Ascomycetes fungi in cotton, soybean, pecan, pomegranate citrus and pistachio in Russia, Greece and Iran. It is characterized by a wet, smelly, rancid, slimy kernel. Kernels with stigmatomycosis can be small, dark green and partially developed with a brown funiculus, well-developed, dark green and rancid or full-sized but abnormal, being white or light yellow and jelly-like, with a lobed appearance. As time passes and the disease develops, necrosis of kernels becomes apparent. The disease is caused by different species of fungi viz. *Ashbya gossypii, Eremothecium coryli* and *Aureobasidium pullulans*. The disease is spread via stink bugs belonging to Pentatomidae and Coreidae families, and the important species acting as vectors are *Thyanta pallidovirens*, *Chlorochroa uhleri*, *C. ligata* and leaf-footed bug *Leptoglossus clypealis*. The fungi causing this disease live within these insects in a symbiotic relationship. The fungi are responsible for the supply of rich nutrition to bugs (just like bacteria, *Buchnera aphidicola*). Ashbya species live in the mouth of insects and produce riboflavin. As these fungi are parasite on milkweeds or oleander plants known for the production of alkaloids the over production of riboflavin by fungi is instrumental in detoxification of alkaloids. With the result, these insects attain the capacity to feed on alkaloid producing plant spcies and this quality of detoxification of alkaloids widens the host range of bugs in nature. Therefore, the fungi are helpful to the insect vectors in respect of supply of nutrients and detoxification of alkaloids and the fungi in turn get benefit of spread and survival via insects.

7.1.3.6 Bugs and trypanosoma (Protozoa)

The monolytic group of single celled eukaryotic plant parasites spread through insects is the *Phytomonas*. The first ever trypanosomatid isolated was from *Euphorbia pilulifera* on the island of Mauritius in 1909 and it was named as *Leptomonas davidi*. The damage due to these organisms is through the depletion of starch content from the latex and parenchyma tissues of plants, in addition to significant erosion of viscosity of latex. In all, the bugs from eleven families are instrumental in the transmission of protozoans, *Phytomonas davidi*, *P. mcgheei*, *P. serpens*, and *P. elmassiani* responsible for causing Phloem necrosis of coffee (*Arabica / Liberica*) (caused by *Phytomonas leptovasorum*) and Hart rot/fatal wilt of coconut palm (*Cocos nucifera*), *Marchitez sorpresiva* (Sudden wilt/slow wilt) in oil palm (*Elaeis guineesis*) (Dilucca et al., 2013; Parthasarathy et al., 1976) in Central and South America (Mitchell, 2004). Both are lethal wilts, first affect the older leaves and causes browning, later on, devour the young leaves. With the further disease development, the inflorescence is captured and the floral parts show rusty spots. The diseased plants bear fewer small sized fruits which finally shed. In the final stage of disease development, the rotting of spear and roots take place causing the death of trees in a couple of months. In addition, *Phytomonas* France has been found inhabiting cassava (*Manihot esculenta*) latex ducts which caused empty root syndrome but it needs further investigation to confirm its etiology. Trypanosoma has recently been found associated with diseases of corn, cassava, *Euphorbia*, *Asclepius* and many fruit plants like mango, bergamot, and annatto. Both the heterogenic and monogenic forms of protozoan have been detected in many insect orders. The presence of protozoan has been confirmed in the alimentary canal, salivary glands and hemolymph but the monogenic forms were detected in the digestive system of insects. The mechanism of transmission is the persistent and propagative type (Dollet, 1984).

bronzing on lower surface of leaves becomes conspicuous. In young seedlings the feeding causes death of growing points. Due to death of growing points, there is an emergence of excessive axillary branches. The lopsided growth of plants makes them highly misshapen and it becomes so difficult to recognize such plant species. The damage is caused both nymphs and adults (Mound, 1997).

7.2.3 Pathogen transmission

The enveloped viruses are included in two families namely Bunyaviridae and Rhabdoviridae. Among these two families the thrip-borne viruses are in the family Bunyaviridae. In all, fourteen species of thrips under the family Thripidae are vectors of twenty viruses (Bunyaviridae: *Tospovirus*) (Ciuffo et al., 2010; Hassani-Mehraban et al., 2010; Pappu et al., 2009; Jones, 2005; Ullman et al., 1997). Of these, eight species are hailing from the genera *Frankliniella* and these are *F. occidentalis, F. schultzei, F. fusca, F. intonsa, F. bispinosa, F. zucchini, F. gemina,* and *F. cephalica.* Besides these species, *Thrips tabaci, T. palmi, T. setosus, Scirtothrips dorsalis, Ceratothripoides claratris,* and *Dictyothrips betae* are vectors of *tospoviruses* (Table 7.2). Besides, *tospoviruses,* thrips are also associated with the spread of viruses belong to Ilarvirus (Tobacco streak virus-TSV), *carmoviruses, alfamoviruses* (Prune dwarf virus-PDV; Prunus necrotic ring spot virus-PNRSV; Apple chlorotic leaf spot virus-ACLSV), machlomoviruses (Maize chlorotic mottle virus-MCMV) and carlaviruses. In addition, further detail about thrips is presented (Table 7.2). The virus vector relationship is worked out for Tomato spotted wilt virus of virus genera *Tospovirus* (Ullman et al., 1997). It is a unique type of relationship as the virus is acquired by early instars in nymphal stage and transmitted only by adults (Jones, 2005). The nymphs are unable to transmit the Tomato spotted wilt virus as the latent period of virus in the body vector is longer than the age of nymphs (Persley et al., 2006; Whitfield et al., 2005; Ullman et al., 1997; Wijkamp and Peters, 1993). Latent period has been determined as 8 days in nymphs with an acquisition access of five minutes (Everth et al., 2013). As the adults and nymphs both feed on virus source, the adults can also pick up the virus but fail to transmit it to healthy plants, it could either be due barriers preventing the entry of virus in to the salivary gland or lack of desired virus titer owning to low concentration of virus in the body of vector. There is no possibility of transovarial transmission in thrips (Wijkamp et al., 1993). The propagative relationship already stands established between thrips and thrip-borne viruses. Another land mark development has taken place in the virus vector relationship aspect of Maize chlorotic mottle virus with *F williamsi* species of thrips recently where semi-persistent relationship has been identified in thrip-borne *Machlomovirus* (Tombusviridae) (Cabanas et al., 2013). Both the nymphs and adults could transmit the virus up to 6 days after acquisition access on virus source.

Table 7.2 Detail of vector species of thrips and viral diseases.

Sl. No.	Virus genera	Virus	Acronym	Vector species	Source
1	*Carmovirus*	Angelonia flower mottle virus/Pelargonium flower break virus	AnFMV/PelFBV	*Frankliniella occidentalis*	Winter et al., 2006
2	*Ilarvirus*	Tobacco streak virus	TSV	*Frankliniella occidentalis*	Sdoodee and Teakle, 2007
3	*Machlomovirus*	Maize chlorotic mottle virus	MCMV	*Frankliniella williamsi*	Lukanda et al., 2014; Cabanas et al., 2013
4	*Sobemovirus/Ilarvirus*	Sowbane mosaic virus, Tobacco streak virus, Prunus necrotic ring spot virus	SoMV, TSV, PNRSV	*Thrips tabaci Lind.*	Hardy and Teakle, 1992
5	*Tospovirus*	Capsicum chlorosis virus	CCV	*Ceratothripoides claratris*	Premachandra et al., 2005a,b
6	*Tospovirus*	Polygonum ring spot virus	PRSV	*Dictyothrips betae*	Ciuffo et al., 2010
7	*Tospovirus*	Ground nut ring spot virus	GRSV	*Frankliniella occidentalis, Frankliniella intonsa, Frankliniella gemina*	Wijkamp et al., 1995; Nagata et al., 2004; de Borbon et al., 1999
8	*Tospovirus*	Impatient necrotic spot virus	INSV	*Frankliniella occidentalis, Frankliniella fusca, Frankliniella intonsa*	De Angelis et al., 1993; Wijkamp et al., 1995; Sakurai et al., 2004; Naidu et al., 2001
9	*Tospovirus*	Tomato chlorotic spot virus	TCSV	*Frankliniella occidentalis, Frankliniella schultzei, Frankliniella intonsa*	Nagata et al., 2004; Whitfield et al., 2005; Wijkamp et al., 1995
10	*Tospovirus*	Tomato spotted wilt virus	TSWV	*Frankliniella occidentalis, Frankliniella schultzei, Frankliniella fusca, Frankliniella intonsa, Frankliniella bispinosa, Frankliniella gemina, Frankliniella cephalica*	Avila et al., 2006; Medeiros et al., 2004; Nagata et al., 2004; Wijkamp et al., 1995; de Borbon et al., 1999; Ohnishi et al., 2006

Table 7.2 contd....

7.2.5.2 Tobacco streak virus disease-TSV (Ilarvirus)

The causal organism is a virus, Tobacco streak virus and has a wide host range. It is economically an important disease and it has caused substantial losses in groundnut and sunflowers crops in India during 2000–01 and 1997, respectively. The virus is quasi-isometric in shape with particles of twenty-seven to thirty-five nm in diameter. The particle contains four ssRNA species. The virus has three nucleoproteins designated as T, M, and B. The virus genome has RNA-1, RNA-2 and RNA-3. Coat protein has been identified as an important component in the life cycle of ilarviruses. The particles are labile and pleomorphic making the crystallization and structural identification difficult. It has two distinct forms and their structure found possessing beta jelly role. The CP was almost similar in Alfalfa mosaic virus and Tobacco streak virus in respect of genome activation (Gulati et al., 2015). In addition to beans, it is known to attack white and yellow clover, Asparagus and cowpea. It is identified by the presence of red discoloration of nodes at a point of attachment of petiole to stems and the plants easily break from this point. Besides, veins and veinlets show red streaks. The streaks are first present on base of the leaf and later on cover the entire leaf lamina. The diseased plants show necrotic lesions on leaves, fruits and inflorescence, in addition to distortion of these parts of a plant. The plant stems are stunted in growth with production of canker and discoloration of bark of stems. The spread of this virus disease is through pollen, seed and four species of thrips (*Frankliniella occidentalis; F. schultzei; Thrips tabaci; Macrocephalathrips abdominalis*). It is widespread in Europe, USA, South Africa, Japan, India, Australia and New Zealand, etc., in all continents.

7.2.5.3 Angelonia flower break mottle virus disease-AnFBMV/ Pelargonium flower break virus disease-PFBV (Carmovirus)

Angelonia angustifolia is an important ornamental plant in USA, Europe and Israel and is attacked by Angelonia flower break virus of *Carmovirus* genus, transmissible through thrips, plant sap and propagation of infected stock. The important hosts are Angelonia, Phlox, Verbena, etc. It is characterized by stunted growth of diseased plants, with mild mottling of leaves. Besides these symptoms, flower mottling and color breaking are apparent symptoms. It is positive ssRNA virus of isometric symmetry with diameter of about thirty nm. It is a new virus disease; therefore, the detailed investigations are still wanting.

7.2.5.4 Maize chlorotic mottle disease-MCMV (Machlomovirus)

It is caused by a Maize chlorotic mottle virus, positive ssRNA, icosahedral spherical geometries with 30 nm in diameter, non-enveloped, linear symmetry and monopartite genome segmentation. The disease is characterized by elongated chlorotic blotches with epinasty of leaves, severe stunting, leaf necrosis with malformation of partially filled ear heads. Male panicles are hard and short and this virus normally infects members of family Gramineae, particularly maize. It is distributed in Peru, Mexico, Argentina and USA. It was first recorded on Peru on maize in 1973. The virus has two distinct strains viz. MCMV-P (Peru) and MCMV-K (Kansas) This virus is unique in the sense that it is transmissible through Chrysomelidae beetles, thrips and seeds. It is transmissible through corn thrips (*Frankliniella williamsi*) in Hawaii. This species of thrips also found on cassava, beans, maize, onion, grasses, rice, pepper, peas, and some weed hosts. Whereas, the natural host of virus is maize crop only. The other areas have hosts belonging to family Poaceae and in all, seventy-three plant species including thirty-five genera have been reported as alternate hosts of virus. The thrip species mentioned above can acquire and inoculate with an access of less than three hours in a semi persistent manner. The longer acquisition access on virus source increases the transmission of this virus. Both nymphs and adult stages are involved in the spread of this virus. Both stages of thrips can continue to transmit virus up to six hours once acquired by these stages of thrips.

7.2.5.5 Sowbane mosaic virus disease-SoMV (Sobemovirus)

It is a disease caused by a Sowbane mosaic virus (isometric, non enveloped RNA virus with a virion diameter of twenty-six to twenty-eight nm, without any capsomere arrangement). It is a synonym of Apple latent virus, *Chenopodium* mosaic virus, *Chenopodium* seed-borne mosaic virus and *Chenopodium* star mottle virus. It has been recorded first from Chenopodiaceae and later on from *Prunus domestica, Atriplex suberecta* and *Vitis.* The characteristic symptom is chlorotic mottling of leaf lamina. It is a first case of *Sobemovirus* transmissible through thrips. Its spread is also through seed, leaf miner, Leafhoppers, leafhoppers and aphids but the transmission through insects is purely mechanical particularly through leaf miner. The mouthparts or the ovipositor become contaminated in the case of leafhopper transmission. The disease is wide spread covering South and Central America, Australia, Bulgaria, Canada, Italy, Czechoslovakia, USA, Yugoslavia and infect plants belonging to three to nine families of plants. The experiments were conducted with infected pollen using *Chenopodium amaranticolor* and *C. quinoa* as hosts . The vectors (five to ten - adults), *Thrips tabaci* were given acquisition access on virus infected hosts for five hours and *C. amaranticolor* and *C. quinoa* were used for inoculation and the transmission efficiency was to the tune of 25% and 83%, respectively. Subsequently, the thrips were tested as vector of this virus without infected pollen. The thrips were able to transmit the virus successfully with a semi persistent manner.

7.3 Thrips and Bacterial Pathogens

Besides plant viruses, the thrips are acting as vectors of bacterial pathogens. Bacterial center rot of onion caused by *Pantoea ananatis* and *P. agglomerates* in Utah (USA) is transmissible by thrips, *Frankliniella fusca* and *Thrips tabaci* and the acquisition access was positively correlated with acquisition access of bacteria, more the access period, more was the transmission of bacteria (Dutta et al., 2015; Gitaitis et al., 2003). The disease is characterised by the water soaked lesions, first appearing on the leaf margins, which subsequently coalesce to produce dieback and wilt-type symptoms. Besides, the larvae of thrips (*Frankliniella moultoni; Taeniothrips* in consequence) are the known vectors of bacteria (*Erwinia amylovora*) causing fire blight of pear. Similarly, *Pseudomonas medicaginis phaseolicola* causing halo blight of beans is spread through thrips, *Hercinothrips femoralis,* thrips (*Frankliniella tritici*) are also known to harbor bacteria (*Candidatus liberibacter solanacearum*) in their bodies (Powell et al., 2015).

7.3.1 Centre rot of onion (Bacterial)

The Centre rot of onion caused by bacteria, *Pantoea ananatis* and *P. agglomerans* and the disease is found in Peru, Poland, South Africa and the USA. It is characterized by whitish to tan lesions with water soaked margins of leaves. The lesions on leaves coalesce and wilting become apparent followed by die back of leaves. The disease progresses towards neck and bulbs, from the leaves and causes light brown discoloration and the plants look like blighted ones. In severe cases, the infection reaches the bulb and causes their rotting. As a result, a foul smell is emitted by bulbs and the rotting becomes conspicuous in the center of the bulb. The spread of bacterium is through rain splashes, water, and thrips.

7.4 Thrips and Fungal Pathogens

Many fungi have been isolated from onion thrips, tested for pathogenicity and found to be pathogenic. The pathogenic fungi isolated from thrips include *Fusarium oxysporum, Phoma medicaginis, Ulocladium* spp., *Penicillium* spp. and *Alternaria* spp. Besides, many other fungi are also isolated from the onion thrips but these fungi are designated non-pathogenic to plants. The fungal pathogens as fungal spores are ingested by thrips as food. Another way of spreading these organisms is through the contaminated body with spores and thereby spreading the fungi as mechanical carriers. The mycophagous thrips (feeding on spores of fungi, viz. *Loyolaia indica, Priesneriana kabandha* and *Elaphrothrips denticollis)* are known to transmit *Fusarium oxysporum* (causal organism of Daming off of *Acacia* and Wilt of *Dalbergia sissoo), Penicillium* (seed fungi of forest trees), *Pestalotia* (Leaf spot causing fungi of Areca and Cashew plantation), *Phomopsis tectonae* (Leaf spot causing fungi on *Tectona grandis*) and *Cystospora* (responsible for disease of *Euclyptus*). The spread of rust fungi *Puccinia graminis tritici* is through *Frankliniella tritici*, Azalia flower spot (*Ovulinia azaleae*) through *Heliothrips haemorrhoidalis* and Coffee rust fungi (*Hemileia vastatrix*) through thrips, *Euphysothrips subramanii* and *Taeniothrips xanthoceros*. It is also transmissible through mechanically transportation of spores.

7.4.1 *Fusarium wilt* (fungus)

It is disease of fungal etiology and it is known to attack cotton, cucurbits, melon, sweet potato, tomato, okra, legumes and banana in nature. It is caused by *Fusarium oxysporum*. There are different sub-species of this fungus known to cause diseases in host plants referred to earlier. These sub-species are *F.o. f sp malvacearum* (cotton). *F.o. f sp batatas* (sweet potato), *F. o. f sp cubense* (banana), *F. o. f sp lycopersici* (tomato) and *F.o. f sp melonis* (melon). The fungus grows in the vascular tissues and as a result of growth of fungus, the nutrient supply along with water is stopped and the above ground plant parts show yellowing and finally wilt. The wilting could be partial or complete. On dissecting the plant, the browning of vascular tissues can easily be seen. There is an arestation of growth and plants remain stunted and bear a lesser number of small malformed fruits. The spread of fungus is through thrips in additions to other traditional method of spread of fungi.

References

Alves e Silva TL, Vasconcellos LRC, Lopes AH and Souto-Padron T (2013). The immune of response Hemocytes of insect *Oncopeltus fasciatus* against the flagellate, *Phytomonas serpens*. PloS one 8: e72076, doi:101371/journal pone 0072076, PubMed 24015207.

Assis-Filho FM, Deom CM and Sherwood JL (2004). Replication of Tomato spotted wilt virus ingested into the alimentary canal of adult thrips by two thrips species. Phytopathology, 94: 333–336.

Avila Y, Stavisky J, Hague S, Funderburk J, Reitz S and Momol T (2006). Evaluation of *Frankliniella bispinosa* (Thysanoptera: Thripidae) as a vector of the Tomato spotted wilt virus in pepper. Florida Entomologist, 89: 204–207.

Bagga HS and Laster ML (1968). Relation of insects to the initiation and development of boll rot of cotton. Journal Economic Entomology, 61: 1141–1142.

Bandla M, Campbell LR, Ullman DE and Sherwood JL (1998). Interaction of Tomato spotted wilt *Tospovirus* (TSWV) glycoproteins with a thrips midgut protein, a potential cellular receptor for TSWV. Phytopathology, 88: 98–104.

Bennett CW and Costa AS (1961). Sowbane mosaic caused by a seed transmitted virus. Phytopathology, 51: 546–550.

Brewer MJ, Anderson DJ, Armstrong JS and Villanueva RT (2012a). Sampling strategies for square and boll feeding plant bugs (Hemiptera; Miridae) occurring in cotton. Journal of Economic Entomology, 105: 896–902.

Brewer MJ, Armstrong JS, Madrano EG and Esquival JF (2012b). Association of Verde plant bug, *Creontiades signatus* (Hemiptera; Miridae) with Cotton boll rot. Journal Cotton Science, 16: 144–151.

Bruton BD, Mitchell F, Fletcher, J, Pair S, Wayadande A, Melcher U, Brady J, Bextine BR and Popham T (2004). *Serratia marcescens*, a phloem colonizing squash bug transmitted bacterium:, is the causal agent of Cucurbit yellowvine virus disease. Plant Disease, 87: 937–944.

Burgess, I, Dueck J and Mckenzie DL (1983). Insect vectors of the yeast *Nematospora coryli* in mustard, *Brassica juncea*, crops in Saskatchewan, Canada. Canadian Entomologist, 115: 25–30.

Cabanas D, Watanabe S, Higashi CHV and Bresan A (2013). Dissecting the mode of Maize chlorotic mottle virus transmission (Tombusviridae: *Machlomovirus*) by *Frankliniella williamsi* (Thysanoptera; Thripidae). Journal Economic. Entomology, 106(1): 16–24.

Camargo EF, Kastelein A and Roitman I (1990). Trypanosomatids parasite of plants (*Phytoplasma*). Parasitology Today, 6: 22–25.

Camargo EP and Wallace FG (1994). Vectors of plant parasites of genus *Phytomonas*. Advances Disease Vector Research 10: 333–359.

Carter W (1973). Insects in relation to plant disease. 2nd edition, New York, Wiley, 750 pp.

Chen CC, Chao CH and Chiu RJ (1996). Studies on host range, transmission and electron microscopy of Peanut chlorotic fan-spot virus in Taiwan. Bulletin of Taichung District Agricultural Improvement Station Pub. 52: 59–68.

Chen CC, Chen TC, Lin YH, Yeh SD and Hsu HT (2005). A Chlorotic spot disease on *Calla lilies* (*Zantedeschia* spp.) is caused by a Tospovirus serologically but distantly related to Watermelon silver mottle virus. Plant Disease, 89: 440–445.

Chen CC, Yeh SD and Hsu HT (2006). Acquisition, inoculation, and transmission of Watermelon silver mottle virus in *Thrips palmi*. Acta Horticulturae, 722: 83–90.

Chen J, Chen J and Xu X (2001). Advances in research in Longan witches broom. Acta Horticulturae, 58: 413–418.

Chu FH, Chao CH, Peng YC, Lin SS, Chen CC and Yeh SD (2001). Serological and molecular characterization of Peanut chlorosis fan spot virus, a new species of the genus *Tospovirus*. Phytopathology, 91: 856–863.

Ciuffo M, Mautino GC, Bosco L, Turina M and Tavella L (2010). Identification of *Dictyothrips betae* as the vector of Polygonum ring spot virus. Annals of Applied Biology, 157: 299–307.

Cortes I, Livieratos IC, Derks A, Peters D and Kormelink R (1998). Molecular and serological characterization of Iris yellow spot virus, a new and distinct *Tospovirus* species. Phytopathology, 88: 1276–1282.

Dammer KH and Grillo H (1990). Verseuchung von *Leptoglossus gonagra* (Fab) mite *Nematospora coryli* Peglion and *Ashbya gossypii* (Ashby et Nowell) Guilliermond in einer Zitrusnlagedee Republik Kuba. Archives of Phytopathol. Pflanzenschutz, 26: 71–78.

Daugherty DM (1967). Pentatomidae as vectors of Leaf spot disease of soybeans. Journal Economic. Entomology, 60: 147–152.

DeAngelis JD, Sether D and Rossignol PA (1993). Survival, development, and reproduction in western flower thrips (Thysanoptera: Thripidae) exposed to Impatiens necrotic spot virus. Environmental Entomology, 22: 1308–1312.

DeBorbon CM, Gracia O and De Santis L (1999). Survey of Thysanoptera occurring on vegetable crops as potential *Tospovirus* vectors in Mendoza, Argentina. Revista de Sociedad Entomologica Argentina, 58: 59–66.

Dilucca AGT, Trinidad chipana EF, Talledo MJ, Davila W, Montoya YC and Zeladdda JLA (2013). Slow wilt another form of Marchitez in oil form associated with Trypanosomatids in Peru. Tropical Plant Pathology, 38: 522–593.

Dollet M (1984). Infection of plants by flagellate protozoa (*Phytomonas*) spp. Trypanosoma. Annual Review of Phytopathology, 22: 115–132.

Dutta, B, Barman AK, Srinivasan R, Avci U, Ullman DE, Langston DB and Gitaitis R (2015). Transmission of *Pantoea ananatis* and *P. agglomomerans*, the causal agent of Center rot of onion, *Allium cepa* by onion thrips (*Thrips tabaci*) through feces. Phytopathology, 104: 812–819.

Everth EE, Ebrath R, Roacio AA, Olga Y, B Omar GG and Walther TA (2013). Tomato spotted wilt virus (TSWV), weeds and thrip vectors with tomato (*Solanum lycopersicum*) in Andean region Cundinamarca (Columbia). Agronomia Columbiana, 31(1): 58–67.

Frazer HL (1944). Observations on the method of transmission of internal boll diseases of cotton by cotton stainer bug. Annals of Applied Biology, 31: 271–290.

German TI, Ullman DE and Moyer JW (1992). *Tospoviruses*: diagnosis, molecular biology, phylogeny, and vector relationships. Annual Review Phytopathology, 30: 315–348.

Gibb KS and Randles JW (1991). Transmission of Velvet tobacco mottle virus and related viruses by mired *Cyrtopeltis nicotianae*. Advances in Disease Vector Research, 7: 1–17.

Gibbs AJ (1957). *Leptomonas serpens n.* sp. *parasitic* in the digestive tract and salivary glands of *Nezara viridula* (Pentatomidae) and in the sap of *Lycopersicon esculentum* (tomato) and other plants. Parasitology, 47: 297–303.

Gitaitis RD, Walcott RR, Wells ML, Diaz JC and Saunders FH (2003). Transmission of Pantoea ananatis causal agent of Center rot of onion by tobacco thrips, *Frankliniella fusca*. Plant Disease, 87: 675–678.

Goldbach R and Peters D (1996). Molecular and physiological aspects of *Tospovirus*. pp 129–157. In: Elliot RM (ed). The Bunyaviridae. Plenum Press, New York.

Goldman V and Czosnek H (2002). Whiteflies (*Bemisia tabaci*) issued from eggs bombarded with infectious DNA clones of Tomato yellow leaf curl virus from Israel (TYLCV) are able to infect tomato plants. Archives in Virology, 147: 787–801.

Golnaraghi AR, Pourrahim R, Farzadfar S, Ohshima K, Shahraeen N and Ahoonmanesh A (2007). Incidence and distribution of Tomato yellow fruit ring virus on soybean in Iran. Journal of Plant Pathology, 6: 14–21.

Gopal K, Reddy MK, Reddy DVR and Muniyappa V (2010). Transmission of Peanut yellow spot virus (PYSV) by thrips, *Scirtothrips dorsalis* Hood in groundnut. Archives of Phytopathology and Plant Protection, 43: 421–429.

Grillo H and Alvarez M (1983). *Nematospora coryli* Peglion (Nematosporaceae: Hemiascomycetidae) y sus transmisores en el cultivo de los citricos. Centro Agricola 10: 13–34.

Gulati A, Alapati K, Murthy A, Savithri H and Murthy MRN (2015). Structural studies on Tobacco streak virus coat protein, insight into the pleomorphic nature of *Ilarviruses*. Journal of Structural Biology, 193: 95–105.

Gutierrez S, Michalakis Y, Van Monster MV and Blanc S (2013). Plant feeding by insect vectors can affect life cycle, population genetics, and evolution of plant viruses. Functional Ecology 27: 610–622.

Hardy VG and Teakle DS (1992). Transmission of Sowbane mosaic virus by *Thrips tabaci* in the presence and absence of carrying pollen. Annals of Applied Biology, 121(2): 315–320.

Hassani-Mehraban A, Botermans M, Verhoeven JTJ, Meekes E, Saaijer J, Peters D, Goldbach R and Kormelink R (2010). A distinct *Tospovirus* causing necrotic streak on *Alstroemeria* sp. in Colombia. Archives of Virology, 155: 423–428.

Hiruki C (1999). Paulownia witches broom disease important in East Asia. Acta. Horticulturae, 496: 63–68.

Hsu CL, Hoepting CA, Fuchs M, Shelton AM and Nault BA (2010). Temporal dynamics of Iris yellow spot virus and its vector, *Thrips tabaci* (Thysanoptera: Thripidae), in seeded and transplanted onion fields. Environmental Entomology, 39: 266–277.

Hunter WB and Ullman DE (1992). Anatomy and ultrastructure of the piercing and sucking mouthparts and paraglossal sensilla of *Frankliniella occidentalis* (Pergande) (Thysanoptera; Thripidae). International Journal of Insect Morphology and Embryology, 21(2): 17–35.

Hunter WB and Ullman DE (1994). Precibarial and cibarial chaemosensilla in western flower thrips, *Frankliniella occidentalis* (Pergande) (Thysanoptera: Thripidae). International Journal of Insect Morphology and Embryology, 23(2): 69–83.

Iwaki M, Honda Y, Hanada K, Tochihara H, Yonaha T, Hokama K and Yokoyama T (1984). Silver mottle disease of watermelon caused by Tomato spotted wilt virus. Plant Disease, 68: 1006–1008.

Table 8.1 Plant viruses transmitted by insects with biting and chewing mouthparts (modified from Harris and Maramorosch, 1980)

Sl. No.	**Virus genera**	**Virus/Acronym**	**Vector (Order)**	**Mechanism (Retention in days)**	**Source**
1	*Comovirus*	Cowpea severe mosaic Virus (CPMV)	Mexican bean beetle, *Epilachna varivestis, Cerotoma ruficornis, Cerotoma aruata, Cerotoma variegata, Diabrotica undecimpunctata, Diabrotica laeta* **(Coleoptera)**	Non-persistent Foregut-borne/days	Fulton and Scott, 1974
2	*Comovirus*	Squash mosaic virus (SqMV)	Western striped cucumber beetle, *Aclymma trivattata, Aclymma vittatum, Diabrotica undecimpunctata, Diabrotica longicornis, Diabrotica virgifera, Epilachna chrysomelina* **(Coleoptera)**	Non-persistent foregut-borne (days)	Freitag, 1956
3	*Comovirus*	Radish mosaic virus (RaMV)	*Phyllotreta striolata, Diabrotica undecimpunctata, Epitrix hirtipennis* **(Coleoptera)**	Non-persistent foregut-borne	Campbell and Colt, 1967
4	*Comovirus*	Bean pod mottle virus (BPMV)	Bean leaf beetle, *Cerotoma trifurcata; Diabrotica undecimpunctata* **(Coleoptera)**	Non-persistent foregut-borne	Walters and Surin, 1973
5	*Comovirus*	Red Clover mottle virus (RCMV)	*Otheca mutabilis* **(Coleoptera)**	Non-persistent foregut-borne	Walters, 1969
6	*Tymovirus*	Turnip yellow mosaic virus (TYMV)	Flea beetles, *Phyllotreta* sp.; *Psylliodes* sp. **(Coleoptera)**	Non-persistent foregut-borne	Hollings and Stone, 1973
7	*Tymovirus*	Egg plant mosaic virus (EPMV)	Flea beetles, *Epitrix* sp. **(Coleoptera)**	Non-persistent foregut-borne	Gibbs and Harrison, 1969
8	*Tymovirus*	Okra mosaic virus (OMV)	*Podagrica sjostedti; Podagrica uniforma; Syagrus calcaratus* **(Coleoptera)**	Non-persistent foregut-borne	Lana et al., 1974
9	*Bromovirus*	Brome mosaic virus (BMV)	Cereal flea beetle, *Oulema melanopus Diabrotica longicornis* **(Coleoptera)**	Persistent circulative?	Gaborjanyi and Szabolcs, 1987

10	*Bromovirus*	Cowpea chlorotic mottle virus (CCMV)	Bean flea beetle, *Cerotoma trifurcata;* Spotted cucumber beetle, *Diabrotica undecimpunctata* **(Coleoptera)**	Persistent circulative?	Walters and Dodd, 1969
11	*Bromovirus*	Bromium mottle virus (BrMV)	Striped cucumber beetle, *Aclymma trivittata; Spotted* cucumber beetle, *Diabrotica undecimpunctata* **(Coleoptera)**	Persistent circulative	Walters and Surin, 1973
12	*Sobemovirus*	Cocksfoot mild mosaic virus (CfMMV)	*Colaspis flavida* **(Coleoptera)**	Non-persistent foregut-borne	Mohamed and Mossop, 1981; Guy, 2006
13	*Sobemovirus*	Rice yellow mosaic virus (RYMoV)	Beetle,*Sesselia pusilla*; Dipterous fly, *Diopsis thoracica* **(Diptera);** grasshoppers, *Stenohippus aequus; Conocephalus merumontanus; Acrida bicolor; Oxya hyla; Paracinema tricolor; Paratettix scaber; Zonocerus variegatus; Acrida sulphuripennis; Acrida turrita; Acrida confusa; Dorsifer* sp. **(Orthoptera)**	Non-persistent foregut-borne	Bakker, 1974; Koudamiloro et al., 2015
14	*Sobemovirus*	Solanum nodiflorum mottle virus (SNMoV)	*Epilachna vigintioctopunctata pardalis (Epilachna sparia); Epilachna vigintisexpunctata doryca; Epilachna guttatopustulata* **(Coleoptera)**	Non-persistent foregut-borne	Greber, 1981
15	*Sobemovirus*	Southern bean mosaic Virus (SBMV)	*Cerotoma trifurcata; E. varivestis D. undecimpunctata howardii* **(Coleoptera)**	Non-persistent foregut-borne	Tremaine and Hamilton, 1983
16	Unassigned	Urdbean leaf crinkle virus (UbLCV)	*Henosepilachna dodecastigma* **(Coleoptera)**	Non-persistent foregut-borne	Bharathan and Beniwal, 1984
17	*Tobamovirus*	Cucumber green mottle Virus (CGMV)	*Raphidopalpa foveicollis* **(Coleoptera)**	Non-persistent foregut-borne	Rao and Varma, 1984

Table 8.1 contd. …

In addition, yellow patches are also seen in the fruit flesh upon dissection. Flesh is mushy, broken and contain cavities.

8.1.4 Beetles and fungi

The beetle vectors of plant pathogens are mainly from families Chrysomelidae, Scolytidae, Curculionidae and Melionidae and spread the fungi in several ways. The beetles ingest the spores of fungi and, thus, carry the fungus internally. The spores of many fungi, being sticky, cling to the body and appendages of beetles and are carried externally. Additionally, beetles also spread the fungi through pollination. Beetles act as predisposing factors for the entry of fungi by creating wounds (feeding and ovipositional punctures). Beetles use some fungi as nutrient rich food and there is a mutual relationship between the two. The potato scab, caused by Actinomyces scabies, is spread by potato flea beetle *Epitrix cucumeris* (Chrysomelidae) in nature. Similarly, under Curculionidae, *Plum curculio Conotrachelus nenuphar* spreads *Sclerotinia fructicola*, a causal organism of brown rot of plum. *Scolytus multistriatus* (Scolytidae) transmits *Ceratocystus ulmi*, causing chestnut blight. The Dutch elm disease (DED) is an important, widespread and destructive disease, caused by fungi, *Ophistoma ulmi* and *O. novo-ulmi*, and is transmitted by beetles. Beetles associated with the spread of fungi are from the family Curculionidae. The killed trees or cut logs of wood are attacked by fungi which remain attached to plants. The bark beetle adults visit the fungus-infested wood for feeding and oviposition. The eggs are laid by females underneath the bark of trees. The hatching of eggs takes place under the bark and the grubs continue to feed on the diseased plants. The fungal spores are sticky in nature and these spores cling to the body of adult beetles which fly away to feed and oviposit on new plants. These beetles while feeding on the bark of new plants, deposit the fungal spores on the wood. The fungi start developing on the trees. The beetles help the fungi travel long distances from diseased to healthy plants and deposit them on suitable substratum for survival and multiplication. As a result of multiplication of fungi, the water and nutrient supply to the aerial parts of plant is blocked and the plant finally dies. Oak wilt disease caused by *Ceratocystis fagacearum* is spread by bark beetles *Pseudopityophthorus minutissimus* Zimm. and *P. pruinosus* Eic and some species of Nitidulidae family. These wood-rotting fungi are also spread by bark beetles such as *Dendroctonus ponderosae* and *Ips pini* (Say). These beetles are responsible for the production of wounds which enable the entry of fungus while the fungi in turn create a congenial microclimate for the feeding and development of beetles. The canker diseases, caused by *Geosmithia* spp., are also spread by walnut twig beetle through *Xyleborus glabratus* and *Sirex noctilio*. The fungus, *Amylostereum areolatum*, also maintains a mutual relationship with the vector in nature. The fungus *Leptographium procerum* and Mycangial fungi both serve as food for the development of young bark beetles. Mechanical spread of *Fusarium* spp. is by *Phylloplatypus pandani* (Curculionidae; Coleoptera) while making mines in the leaves of *Pandanus boninensis* (Sugiura and Masuya, 2010). Black stem rot of pines. caused by fungus *Leptographium wageneri*, is vectored by root feeding beetle *Hilastes nigrinus*, root weevil *Steremnius carinatus* and crown weevil *Pissodes fasciatus*. The transmission mechanism is not very complex but the spread of fungus is through the externally contaminated body and appendages. Wood fungi like *Ceratocytis* and *Ophistoma* create a congenial micro climate for the development of beetle *Dendroctonus ponderosae*, while the beetle causes wounds for infection of fungus. Similarly, Wilt diseases caused by *Cephalosporium diospyri* in persimmon (vectored by *Xylobiops basilaris* and *Oncideres cingulatus*), *Diplodia recifensis* in mango (vectored by *X. affinis*) and *Ceratocystis fimbriata* (vectored by *Hypocryphalus mangiferae*) are spread by carrying the sticky spores of these fungi. The Cucurbit anthracnose caused by *Colletotrichum lagenarium* is spread through *D. undecimpunctata*. Hence, the spread of pathogens through beetles is both internal as well as external.

8.1.4.1 Dutch elm disease. The Dutch elm disease (DED) is an important, widespread and destructive disease caused by fungi, *Ophistoma ulmi* and *O. novo-ulmi*, and is transmitted by beetles. It was first described in Holland, hence the name "Dutch elm disease". In America, it is transmitted by *Hylurgopinus rufipes* (Eichh.) and *Scolytus multistriatus*, while in Europe, its spread is through *S. Scolytus, S. multistriatus, S. pygmaeus, S. kirschii* species of bark beetles. The disease is characterised by flagging and wilting of leaves, which later on turn yellowish. Initially, one to two limbs are affected but the plant dies within 1–3 years after infection of the pathogen. As a result of fungal infection, the trees start producing a gum that causes the trees to wilt and finally collapse. The beetles carry the spores of fungus externally on their bodies and the spread of fungus takes place in this manner. The feeding wounds created by beetles are the target sites for deposition of spores. Over time, the wilting of plants can be seen. In this case, the cause of wilting of plants is the growth of fungus in the xylem. The fungal infestation of elm trees creates conditions for breeding of beetles while fungi, in turn, benefits from spread by beetles.

8.1.4.2 Laurel wilt disease. This is a disease caused by one of the symbionts (*Raffaelea lauricola*) of Red bay ambrosia beetle, *Xyleborus glabratus* (Coleoptera; Curculionidae: Scolytidae), in Red bay in Southern USA (Fraedrich et al., 2008; Crane et al., 2008). In addition to this species, there are many more species but all of them feed on dead wood and already attacked trees. The *X. glabratus* is the only species known to attack healthy red bay. The amber colored adult beetles can be identified by V shaped pointed snout representing mandibles, elongated cylindrical body and their *musarum* causing Anthracnose of *Musa species* is through the bees (*Trigona corvine*), wasp (*Polybia occidentalis*) and *Synoeca surinama*. Bees and wasps spread the azalea flower spot by carrying the fungus externally. Besides bees and wasps, ants (*Crematogaster striatula*) are also instrumental in the spread of *Phytophthora palmivora*, which causes Black rot of cacao, because the fungus is soil-borne. In order to build their nests, the ants must carry spore-covered soil particles from the substratum onto the cacao tree. Wasps are also associated with the spread of another disease caused by a fungus in birch. In the birch constriction disease, the lower part of the tree, where the wood wasp *Xiphydria betulae* feeds, becomes constricted. At that point of constriction, the leaves wither and are found clinging to the branches. The tip of the withered birch tree is soon devoured by anthracnose fungi, *Melanconium bicolor.*

8.3.2.1 Endosepsis of fig disease. The disease is caused by *Fusarium verticillioides* (=*Fusarium moniliforme*). It is spread through tiny wasp *Blastophaga psenes* and through thrips. The wasp inhabits the cavity called a syconium and its body gets contaminated with propagules of fungus. While pollinating, the fungus is carried from pollination fig to edible fig of spring crop; while doing so it also transmits fungus. Thrips can also enter the syconium and their bodies also become contaminated. These thrips contaminated with fungus also spread the pathogen, but only to nearby flowers on the same tree; long distance spread is through wasps only.

8.3.3 Hymenoptera and bacteria

Honey bees are also a vector of bacterial pathogens. Fire blight of apple and pear is a disease caused by a bacterium *Erwinia amylovora*, prevalent in North America and Europe and transmissible through honey bees *Apis malleifera*. The disease symptoms are apparent in the whole plant viz. blossom, fruit, shoot, and twigs. The diseased blossom appears gray initially but subsequently becomes black, twigs bend down, fruits shrivel and appear dark green and canker on branches becomes conspicuous. The blighted parts continue to hang on the trees throughout the winter season. The cracking of bark takes place and the wood under the bark appears red and discolored. The spread of disease is via air currents and honey bees. The droplets of sweet, honey-coloured bacterial ooze runs over the twigs and shoots and is spread by insects like bumble bees and wasps as the ooze sticks to their body parts. Since the bacterium causes blossom blight and the pollinating insects visit these diseased flowers, their body and appendages are smeared with bacterial ooze. The ooze-containing bacterial cells are carried externally by the pollinating bees as these insects visit many trees in order to collect nectar (Hildebrand et al., 2000). Another disease, Southern wilt of solanaceous crops, caused by *Ralstonia solanacearum* bacteria, is spread through Hymenoptera. The bees and wasps spread plant pathogens via pollination, transporting the pathogens by carrying them externally on their body and appendages, and via the symbiotic relationship between pathogen and vector.

8.3.3.1 Fire blight of apple and pear. A serious disease of apple and pear trees, caused by *Erwinia amylovora* bacterium. The bacterium overwinters in the cankers formed on the stem; with the onset of the spring season the bacteria start multiplying and produce amylovoran, this causes plugging, which is further responsible for wilting of trees (Oh and Beer, 2005). Unlike other bacterial species, this bacterium overwinters in cankers. With the cracking of cankers due to freezing temperatures, the bacteria start oozing out from the cankers in a sticky, viscous liquid. Due to rain splashing the ooze-containing bacteria onto other parts of the tree, the bacteria often spreads to blossoms. In addition to rain splashes, honey bees also carry the bacteria to blossoms (Rezzonio and Duffy, 2007). Once the bacteria are in the inflorescence, they enter through insect wounds, caused by psylla, plant bugs or leafhoppers. As a result of infection, the wilting of inflorescence occurs within one to two weeks. Initially, blossoms appear water soaked an gray-green but they soon turn black or brown. This is followed by blighting of the entire cluster and death. Another stage of the disease is the blight of young twigs and shoots, in which the petals and leaves fall and the young stem bends in the shape of a shepherd's crook. The bacterial ooze continues to seep from the affected parts. In severe cases, the entire tree looks scorched. The affected fruits show sunken lesions and ooze is exuded from these affected parts as well.

8.4 Diptera as Vector

Of the three families, namely Agromyziidae Tephritidae and Anthomyiidae, associated in the spread of plant pathogens, Apple maggots *Rhagoletis pomonella* Walish belong to the family Tephritidae. They are involved in the spread of *Pseudomonas melophthora,* which causes bacterial rot of apple. Also, the spread of black leg disease caused by *Phoma lingam* is via the cabbage maggot, *Delia radicum* (Anthomyiidae).

8.4.1 Flies and fungi

The dipterous flies are not vectors of plant viruses, but they are known to spread fungal pathogens through mutual relationships, externally through the contaminated body and internally by ingesting spores. In this context, shore flies and fungus gnats are considered as major spreaders of fungi. The needle blight of red palm caused by fungus *Pullularia pullulans* is spread by a gall midge belonging to the family Cecidomyiidae. The fungus

Claviceps purpurea (Ergot of cereal) is spread by dipterous flies who carry the sticky spores. The transmission of fungi causing rots of fleshy fruits is through apple maggot (*Rhagoletis pomonella*), Mediterranean fruit fly (*Ceratitis capitata*) and house fly (*Musca domestica*), and the spread of root infecting fungus also takes place through Hessian fly *Phytophaga destructor* (Say). Fungus gnats ingest the zoospores of *Pythium aphanidermatum* (Jarvis et al., 1993) which pass through the digestive tract of flies; therefore, the spread is through carrying the pathogens internally (El-Hamalawi and Stanghellini, 2005). Fungus gnats also carry the macro conidia of *Fusarium avenaceum* internally. The spores of many different fungi are ingested by flies and spread from one location to another *(Botrytis cinerea* - gray mildew; *Fusarium oxysporum fsp. radicis - lycopersici; Thielaviopsis basicola* - black rot fungi; *Verticillium albo - atrium)* (Stanghellini et al., 1998; James et al., 1995; Gillespie and menzies, 1993; Kalib and Millar, 1986). In addition to fungus gnats, shore flies are also instrumental in the spread of fungi. Like the gnats, the shore flies also ingest spores of *Pythium aphanidermatum* and *F. avenacearum.* These are carried in the alimentary canal and passed along with excreta by both adults and maggots. Fungus namely *B. basicola* is also spread by shore flies; the flies carry the spores internally from place to place.

8.4.1.1 Needle blight of red palm. This disease is caused by fungus *Dothistroma septosporum* in red palms and is prevalent throughout all the continents. It is characterized by the presence of reddish-brown spots which later turn into bands. Eventually, the needle tips become completely covered with this discolouration. The bands then turn brown and develop minute black dots; these dots are spores of the fungus. Since the gall midge maggots feed on needles, these insects ingest spores and carry them internally. The body of the maggot is also smeared with fungal spores. The needle blight is also caused by *Pullularia pullulans*, the spread of this fungus is also through gall midges belonging to family Cecidomyiidae in the order Diptera.

8.4.2 Flies and bacteria

The spread of bacteria in nature is brought about by dipterous flies. The olive knot disease of olive, prevalent in the Mediterranean region and California, USA, is caused by a bacterium *Pseudomonas savastanoi.* The diseased trees show the presence of galls and wounds on all plant parts. It is transmitted by olive fly *Bactrocera oleae* which has a symbiotic relationship with the bacterium. The olive fly is responsible for the spread of pathogen through feeding and ovipositional wounds as the bacteria lives in the digestive tract and eggs of the fly. The bacteria enrich the food and the olive fly helps in the bacterium spread and multiply. The maggot's body is smeared with bacteria during egg hatching, Southern wilt of Solanaceous plants, caused by *Ralstonia solanacearum* in tropical countries, is also spread by *Drosophila melanogaster* in addition to bees and wasps. These flies visit the bacterial ooze and their body and appendages are smeared with bacterial spores and, thus, the bacteria is spread to other plants. Soft rot of vegetable crops, fruits, stored potatoes and ornamentals is caused by bacteria like *Erwinia carotovora*, and *Pseudomonas fluorescens Pectobacterium carotovorum* (*Erwinia carotovora*) causes Black leg of potato (De Boer, 2002) and it produces pectolytic enzymes responsible for the dissolution of plant cell walls. Besides causing black leg in potato, the bacterium also uses dipterous fly (*Drosophila melanogaster*) as a secondary host for horizontal spread between plants. It has been well established that colonization of bacterium in *Drosophila* is under the control of genes (Basset et al., 2003). The evf gene has been identified as being responsible for persistence and survival of bacterium in the gut of the fly and production of toxin. The above species, responsible for production of pectolytic enzymes, are now put into a separate species, *Pectobacterium atrosepticum* and are known to liquefy pectates. The spread of these bacteria is generally through dipterous flies, namely *Hylemia platura* (seed corn maggot), *H. florilega* (bean seed maggot), *Delia radicum* (cabbage maggot), *D. antiqua* (onion maggot), *Drosophila buscki, Tritoxa flexa* (onion black fly), and *Eumerus strigatus* (onion bulb fly). Black leg bacterium remains in the intestinal tract of these maggots. While feeding, these maggots place the bacterium on cut tubers. The body of these insects also becomes smeared with bacteria and they deposit these bacteria on the feeding wounds. The major spread of black leg is through contaminated potato tubers during planting in the soil. The diseased tubers rot in the soil and are responsible for infection of developing progeny in the soil.

8.4.2.1 Bacterial soft rot of vegetables. The soft rot of vegetables is caused by bacteria *Erwinia carotovora sub-species carotovora* and *E.c. sub-species atroseptica* and *Pseudomonas flluorescens* in cabbage, cauliflower, carrot, radish, spinach, rape, cucumber, asparagus, gladiolus including many fruits and ornamentals. Insects are predisposing factors for the entry of bacteria through feeding wounds. Initially, a minute water-soaked lesion is produced at the entry point, followed by rapid rotting of the entire piece of produce. After the infection, a foul smell is emitted from that lot of produce in the storage, transient or in the market. These species of bacteria produce proteolytic enzymes that are responsible for maceration of tissues. The cabbage maggots *Delia platura* and *Hylemya florilega* are both dipterous insects known to act as creators of suitable substratum for the entry of bacteria. In addition, onion rot is inflicted by a bacterium, *Erwinia caratovora*, and this bacterium is spread via *Tritoxa flexa, Hylemya antiqua* and *H. platura*. Another important disease called Olive knot, caused by *Pseudomonas savastanoi* (Smith), is common in California and spreads through splashing rain water. Later on, it was also demonstrated that *Dacus oleae* (the olive fly) is also responsible for its spread. The olive fly carries the

8.6.2 Diseases

8.6.2.1 Apple brown rot disease. Brown rot is a fungal disease caused by *Monolinia* (=*Sclerotinia*) *fructigena*, *M. fructicola* and *M. laxa*. It is also known as European brown rot in Europe. It is a disease of fruits such as apple, pear, plum, and cherries and is prevalent in Australia, North America, New Zealand, Java, Brazil, Europe, Africa and Chile. The disease appears at two stages, viz. blossom and twig damage stage, and fruit damage stage. In early spring, the infection of flowers is generally from the adjoining orchards. Aeciospores are carried by the wind from a nearby source and cause infection of floral tube, ovary, and peduncle of the flower. As a result, the inflorescence wilts and hangs from the twigs. The fruit damage is the next stage of the disease. The bird damage or insect damage (Earwigs or codling moth) are the predisposing factors for the entry of fungi through feeding and ovipositional wounds. At the point of entry, there is a formation of circular rings which are easy to identify. Besides, the infected fruits turn brown on the tree prior to ripening and fall to the ground as rotten material.

8.7 Cockroaches

These insects belong to insect order Dictyoptera (Family: Blattidae) and have biting and chewing mouthparts. The cockroaches are generally pests of stored food but have recently been identified as pests of greenhouse plantations, particularly among orchids. The orchids are known to suffer from Cymbidium mosaic virus disease in greenhouses. Cockroaches have recently been discovered as a vector of this virus and *Periplaneta australassiae* has been identified specifically as a vector of Cymbidium mosaic virus in greenhouses (Allen, 2010), though the rate of transmission was relatively low.

8.7.1 Cockroaches and plant viruses

8.7.1.1 Cymbidium mosaic virus disease (Potexvirus)

It is a common disease caused by a virus called Cymbidium mosaic virus (CymMV) of genera *Potexvirus* and family Alphaflexiviridae. It is ssRNA, filamentous and non-enveloped, with a particle size of 480 x 13 nm. The genome contains 6227 nucleotides with five ORF. The spread of the virus is via contaminated tools, man or contaminated water but also through the feeding behaviour of cockroaches in the greenhouses. The disease was recorded by DD Jensen in 1951 as black necrotic spots on leaves along with distortion of leaves and overall stunted growth of plants.

References

Agrios GN (2008). Transmission of plant diseases by insects. pp. 3853–3885. In: Capinera JL (ed). Encyclopedia of Entomology. 2nd Edition, Dordrecht: kluwer Academic.

Allen C (2010). Virus transmission in orchids through the feeding damage of Australian cockroach, *Periplaneta australassiae*. Acta Horticulturae, 878: 375–379.

Bharathan N and Beniwal SPS (1984). Transmission characteristics of Urd bean leaf crinkle disease by Epilachna beetle, *Henosepilachna dodecastigma*. Indian Phytopathology, 37: 660–664.

Bakker W (1974). Characterization and ecological aspects of Rice yellow mottle virus (RYMV) genus *Sobemovius*, a continental problem in Africa. Plant Protection Science, 40: 26–35.

Basset A, Tbou P, Lemaitre B and Boccard F (2003). A single gene that promotes interaction of phytopathogenic bacteria with its insect vector, *Drosophila melanogaster*. EMBO Report, 4: 205–209.

Bercks R (1973). Scrophularia mottle virus CMI/AAB Descriptions of plant viruses, No 113.

Bock KR (1971). Notes on East African plant virus diseases I Cowpea Mosaic Virus. East African Agricultural and Forestry Journal, 37: 60.

Boylan-Pett W, Ramsdell DC, Hoopingarner RA and Hancock JF (1992). Honey bees foraging behavior and the transmission of the Pollen borne blue leaf mottle virus in high bush blueberry. Acta Horticulturae, 308: 99–108.

Broadbent L and Heathcote GD (1958). Properties and host range of Turnip crinkle rosette and yellow mosaic viruses. Annals of Applied Biology, 46: 585–592.

Cabanas D, Watanabe S, Higashi CHV and Bressan A (2013). Dissecting the mode of Maize chlorotic mottle virus transmission (Tombusviridae: *Machlomovirus*) by *Frankliniella williamsi* (Thysanoptera; Thripidae). Journal of Economic Entomology, 106(1): 16–24.

Campbell RN and Colt WM (1967). Transmission of Radish mosaic virus. Phytopathology, 57: 502–504.

Cochran DG (1999). Cockroaches; their biology, distribution, and control. World Health Organization, Communicable diseases, prevention and control,World Health Organization) Pesticide Evaluation Scheme (WHOPES) 1999.

Costa AS, DeSilva DM and Duffus JE (1958). Plant virus transmission by leaf-miner fly. Virology, 5: 145–149.

Crane JH, Pena J, Ploetz RC and Osborne JL (2008). Redbay Ambrosia beetle-laurel wilt pathogen; a potential problem for Florida Avocado Industry. EDIS http/edis.ifas.ufl.edu.

Croxall HE, Collingwood CA and Jenkins JEE (2008). Observations on brown rot (*Sclerotinia fructigena*) of apples in relation to injury caused by earwigs (*Forficula auricularidae*). Annals of Applied Biology, 38(6): 833–843.

Dale WT (1954). Sap-transmissible mosaic diseases of solanaceous crops in Trinidad. Annals of Applied Biology, 41: 240–247.

De Boer SH (2002). Relative incidence of *Erwinia carotovora* sub-species *atroseptica* in stolen end and peridermal tissue of potato tubers in Canada. Plant Disease, 86: 960–964.

El-Hamalawi ZA and Stanghellini ME (2005). Disease development on *Lisianthus* following aerial transmission of *Fusarium avenaceum* by adult shore flies, fungus gnats, and moth flies. Plant Disease, 89(6): 619–623.

Feldman TS, Obrien HE and Arnold AE (2008). Moth that vector a plant pathogen also transport endophytic and mycoparasitic antagonists. Microbiology Ecology, 56: 742–750.

Fraedrich SW, Harrington TC, Rabaglia RJ, Ulysben MD, Mayfield AE iii, Hanula JL, Eickwort JM and Miller DR (2008). A fungal symbiont of the redbay Ambrosia beetle causes a lethal wilt in redbay and other Lauraceae in the Southern United States of America. Plant Disease, 92: 215–224.

Freitag JH (1956). Beetle transmission, host range, and properties of Squash mosaic virus. Phytopathology, 46: 73–81.

QUESTIONS (EXERCISE)

Q 1. Discuss in detail the mechanism of spread of endosepsis of fig in nature.

Q 2. How does the spread of fungi take place through the involvement of insects?

a) Feeding punctures
b) Oviposition wounds
c) Contaminated mouthparts

Q 3. Cockroaches and grasshoppers are vectors of plant viruses. Support your answer with suitable examples.

Q 4 Discuss the mechanism of feeding and virus transmission through beetles.

Q 5 How are the following insects involved in the transmission of plant pathogens?

a) Plant bugs
b) Dipterous flies
c) Moths

CHAPTER 9

Mites

9.1 Identification and Brief Biology

Arthropods are known vectors of plant pathogens. Insects are considered as the main agents of the spread of plant pathogens among arthropods. Besides insects, mites are another group, Acarina (Class: Arachnida), considered as vectors of plant pathogens. So far, 48,200 species of mites have been identified. These mites are tiny, elongated, spindle-shaped and translucent, with transverse rings on the abdomen. They can be orange, white or yellow in colour, and measure 0.1 to 30 mm in length, with two (Eriophyidae) or four pairs of legs (Tetranychidae). The body of a mite is divided into Gnathosoma and Idiosoma. Gnathosoma encloses the mouthparts while the rest of the body is idiosoma. They live in congregations on the underside of leaves and suck sap with piercing and sucking mouthparts. As a result of feeding, conspicuous specks of yellow color on the upper leaf surface become apparent. The leaves show downward curling and are brittle in texture. However, these creatures are more destructive as vectors of plant pathogens. Mites prefer hot and humid conditions for development. After mating, the female lays around 300 straw colored/glossy and sphere-shaped stalked eggs in webs or on bark or leaf near the dormant buds. Eggs hatch in three days and, on hatching, six-legged larvae are produced which develop into 8-legged nymphs that closly resemble adults of the species. It goes through proto-nymph and deutonymph stages before reaching the adult stage. The adult mite has four pairs of legs (except eriophyids with two pairs of legs) which sets it apart from other insect vectors. Under suitable conditions, the life cycle is completed in about seven to ten days.

in diameter and is bound by double layer. It has thirty-two kDa special proteins associated with nucleic acid. It can be identified by mild mosaic, chlorosis or necrosis symptoms and reddish purple band of 0.25–0.5 width running in the leaves parallel to veins. It is transmissible by *A. tosichella* in a non-persistent manner.

9.6.5 Trichovirus (Cherry mottle leaf virus-CMoLV)

A virus belonging to *Trichovirus* genera (family: Betflexiviridae) of plant viruses and was first recorded in Washington, USA in 1917. It is transmissible via mite *Eriophyes inaequalis* (Ma et al., 2014). It causes irregular chlorotic leaf mottling and distortion of leaves with puckering of leaf lamina. The causal organism is responsible for shortening of intermodal length and further instrumental in reducing fruit flavor. The details regarding this pathogen are still incomplete.

9.6.6 Rymovirus (Rye grass mosaic virus-RGMV)

It is characterised by chlorotic streaks on the leaves. It is an ssRNA filamentous virus whose particle length is 700 nm. The total genome size is 8.5 kb. The virion has 4700 coat proteins (M). It causes light green to yellow mottling and streaking in ryegrass. The virus is closely related *to* Rye grass mosaic virus, Agropyron mosaic virus-AgMV and Oat necrotic mosaic virus-ONMV and is likely to be merged with genera *Potyvirus* in the near future. It is transmissible by mite *Abacarus hystrix* in a mechanical manner. The acquisition access is two hours and with increase in access period, the transmission rates increases. Adult mites lose the virus after twenty-four hours. All nymphal instars and adults transmit the virus. The virus is sap-transmissible as well.

9.6.7 Tritimovirus (Wheat streak mosaic virus-WSMV)

The virus falls under the family Potyviridae and is sap-transmissible. It is known to infect wheat, maize, etc., and it was first detected in 1922 in Nebraska in the USA where is caused enormous losses in crop yields until 1960. The diseased plants are dwarfed/stunted and show a mosaic pattern and necrotic lesions in the leaves. The root system is also poorly developed in diseased plants. The early maturity is common and causes shriveling of grains. It is an ssRNA virus with filamentous particles and is transmissible by *A. tosichella*. The virus was detected in midgut without degradation throughout the life of mite (six to nine days). The virus was also detected in the body cavity and salivary glands of mite, this demonstrated the circulative relationship mechanism of virus spread (Paliwal, 1980). All stages are known to transmit the virus except the egg stage. The virus is acquired in larval stages and transmitted in adult stages.

9.6.8 Potyvirus (Potato virus Y-PVY)

This is the only virus (other than the Eriophyidae family of mites) that spreads through the *Tetranychus telarius* (Tetranychidae). The mechanism of transmission is stylet-borne as the virus is acquired in five minutes and inoculated in five minutes with 40% transmission.

9.6.9 Nepovirus (Black currant reversion virus-BCRV)

It is first virus of the family Comoviridae to be identified as vectored by mite. It is an icosahedral ssRNA virus found only on black currants (*Ribes nigrum*). The Eriophyid mite *Cecidophyopsis ribis* is a vector of this virus. All stages of mite can transmit virus except the egg stage. The acquisition access period is three hours and the virus retention is for twenty-five days. The virus exists in two forms viz. European and Russian. The disease is characterised by scales on stems, leaves curled and rolled and marked with yellow/black spots or flecks and the presence of bubble like blisters. The buds are smooth and without pubescence and the fruits are covered with black markings.

9.6.10 Cytorhabdovirus (Barley yellow streak mosaic virus-BYSMV)

It is one of the serious diseases of barley which is a filamentous, ssRNA, enveloped virus with a particles size of 64 x 127–4000 nm. The disease is characterized by yellow streaks in the leaves, stunted growth of plants and the symptoms first appear on the edges of the field. The rosetting of flowers can also be seen in diseased plants. It is transmitted by a mite species named *Petrobia latens* (Muller) and is transovarially transmitted through the eggs of mites. The relationship of the virus with the mite is of persistent circulative type (Smidansky and Carrol, 1996; Robertson and Carroll, 1988).

9.6.11 Allexivirus (Shallot virus X-ShV-X)

This virus from the family Alexiviridae is ssRNA and has filamentous helical particles with a diameter of twelve nm and virion length 800 nm. It is known to attack *Allium cepa* belonging to family Liliaceae (Hamed et al., 2012). It was first recorded in the Netherlands as showing mild mottling in onions. The virus is carried through mite *Aceria tulipae* in semi–persistent or foregut-borne non-persistent manner.

Smidansky ED and Carroll TW (1996). Factors influencing the outcome of Barley yellow streak mosaic virus–brown mite-barley interactions. Plant Disease, 80: 186–193.

Skare JM, Wijkamp I, Denham I, Rezende J, Kitajima EW, Park JW, Desvoyes B, Rush CM, Michels G, Schlthof KB and Scholthof KBG (2006). A new eriophyid mite-borne membrane–enveloped virus like complex isolated from plants. Virology, 347: 343–353.

Susi P (2004). Black currant reversion virus, a mite transmitted *Nepovirus*. Molecular Plant Pathology, 5: 167–173.

Vacante V (2013). The mites as vectors of pathogens of the plant. Protezione Delle Colture, 2: 2–8.

Van Der, GLPS, de Navia MGYD and Tanzini MR (2002). New records of pathogenic fungi in mites (Arachnida; Acari) from Brazil. Neotropical Entomology, 31: 493–495.

QUESTIONS (EXERCISE)

Q 1. Name the mite vectors of following viruses:

a) Barley yellow streak mosaic virus
b) Shallot virus X
c) Peach mosaic virus
d) Cherry mottle leaf virus
e) Orchid fleck virus

Q 2. Describe the feeding mechanism of mites.

Q 3. Discuss the mechanism of spread of a mite-borne propagative virus.

Q4. How can the damage caused by mites be differentiated from viral damage?

Q5. The specificity is more pronounced in mites, Support your answer with suitable examples.

CHAPTER 10

Nematodes

10.1 Identification and Brief Biology

The nematodes are a microscopic 1.3–3.0 mm in length. They are poikilothermic animals under the phylum Nematoda. The development is dependent on ambient temperature, as their body temperature does not change with change in temperature. These are ectoparasities of plants, divided into three groups viz. dagger nematodes (*Xiphinema*), needle nematodes (*Longidorus*) and stubby root nematodes (*Paratrichodorus*) and these are known vectors of plant pathogens. The nematode vectors of plant pathogens are from two ectoparasitic orders, namely Dorylaimida (sub-order: Dorylaimida; Family: Longidoridae) and Trilonchida (sub-order: Dipherophorina; Family: Trichodoridae). The sub-order Dorylaimida contains nematodes belonging to genera *Xiphinema* and *Longidorus,* whereas Dipherophorina contains large-sized *Trichodorus* and *Paratrichodorus* (2–12 mm), medium-sized *Longidorus* and *Paralongidorus* (1.6–6 mm) and small-sized *Xiphinema* (0.5–1.5 mm) nematodes (Table 10.1; 10.2). The females are C shaped and taper at both ends. They have a spear like odontostylet, similar to hemipteran insects, which penetrates deep into the root tip. The odontostylet is forked at the tip and has two rings at its base, the posterior one being more prominent. The posterior part of esophagus is swollen, connected to the guiding ring via odontophore. The odontotostyle (100 ug) is attached to the guiding ring. There is a bulb, eighty ug long and twenty ug wide, present between the esophagus and intestine. In addition, a tail with two to three pores on the cauda is provided. The male is also similar in structure but slightly smaller in size. The female is always found in the soil in thin films of water. The female lays the eggs in the soil in spring or in early summer, to give rise to juveniles (three to four) which later on turn into adults. The first juvenile enters the soil after hatching and undergoes molts. After molting three to four times, the juvenile becomes an adult.

...Table 10.1 contd.

Sl. No.	Nematode species	Virus	Acronym	Source
41	*Xiphinema californicum*	Cherry rasp leaf virus	CRLV	Brown and Halbrendt, 1992
42	*Xiphinema californicum*	Tobacco ring spot virus	TRSV	Brown and Halbrendt, 1992
43	*Xiphinema californicum*	Tomato ring spot virus	TomRSV	Brown and Halbrendt, 1992
44	*Xiphinema coxi*	Arabis mosaic virus	ArMV	Raski, 1988
45	*Xiphinema coxi*	Cherry leaf roll virus	CLRV	Jones et al., 1981
46	*Xiphinema diversicaudatum*	Arabis mosaic virus	ArMV	Jha and Posnette, 1961
47	*Xiphinema diversicaudatum*	Strawberry latent ring spot virus	SLRSV	Lister, 1964
48	*Xiphinema diversicaudatum*	Carnation ring spot virus	CRSV	Brown and Trudgill, 1984
49	*Xiphinema diversicaudatum*	Brome mosaic virus	BMV	Mojtahedi et al., 2002
50	*Xiphinema diversicaudatum*	Weidelgrass mosaic virus	WGMV	Schmidt et al., 1963
51	*Xiphinema brevicolle*	Tomato ring spot virus	TomRSV	Taylor and Murant, 1969
52	*Xiphinema index*	Grapevine fan leaf virus	GFLV	Hewitt et al., 1958
53	*Xiphinema rivesi*	Tobacco ring spot virus	TRSV	Brown and Halbrendt, 1992
54	*Xiphinema rivesi*	Tomato ring spot virus	TomRSV	Forer et al., 1984
55	*Xiphinema vuittenezi*	Cherry rasp leaf virus	CRLV	Van Hoof, 1972
56	*Xiphinema vuittenezi*	Bulgarian latent virus	BLV	Stellbach and Goheen, 1988; Raski, 1988

Table 10.2 Nematode vectors of tubular shaped plant viruses.

Sl. No.	Nematode species	Virus	Acronym	Source
1	*Longidorus attenuates*	Docking disorder virus	DDV (TRV)	Mojtahedi et al., 2002
2	*Longidorus attenuates*	Potato corky ring spot virus	PCRSV	Gudmestad et al., 2008
3	*Longidorus elongatus*	Docking disorder virus	DDV (TRV)	Mojtahedi et al., 2002
4	*Longidorus elongatus*	Potato corky ring spot virus	PCRSV	Gudmestad et al., 2008
5	*Longidorus leptocephalus*	Docking disorder virus	DDV (TRV)	Mojtahedi et al., 2002
6	*Longidorus leptocephalus*	Potato corky ring spot	PCRSV	Gudmestad et al., 2008
7	*Paratrichodorus porosus*	Tobacco rattle virus	TRV	Van Hoof et al., 1966
8	*Paratrichodorus christiei*	Tobacco rattle virus	TRV	Gibbs and Harrison, 1964
9	*Paratrichodorus anemones*	Pea early browning virus	PEBV	Van Hoof, 1962
10	*Paratrichodorus pachydermus*	Pea early browning virus	PEBV	Van Hoof, 1962
11	*Paratrichodorus pachydermus*	Tobacco rattle virus	TRV	Van Hoof, 1968; Gibbs and Harrison, 1964
12	*Paratrichodorus allius (=tunisiensis)*	Potato corky ring spot virus	PCRSV	Schneider and Ferris, 1987
13	*Paratrichodorus anemones*	Pea early browning	PEBV	Harrison, 1967
14	*Paratrichodorus christiei*	Tobacco rattle virus	TRV	Mojtahedi et al., 2002
15	*Paratrichodorus hispanus*	Tobacco rattle virus	TRV	Van Hoof, 1962; Van Hoof, 1964
16	*Paratrichodorus minor (=christiei)*	Pea early browning	PEBV	Van Hoof et al., 1966
17	*Paratrichodorus nanus*	Tobacco rattle virus	TRV	Cooper and Thomas, 1970
18	*Paratrichodorus pachydermus*	Tobacco rattle virus	TRV	Van Hoof et al., 1966; Van Hoof, 1968

Table 10.2 contd. …

...Table 10.2 contd.

Sl. No.	Nematode species	Virus	Acronym	Source
19	*Paratrichodorus porosus*	Tobacco rattle virus	TRV	Gibbs and Harrison, 1964
20	*Paratrichodorus teres*	Pea early browning virus	PEBV	Van Hoof, 1962; Gibbs and Harrison, 1964
21	*Paratrichodorus teres*	Tobacco rattle virus	TRV	Gibbs and Harrison, 1964
22	*Paratrichodorus tunisiensis (=allius)*	Tobacco rattle virus	TRV	Mojtahedi et al., 2000; Cooper and Thomas, 1970
23	*Paratrichodorus tunisiensis*	Tobacco rattle virus	TRV	Cooper and Thomas, 1970
24	*Trichodorus minor (=christiei)*	Tobacco rattle virus	TRV	Gibbs and Harrison, 1964
25	*Trichodorus cylindricus*	Tobacco rattle virus	TRV	Van Hoof, 1968
26	*Trichodorus primitivus*	Pea early browning virus	PEBV	Gibbs and Harrison, 1964
27	*Trichodorus primitivus*	Tobacco rattle virus	TRV	Gibbs and Harrison, 1964
28	*Trichodorus Primitivus*	Pea early browning	PEBV	Van Hoof, 1962
29	*Trichodorus similis*	Tobacco rattle virus	TRV	Cremer and Schenk, 1967
30	*Trichodorus viruliferus*	Pea early browning	PEBV	Gibbs and Harrison, 1964
31	*Trichodorus viruliferus*	Pea early browning	TRV	Van Hoof, 1962
32	*Trichodorus viruliferus*	Pea early browning	PEBV	Van Hoof, 1962

Normally, the life cycle is completed in about seven months and the adult longevity is three to five years. The nematodes complete all the stages in the soil, particularly the ectoparasites belonging to the *Xiphinema. Xiphinema index* is widely distributed in Mediterranean areas while *X. diversicaudatum* is prevalent in Europe. *Longidorus* is mostly localized in Northern Europe. The *Trichodorus* are commonly met with in North America and Europe. With respect to soil conditions, *X. diversicaudatum* prefers clay soils. *Longidorus* and *Xiphinema* are more adapted to light soils. *Trichodorus* nematodes enjoy free drained sandy soils. *P. pachydermus* is found in abundance in sandy soils rather than in clay conditions. The symptoms of damage are quite distinct; the nematode damage can be identified in the field by the yellow patches, start of symptoms from the outer periphery and stunted plant growth. The infection persists even in the absence of crop in the field.

10.2 Feeding Apparatus

These nematodes are obligate ectoparasites and always require living organisms to feed and thrive. They remain confined to the roots which they consume. The total life cycle of vector nematode is completed outside the root. While feeding on roots, half of the body remains outside the root. Some species of nematodes feed entirely on apical meristems of roots, such as *Longidorus* and *Paratrichodorus*, while *Xiphinema* devours the portion of root behind the tips. *Trichodorids* feed on the epidermis of root cells. These nematodes feed on epidermal cells or even deeper layers. These nematodes possess solid and vertically curved teeth or the odontostyle. The nematodes puncture the tissues with odontostyle, along with penetration of pharyngeal bulb, and secrete enzymes to induce symptoms like hypotrophy and thickening of cells. Before feeding, the nematodes form a feeding tube for secreting saliva that solidifies into a sheath; the feeding occurs within this sheath. The cell contents are pumped through this into the mouth from the cytoplasm of cells. The feeding occurs when the odontostyle penetrates into the root tissues with a thrust and moves to deeper layers with pulsation. The esophageal bulb triggers the ingestion of cell sap with staggered pulsatory movements. While feeding, the nematodes destroy the cells of the root in the region around the point of feeding. As a result of the destruction of cells, the supply of water and nutrients is obstructed, and the disease symptoms start appearing. The plants remain stunted in growth, discoloured (yellowish), wilted and the roots show swellings along with the development of lateral roots called galls, depending on the species of nematodes. The feeding of nematodes may last for hours to days. *Xiphinema* and *Longidorus* have a wide host range, the former mainly devours the herbaceous crops and woody plants and weeds while the latter is known to infect rye grass, strawberry, sugar beet, black currant, chickweed, etc.

10.3 Mechanism of Pathogen Transmission

The ectoparasites belonging to two families, Longidoridae (*Longidorus, Paralongidorus,* and *Xiphinema*) and Trichodoridae (*Trichodorus* and *Paratrichodorus*) and are vectors of plant pathogens. The *Trichodorids* are known to transmit *Tobravirus* (tubular/rod shaped viruses) (Tobacco rattle virus) while *Longidorids* spread nepoviruses (polyhedral shape viruses) (Tobacco ring spot virus). The tobraviruses are straight, tubular particles (helical symmetry) with ssRNA, bipartite, with a size of 180–210 x 45–115 nm, while nepoviruses are of isometric symmetry (twenty eight nm in diameter), with a bipartite genome and two single stranded RNA. In all, there are thirty-six plant viruses which are spread by nematodes belonging to two families, namely Longidoridae (twenty two) and Trichodoridae (14) (Lamberti and Roca, 1987). Already, a mention of ten top plant parasitic viruses transmitted by nematodes has been made (Jones et al., 2013). The nematode *Xiphinema index* was identified as a vector of Fan leaf of grapevine virus for the first time in 1958 (Hewitt et al., 1958). The virus transmission is passive in nature, thus, it requires an injury which is created by nematodes while feeding. The nematodes explore the roots and thrust the odontostyle into the tissues for feeding. If the host is a diseased one, the causal virus is picked up along with the sap. The morphology of odontophore varies among the different species of nematodes. *Longidorus* species have short cheilostome while in *Xiphinema* it is much longer and the odontophore is more sclerotized. The viruses have specific sites where they become attached to the body of vector nematodes. Normally, the viruses are adsorbed on the odontostylet. These viruses have different specific sites for retention in the nematode vectors. Therefore, the viruses do not compete for the sites of retention in the vector body. For example, the *Xiphinema americanum* retains the virus in the cuticle linings of esophageal lumen or stylet lumen wall and the odontophore along with the sap that passes from plant to nematode. In *Longidorus,* the virus is carried and retained between the guiding sheath/ odontostyle area that covers the upper region of odontostyle or attaches to the lumen wall. While in *Trichodorus* and *Paratrichodorus,* the virus is retained on the pharyngeal wall in the linings of esophageal lumen and onchiostyle region. The transmission process of the virus through nematodes involves adsorption of virions at the site and their subsequent detachment. The detachment of virus takes place with saliva released by oesophageal glands being passed on to the plant cell via the lumen of the oesophagus. The salivary secretions alter the pH, which subsequently changes the charge of virus particles and triggers detachment of particles from the site of retention. The detachment of viruses is under the interaction of coat protein of virus and sites of virus location in the vector. During feeding, the saliva is secreted by salivary glands and causes detachment of virus particles. The saliva contains enzymes that interact with material meant for binding of the virion. The virus, after having been detached, is deposited in the wound and produces the disease. The detachment of the virus is a slow process that takes a significant amount of time (weeks). The nematode loses it if kept on non-hosts for a longer period of time. The species belonging to *Xiphinema* and *Trichodorus* normally retain the virus for life, while *Longidorus* spp. lose the virus within eight weeks of acquisition. Arabis mosaic virus in strawberry is transmissible by *X. diversicaudatum*. The Raspberry ring spot virus in cherry is transmissible through *L. macrosoma*. Both the adult and larval stages are able to transmit the viral pathogens and these stages can acquire the viruses within an acquisition access of less than an hour. There is no trans-stadial or transovarial transmission in nematodes. The transmission efficiency increases with increase in the acquisition access period on virus source. The acquisition access of Grapevine fan leaf virus by *X. index* is twenty-four hours (Van-Zyl et al., 2012) while it is one hour for *X. americanum*. Both the species of *Xiphinema* can retain the Grapevine fan leaf virus for several months, while the other strain of this virus may be lost within a month. The adult and larval stages of many nematodes can transmit plant viruses with the same efficiency as *X. index* (Fan leaf of grapevine), *X. diversicaudatum* (Arabis mosaic virus-ArMV and Strawberry latent ringspot virus-SLRSV), *X. americanum* (Tomato ring spot virus-TomRSV) and *T. pachydermus* (Tobacco rattle virus). However, *L. elongatus* adults failed in the transmission of Tomato black ring virus-ToBRV, while the larvae transmitted it successfully. The virus-vector study has indicated that both the larval and adult stages can transmit plant viruses (Grapevine fan leaf virus; Tomato ring spot-TomRSV; Strawberry ring spot virus-SRSV; Arabis mosaic virus-ArMV, etc.). The acquisition threshold period is estimated to be five to fifteen min in the case of *X. index* transmitting GFLV; the inoculation threshold period is less than twenty-four hours for almost all viruses. The persistence of viruses in their respective vectors is for 240 days (GFLV in *X. index)*, fifty-six days (TomRSV/RpRSV in *L. elongatus)* and thirty-six days (TRV in *T pachydermus*) depending upon virus and nematode vector (Ayala and Allen, 1968; Taylor and Ruski, 1964).

10.4 Specificity in Transmission

All viruses are not transmissible through all the species of nematodes; this confirms the concept of specificity. There is a greater specificity in the transmission of NETU (Table 10.2) viruses as compared to NEPO (Table 10.1) viruses transmissible through mites. Arabis mosaic virus, transmitted by *X. diversicaudatum* but not by *X. index*, is an example. Similarly, the viruses like Brome mosaic virus-BMV (*X. diversicaudatum* and *L. macrosoma*), Carnation ring spot virus-CRSV (*X. diversicaudatum* and *L. elongatus*) and Prunus ring spot virus-PRSV (*L. macrosoma*) have different vectors. Besides these, different strains of viruses are also spread through different species of nematodes. In the case of pea early browning virus, the English strain is vectored by *P. anemones, T. primitivus* and *T. viruliferus* and the Dutch strain is vectored through *P. pachydermus* and *P teres*. Likewise, in the case of TRV, the European strain is carried by *P. anemones* and *P. manus* while the American strain is carried by *P. allius, P. christiei* and *P. porosus*. Thus, there is high specificity in nematode vectors. Initially, it was thought that there is a low specificity in the spread of plant viruses, as the strain of RpRSV has several species of *Longidorus* as potential vectors. Now, the importance of the chemical nature of sites of virus location and virus determinants like coat protein has been demonstrated in nematodes. In addition, the presence of genes responsible for differential transmission of plant viruses is also a factor. Further, the chemical composition of salivary secretions in different species of nematodes varies in nature; this results in differential release of virus from the site of location of virus. The detachment of *X. americanum* virus from the lumen of esophagus is dependent on interaction of viral coat protein and retention site in nematode. There are five genera, namely *Xiphinema* (GFLV, ArMV, CLRV, TomRSV, Tobacco ring spot virus-TRSV, Peach rosette mosaic virus-PRMV, CRLV, Peach yellow, Bud mosaic virus, Grapevine yellow vein virus-GYVV, Raspberry ring spot virus-RpRSV), *Trichodorus* (Tobacco rattle virus-TRV, Pepper ring spot virus-PRSV, Pea early browning virus-PEBV, Docky disorder of sugar beet virus-SBDDV), *Paratrichodorus* (TRV, PEBV, Corky ring spot virus of potato-PCRSV), *Longidorus* (TBRV, RpRSV, Docky disorder of sugar beet virus-SBDDV, ToBRV, Cherry rasp red leaf virus-CRRLV) and *Paralongidorus* (ArMV, RpRSV, Strawberry latent ring spot virus-SLRSV), acting as vectors of plant pathogenic viruses. To further strengthen the concept of specificity, the transmission efficiencies of *Trichodorus* and *Paratrichodorus* species were compared and it was found that when the nematode was given access to the virus isolate with which it was associated in the field, the nematode transmitted the virus. There was no transmission of virus in situations where the virus isolate was not associated with the vector in the field, as in the case of Tobacco rattle virus in the Netherlands (Van Hoof, 1968).

10.5 Diseases

10.5.1 Fan leaf of grapevine virus disease-FLGV (Nepovirus) (Xiphinema)

Fan leaf is a disease caused by a virus belonging to *Nepovirus* (Sequiviridae) and is transmissible by *Xiphinema index.* The nematode vector can acquire the virus with an acquisition access of five to fifteen minutes on virus source and can retain it for 240 days. It is characterized by chlorosis of leaves with yellow vein bands, shortening of internodes, and undersized berries. It is an RNA virus and contains 2 RNA molecules, icosahedral, twenty eight nm in size, with sixty identical protein sub units. It is also a seed-borne and pollen-borne virus. The fan leaf-like symptoms are caused by number of viruses in grapevine and these includes ArMV, Artichoke Italian latent-AILV, Blueberry mottle, Cherry leaf roll, Grapevine Anatolian ring spot, Bulgarian ring spot, Grapevine chrome mosaic, Grapevine deformation, Fan leaf of Grapevine, Grapevine Tunisian ring spot, Peach mosaic mottle, Raspberry ring spot, Strawberry ring spot, Tobacco ring spot, Tomato black ring and Tomato ring spot viruses. It is widespread in the USA in all the grape cultivated areas.

10.5.2 Tobacco rattle virus disease-TRV (Tobravirus) (Trichodorus)

A disease of viral etiology, it is known to cause Corky disease in potato. It is prevalent in Europe, Japan, New Zealand and North America. The virus is transmissible via *Trichodorus allius,* nematodes and Cuscuta. The disease is characterised by ring spots, line patterns, leaf and stem mottling and ring formation on potato tubers. It is known to attack plants of more than 400 species from mono- and dicotyledonous families. The virus is +ve RNA tubular rods and exists in two sizes viz. 185 x 196 nm and 50 x 115 nm. It has two isolates and of these two, one contains two RNA (non-transmissible by nematodes) and the other single RNA (nematode transmissible).

10.5.3 Pea early browning virus disease-PEBV (Tobravirus) (Paratrichodorus)

It is *Tobravirus* (Virgaviridae), +ve ssRNA with helical symmetry and a virus with linear genome arrangement. It has particles of size twenty-one nm with two possible lengths viz. 105 nm and 215 nm. It has two strains viz. Dutch and British, transmissible by different species of nematodes. The Broad bean yellow band virus is a synonym of Pea early browning. It prevails in Western Europe, The Netherlands, and England. The important symptoms of damage are early browning of leaves and large necrotic spots in the stipules and leaflets. The necrotic spots are present on stems or sometimes on pods. It is spread through different species of nematodes (*Trichodorus teres, T. pachydermus, T. anemones, T. primitivus* and *T. similis*). It is also seed-borne in nature.

10.5.4 Cherry rasp leaf virus disease-CRLV (Cheravirus) (Xiphinema)

CRLV is of viral origin; It has +ve ssRNA, contains two isometric, hexagonal particles, measuring thirty nm in diameter. The disease is prevalent in Asia, Europe, North America and Africa and is transmissible through *Xiphinema americanum* and *X. rivesi*. In nature, cherries and peaches are attacked by this virus. It can be identified from the leaf enations, abnormal fruit formation, canker, dieback, leaf rolling, stunting and rosetting of plants.

10.5.5 Arabis mosaic virus disease-ArMV (Nepovirus) (Xiphinema)

It is a virus that is responsible for disease in peaches, plum, raspberry, strawberry, rose, cherry, etc. It is prevalent in Europe and Canada and belongs to *Nepovirus* (Sequiviridae). It is a +ve ssRNA virus measuring thirty nm in size. It is characterized by mottling and flecking of leaves with conspicuous deformities and enations in affected plants. The presence of streaks/lines or rings is also possible in this disease. It is transmissible through nematodes, *Xiphinema diversicaudatum.*

10.6 Nematodes and Bacteria

In addition to viruses, nematodes are also carriers of plant pathogenic bacteria as well. The bacteria are carried internally by nematodes and exuded as fecal matter in the environment where they are known to cause diseases. Besides, the free-living ectoparasitic nematodes feed on bacterial cells present in wounded plant tissues. It is very well established in the case of tundu disease of wheat, annual rye grass toxicity and strawberry cauliflower disease complex. The association of nematodes with bacteria has been demonstrated in all three situations. The nematode, *Anguina tritici* (Vector of *Clavibacter tritici* causal organism of yellow ear rot in wheat) (Pathak and Swarup, 1984), *A funesta* (vector of *Clavibacter* sp. causal organism of annual rye grass toxicity disease) (Riley and McKay, 1990) and *Aphelenchoides fragariae* (Vector of *Rhodacoccus fasciens* causing cauliflower disease in strawberry) (Crosse and Pitcher, 2008) are associated with the spread of bacterial diseases (Riley and Beasdon, 1995; Riley, 1992; Jensen, 1978). In addition, nematodes *Dylenchus dipsaci* (vector of *Clavibacter michiganensis* causing wilt disease in alfalfa) (Hawn, 1971), *Pseudomonas fluorescens* (causing Café au Lait bacteriosis in garlic) (Crubel and Sampson, 1984) and *Bacterium rhaponticum* (causing crown rot of Rhubarb) (Metcalfe, 1940) are also vectors of bacteria. Another disease, wilt of potato, is caused by *Pseudomonas solanacearum* bacteria vectored by *Globodera pallida* nematode (Jensen, 1978).

10.6.1 Disease

10.6.1.1 Tundu disease of wheat. This disease of wheat, also known as yellow ear rot, is caused by *Anguina tritici* nematode and bacteria *Clavibacter tritici* by association. It is widespread in India, Ethiopia, Syria, Romania, Yugoslavia and Australia. The bacterium is unable to cause disease without nematode. The bacterium is carried externally on the surface of the nematode and in the inflorescence. The gall formation is caused by the nematode while inhabiting this outgrowth. As a result of infection, the plant shows rolling, curling and spiraling of leaves, delayed maturity and reduced yield (up to

50% in infected plants). The interaction of both is responsible for infection of wheat that develops from the oozing bacteria in the grain. The juvenile nematodes penetrate into the primordia of flower, undergo molting and turn into adults. The galls contain adults which mate inside these galls. The egg-laying takes place within the galls. A gravid female lays about 2000 eggs within a period of about one week. These galls are harvested along with wheat grains and the nematodes remain in these galls. On absorption of water, the juveniles become active but they do not enter into the roots; instead, they feed externally as ectoparasites until flowers are formed.

10.7 Nematodes and Fungi

As carriers of fungal pathogens, the nematodes are less active than bacteria. The association of *Meloidogyne* species with *Fusarium oxysporum* in cotton crop has been apparent since 1892. The nematodes are predisposing factors for the entry of fungi. The nematodes alter the nutritional status, suppress host toxins, antifungal metabolites, growth factors and modify their predisposition (Riedel, 1988). The pine wood nematode is known to carry a large number of fungi; several genera have been isolated, these include *Mucor, Penicillium, Ophistoma, Bionectria, Botrytis, Fusarium, Hypocrea, Nectria* and *Trichoderma camerops* (Hyun et al., 2007).

References

Argelis A (1967). Present situation of Grapevine virus diseases with reference to the problems which they cause in Greek vine yards. pp. 309–312. In: Integrated Pest Control in Viticulture. Rotherdam Netherlands, 395 p.

Auger J, Leal G, Magunacelaya JC and Esterio M (2009). *Xiphinema rivesi* from Chile transmits Tomato ring spot virus to cucumber. Plant Disease 93: 971.

Ayala A and Allen MW (1968). Transmission of the California tobacco rattle virus by three species of the nematode, genus *Trichodorus*. Journal of Agricultural University of Puerto Rico, 52: 101–125.

Brown DJF and Boag B (1977). *Longidorus attenuates* C.I.H. Descriptions of Plant-Parasitic Nematodes set 7 No 101.4 pp.

Brown DJF and Trudgill DL (1984). The spread of Carnation ring spot virus in soil with or without nematodes. Nematologica 30: 102–105.

Brown DJF and Halbrendt JM (1992). The virus vector potential of *Xiphinema americanum* and related species. Journal of Nematology, 24: 584.

Brown DJF, Grunder J, Hooper DJ, Klingler J and Kunze P (1994). *Longidorus arthensis* sp.n. (Nematoda; Longidoridae) the vector of Cherry rosette disease caused by a new *Nepovirus* in cherry trees in Switzerland. Nematologica, 40: 133–149.

Caubel G and Samson R (1984). Effect of the nematode *Ditylenchus dipsaci* on the development of Café au Lait Bacteriosis in garlic (*Allium sativum*) caused by *Pseudomonas fluorescens*. Agronomie, 4: 311–313.

Cohan E, Tanne E and Nitzany FE (1970). *Xiphinema italiae*, a new host of Grape fan leaf virus. Phytopathology, 60: 181–182.

Cooper JL and Thomas PR (1970). *Trichodorus nanus*, a vector of Tobacco rattle virus in Scotland. Plant Pathology, 19: 197.

Cremer MG and Schenk PK (1967). Notched leaf in *Gladiolus* spp. caused by viruses of Tobacco rattle virus group. Netherlands Journal of Plant Pathology, 73: 33–48.
Crosse JE and Pitcher RS (2008). Studies on the relationship of eelworm and bacteria to certain plant disease. II. The etiology of strawberry cauliflower disease. Annals of Applied Biology 39: 475–486.
Eveleigh ES and Allen WR (1982). Description of *Longidorus diadecturus* n sp. (Nematoda; Longidoridae) a view of Rose rosette mosaic virus in peach orchards in Ontario, Canada. Canadian Journal Zoology, 60: 112–115.
Forer LB, Powell CA and Stouffer RF (1984). Transmission of Tomato ring spot virus to apple root stock cuttings to cherry, peach seedlings by *Xiphinema rivesi*. Plant Disease, 68: 1052–1054
Gibbs AJ and Harrison BD (1964). A form of Pea early browning virus found in Great Britain. Annals of Applied Biology, 54: 1–11.
Gonsalves D (1988). Tomato ring spot virus decline. pp. 49–50. In: Pearson RC and Goheen AC (eds). Compendium of Grape Diseases. American Phytopathological Society Press St Paul Minnesota 93 p.
Griesbach JA and Maggenti AR (1989). Vector capability of *Xiphinema americanum* sesulta lato in California. Revue de Nematologie, 13: 93–103.
Gudmestad NC, Mallikc I, Basche JS and Crosslin JM (2008). First report of Tobacco rattle virus causing corky ring spot in potatoes grown in Minnesota and Wisconsin. Plant Disease, 92: 1254.
Han L-J and Liu W-H (2007). Studies on Prunus necrotic ring spot virus (PNRSV) occurring on the lily. Agricultural Sciences China, 6: 1201–1208.
Harrison BD (1967). Pea early browning virus (PEBV) Report of Rothamsted Experimental Station for 1966: 115.
Harrison BD (1964). Specific nematode vectors for serologically distinctive forms of Raspberry ring spot and tomato black ring viruses. Virology, 22: 544–550.
Hawn EJ (1971). Mode of transmission of *Corynebacterium insidiosum* by *Ditylenchus dipsaci*. Journal of Nematology, 3:420–421.
Hewitt WB, Raski DJ and Goheen AC (1958). Nematode vector of soil-borne Fan leaf virus of grapevines. Phytopathology, 48: 586–595.
Hyun MW, Kim JH, Suh D, Lee SK and Kim SH (2007). Fungi isolated from pine wood nematode, its vector, Japanese pine swayer and the nematode infected Japanese black pine wood in Korea. Microbiology, 35: 159–161.
Jensen HJ (1978). Interrelations of nematodes and other organisms in disease complex. International Potato Centre Report of the Second Planning Conference on the Developments in the Control of Nematode Pests of Potatoes. CIP, Lima, Peru.
Jha A and Posnette AF (1961). Transmission of Arabis mosaic virus by nematode *Xiphinema diversicaudatum* (Micol). Virology, 13: 119–123
Jones AT, McElroy FD and Brown DJF (1981). Tests for transmission of Cherry leaf roll virus using *Longidorus, Paralongidorus,* and *Xiphinema* nematodes. Annals of Applied Biology, 99: 143–150.
Jones AT, Brown DJF, McGavin WJ, Rudel M and Altmayer B (1994). Properties of an unusual isolate of Raspberry ring spot virus from grapevine in Germany and evidence for its possible transmission by *Paralongidorus maximus*. Annals of Applied Biology, 124: 283–300.
Jones JT, Haegeman A, Danchin EGJ, Gaur HS, Helder J, Jones MGK, Kikuchi T, Manzanilla-Lopez R, Palomares-Rius JE, Wesemael WML and Perry RN (2013). Top 10 plant parasitic nematodes in molecular plant pathology. Molecular Plant Pathology, 4: 946–961.
Klos EJ, Finck F, Knierin JA and Cation D (1995). Peach rosette mosaic transmission and control studies. Quarterly Bulletin Michigan State Agricultural Experimental Station, 49: 287–293.
Kyriakopoulou PEC (2008). Artichoke Italian latent virus causes artichoke patchy chlorotic stunting disease. Annals of Applied Biology, 127: 489–497.

Lamberti F and Roca F (1987). Present status of nematodes as vectors of plant viruses. pp. 321–328. In: Veech JA and Dickson DTW (eds). Vistas on Nematology. Society of Nematologists, Hyattsville, Maryland, 509 p.
Lister RM (1964). Strawberry latent ring spot virus: a new nematode-borne virus. Annals of Applied Biology, 54: 167–176.
Metcalfe G (1940). *Bacterium rhaponticum* (Millard) Dawson, a cause of crown rot disease of rhubarb. Annals of Applied Biology, 27: 502–508.
Mojtahedi H, Santo GS, Crosslin JM, Brown CR and Thomas PE (2000). Corcky ring spot disease.A review of the current situations. pp. 9–13. In: Moses Lake WA (ed). Proceedings 39th Washington State, Potato Conference and Trade Fair.
Mojtahedi H, Santo GS, Thomas PE, Crosslin JM and Boydston RA (2002). Eliminating Tobacco rattle virus from viruliferous *Paratrichodorus allius*, establishing a new vector combination. Journal of Nematology, 34: 66–69.
Nyland G, Lownsbery BF, Lowe BK and Mitchell JF (1969). The transmission of Cherry rasp leaf virus by *Xiphinema americanum*. Phytopathology, 59: 111–112.
Pathak KN and Swarup G (1984). Incidence of *Corynebacterium michiganense pv. tritici* in the ear-cockle nematode (*Anguina tritici*) galls and pathogenicity. Indian Phytopathology, 37: 267–270.
Ramsdell DC (1988). Peach rosette mosaic virus decline. pp 51–52. In: Pearson RC and Goheen AC (eds). Compendium of Grape Diseases. American Phytopathological Society Press St Paul Minnesota, 93p.
Raski DJ (1988). Dagger and needle nematodes. pp 56–59. In: Pearson RC and Goheen AC (eds). Compendium of Grape Diseases American Phytopathological Society Press St Paul Minnesota, 93p.
Riedel RM (1988). Interactions of plant parasitic nematodes with soil-borne plant pathogens. Agriculture, Ecosystem and Environment, 24: 281–292.
Riley IT and McKay AC (1990). Specificity of the adhesion of some plant pathogenic microorganisms to the cuticle of nematodes in genus Anguina (Nematoda; Anguinidae). Nematologica, 35: 90–103.
Riley IT (1992). *Anguina tritici* is a potential vector of *Clavibacter toxicus.* Australasian Plant Pathology, 21: 147–149.
Riley IT and Reardon TB (1995). Isolation and characterization of *Clavibacter tritici* associated with *Anguina tritici* in wheat from Western Australia. Plant Pathology, 44: 805–810.
Roca F, Rana GL and Kyriakopoulou PE (1982). *Longidorus fasciatus,* Roca and Lamberti, a vector of a serologically strain Artichoke Italian latent virus in Greece. Nematologia Mediterranea, 10: 65–69.
Salomao TA (1973). Soil Transmission of Artichoke Yellow Band Virus Atti Second Conference International Studies Carciofo Bari, Italy, 7975: 831–854.
Schmidt HB, Fritzsche R and Lehmann W (1963). Die ubertragung des weidel-grasmosaik-virus durchnematoden. Naturwissenschaften, 50: 386.
Schneider SM and Ferris H (1987). Stage-specific population development and fecundity of *Paratrichodorus minor.* Journal of Nematology, 19: 267–394.
Stellbach G and Goheen AC (1988). Other virus and virus like diseases. pp 53–54. In: Pearson RC and Goheen AC (eds). Compendium of Grape Diseases. American Phytopathological Society Press St Paul Minnesota, 93p.
Taylor CE (1962). Transmission of Raspberry ring spot virus by *Longidorus elongatus* (de Man) (Nematoda; Dorylaimidae). Virology, 67: 493–494.
Taylor CE and Ruski DJ (1964). On transmission of Grape fan leaf by *Xiphinema index.* Nematologica 10: 489–495.
Taylor CE and Murant AF (1969). Transmission of strains of Raspberry ring spot and tomato black ring viruses by *Longidorus elongatus* (de Man). Annals of Applied Biology, 66: 43–48.
Taylor CE and Robertson WM (2008). The location of Raspberry ring spot and Tomato black ring viruses in the nematode *Longidorus elongatus* (de Man). Annals of Applied Biology, 64: 233–237.

Van Hoof HA (1962). *Trichodorus pachydermus* and *T. teres*, vectors of Pea early browning of peas. Tijdschr Plantenzieten, 68: 391–396.

Van Hoof HA (1964). *Trichodorus teres*, a vector of Tobacco rattle virus. Netherland Journal of Plant Pathology, 70: 187.

Van Hoof HA (1968). Transmission of Tobacco rattle virus by *Trichodorus* species. Nematologica 14: 20–24.

Van Hoof HA (1972). Viruses transmitted by *Xiphinema* species in The Netherlands. Netherlands Journal of Plant Pathology, 77: 30.

Van-Zyl S, Vivier MA and Walker MA (2012). *Xiphinema index* and its relationship to grapevines: A review. South African Journal of Enology & Viticulture, 33: 21–32.

Van Hoof HA, Maat DZ and Seinhorst JW (1966). Viruses of tobacco virus group in North Italy: the vector and serological relationships. Netherland Journal of Plant Pathology, 72: 253.

Yagita H and Kumuro Y (1972). Transmission of Mulberry ring spot virus by *Longidorus martini* Meryl. Annals of Phytopathological Society of Japan, 38: 275–283.

QUESTIONS (EXERCISE)

Q 1. List the vectors of the following nematode-borne viruses:

a) Cherry leaf rasp virus
b) Fan leaf of grapevine virus
c) Carnation ring spot virus
d) Arabis mosaic virus
e) Peach rosette virus

Q 2. The location of viruses is different in nematode vectors. Discuss with suitable examples.

Q 3. Describe the feeding mechanism of nematodes.

Q 4. What is the difference between NEPU and NETU viruses? List three virus diseases along with their vectors belonging to these categories.

Q 5. Briefly explain the mechanism of spread of plant viruses through nematodes.

CHAPTER 11

Fungi

11.1 Introduction

Plants are sessile and are unable to move from one place to another; they can only spread via pollen or seed. Likewise, most of the plant pathogens are dependent on various arthropods, nematodes, fungi and Plasmodiophorids for their spread in nature. Among the fungi, not all are involved in the spread of plant viruses due to specificity in fungal vectors. The specificity is governed by determinants, and their interaction with the vectors that are provided with specific sites that mediate their recognition. Of the determinants, coat protein (CP) is the one that is considered to be most vital in the transmission of the virus. In coat protein, it is the specific amino acid that determines the spread of a particular virus through selective attachment on a particular site in the body of the vector. So far, thirty plant viruses have been identified as soil-borne and vectored by fungi (Campbell, 1996). Of these, ten are polyhedral in shape (Tombusviridae) and eighteen are rod-shaped (furoviruses and bymoviruses) (Kulne, 2009; Shukla et al., 1998). The furoviruses are Soil-borne wheat mosaic virus-SBWMV, Soil-borne cereal mosaic virus-SBCMV, Chinese wheat mosaic virus-CWMV and Oat mosaic virus-OMV, while genera *Bymovirus* includes Barley yellow mosaic virus-BaYMV, Barley mild mosaic virus-BaMMV, Wheat spindle streak virus-WSStV, Wheat yellow mosaic virus-WYMV and Oat golden stripe mosaic virus-OGStMV. These viruses are carried either internally (*Polymyxa* species/*Spongospora*) or externally by the vector (*Olpidium* species) through zoospores. These viruses are adsorbed on the zoospores' plasmalemma. These viruses are released into the soil with the disintegration of roots and are picked up by the zoospores (Campbell, 1996).

11.2 Characteristic Features Common in Vector Species

The features common in vector fungi are a high level of specificity (transmission through one species of vector), multiplication in plant cells, limited genera (*Olpidium, Synchytrium* from Chytridales and *Polymyxa, Spongospora* from Plasmodiophorales) acting as vectors of viruses, root-inhabiting obligate parasites and viruses being carried by zoospores both externally (adsorbed on the surface of zoospores) and internally (inside the zoospores). Plasmodiophorids also share some characteristic features (Braselton, 1995). These features include an unusual cruciform type of nuclear division, unequal length of biflagellate, multinucleated plasmodia, in addition to features common in four genera. Of these four genera, *Synchytrium endobioticum* is in itself pathogenic and causes Potato wart disease, as well as acting as vector of Potato virus X. The virus is acquired *in vivo* but not *in vitro* and is carried internally in resting spores (untested). These fungal-borne viruses are double-stranded RNA viruses. The viruses are transmitted by fungi belonging to genera *Benyvirus, Bymovirus, Furovirus, Pecluravirus* and *Pomovirus*. *Olpidium virulentus* is a vector of Tobacco stunt virus-TSV (*Ophiovirus*), Pepino mild mosaic virus-PepMMV, Mirafiori lettuce big vein virus-MiLBVV (*Ophiovirus*) and Lettuce big vein associated virus-LBVaV (Lettuce big vein virus-LBVV) (*Varicosavirus*) (Table 11.1). Similarly, Lettuce ring necrosis virus-LRNV (*Varicosavirus*), Fereesia leaf necrosis virus-FLNV (*Varicosavirus*), Chenopodium necrosis virus-ChNV (*Necrovirus*), Lisianthus necrosis virus-LiNV (*Necrovirus*), Tobacco necrosis virus-A-TNV-A (*Necrovirus*), Tobacco necrosis virus-D-TNV-D (*Necrovirus*), Tulip mild mottle mosaic virus-TuMMV and Olive mild mottle virus-OMMV are all transmissible by *O. brassicae*. *O. bornovanus* is known to transmit Cucumber necrosis virus-CNV (*Tombusvirus*), Cucumber leaf spot virus-CLSV (*Aureovirus*), Cucumber soil-borne virus-CSBV (*Carmovirus*), Melon necrotic spot virus-MNSV (*Carmovirus*), Squash necrosis virus-SqNV (Carmovirus) and Red clover necrotic mosaic virus-RCNMV (*Dianthovirus*) (Rochon, 2009; Rochon et al., 2004). *Polymyxa graminis* transmits Barley yellow mosaic virus-BaYMV, Barley mild mosaic virus-BaMMV, Chinese wheat mosaic virus-CWMV, Soil-borne wheat mosaic virus-SBWMV, Wheat spindle sheath mosaic virus-WSSMV, Oat mosaic virus-OMV, Wheat yellow mosaic virus-WYMV, Aubean wheat mosaic virus-AuWMV, Soil-borne cereal mosaic virus-SBCMV, Oat golden stripe virus-OGStV, Sorghum chlorotic necrosis virus-SCNV, Rice necrosis mosaic virus-RNMV, Rice stripe necrosis virus-RStNV, Broadbean necrotic virus-BBNV and Peanut clump virus-PeCV (Kanyuka et al., 2003). *O. betae* is a known vector of Beet necrotic yellow vein virus-BNYVV. *Spongospora subterranea* transmits Potato mop top virus-PMTV and Melon necrotic spot virus-MNSV is transmissible through *O. radicale*.

Table 11.1 Fungi as vector of plant viruses (added new information from Mukhopadhyay, 2011; Yilmaz et al., 2003; Dijkstra and Khan, 2002).

Virus (Acronym)	**Genus/family**	**Characters**	**Vectors**	**Spread**	**Source**
Barley mild mosaic Virus (BaMMV)	*Bymovirus*/Potyviridae	Filamentous, ssRNA	*Polymyxa graminis*	Internal	Jianping et al., 1991
Beet necrotic yellow vein virus (BNYVV)	*Benyvirus*/Unassigned	Rod-shaped, ssRNA	*Polymyxa betae*	Internal	Hugo et al., 1996
Lettuce big vein virus (LBVV)	*Varicosavirus*/ Unassigned	Rod-shaped, ssRNA	*Olpidium brassicae*	Internal	Tomlinson and Garrett, 1964
Melon necrotic spot virus (MNSV), Cucumber leaf spot virus (CLSV)	*Carmovirus*/ Tombusviridae	Icosahedral, ssRNA	*Olpidium bornovanus; O. radicale*	External	Ohki et al., 2010; Campbell et al., 1991
Potato virus X (PVX)	*Potexvirus*/Potyviridae	Filamentous ssRNA, Monopartite	*Synchytrium endobioticum*	Internal	Lange and Olson, 2011
Potato mop top virus (PMTV)	*Pomovirus*/unassigned	Rod-shaped, ssRNA	*Spongospora subterranea*	Internal	Montera–Astua et al., 2008
Pathos latent virus (PLV)	*Aureovirus*/ Tombusviridae	Icosahedral, ssRNA	*Olpidium bornovanus*	External	Chen et al., 2016
Peanut clump virus (PeCV)	*Pecluravirus* unassigned	Rod-shaped, ssRNA	*Polymyxa graminis*	Internal	Dieryck et al., 2008
Soil-borne wheat mosaic virus (SBWMV)	*Furovirus* /Unassigned	Rod-shaped, ssRNA	*Polymyxa graminis*	Internal	Driskel et al., 2004
Tobacco necrosis virus (TNV)	*Necrovirus*/ Tombusviridae	Icosahedral, ssRNA	*Olpidium brassicae*	External	Kassanis and Mcfarlane, 1964

11.3 Mechanism of Transmission

The virus vector fungus belongs to chytrids (Family: Olpidiaceae) and Plasmodiophorids (Family: Plasmodiophoraceae). The transmission of viruses can be both vertical and horizontal. In vertical transmission, the movement of the virus is to the reproductive region of fungus via cytoplasm during the fusion of hyphae and further to the healthy mycelium. The relationship of the virus with the vector is of two categories viz. non-persistent and persistent (Adams, 1991). Zoospores are released from the plant part (root) and the virus particles are adsorbed on the surface of zoospores. In the non-persistent category, the virus is adsorbed on the plasma membrane of zoospores and carried externally. Once the virus acquisition by the fungus takes place *in vitro* or in soil water, the virus is picked up by zoospores' protoplasm during the process of encystment or infection. The virus does not enter into the thallus as it grows and produces zoospores or resting spores, therefore, it persists in infected roots, root residues or in the soil as in the case of Tobacco necrosis virus vectored by *O. brassicae*. In the persistent category, the virus, under genetic control, is carried internally; the coat protein plays a major role in this process. The acquisition of virus occurs *in vivo* in roots of infected plants and the virus is carried internally in zoospores. The virus is carried by protoplast of zoospores as it infects the host cell. It starts multiplying within the thallus as resting zoospores and passes on from one season to the next, as in the case of Soil-borne wheat mosaic virus transmitted by fungus. Zoospores are motile propagule two to fourteen um long and two to six um in diameter, lack cell walls but possess 1 (Chytridales) or 2 (Plasmodiophorales) flagella. These fungi are obligate parasites that live on roots and develop in the protoplast of host plants as thallus after 36 hours; during this period, the exchange of virus takes place. The process continues after infection due to the obligate relationship of the pathogen (Campbell, 1996). The fungi further develop and produce sporangia and more zoospores. These are thickened spores called resting spores and are helpful to the fungi for surviving during the off-season when the host is not available in the field. Zoospores are, thus, an important link in virus transmission. With respect to transmission, Tobacco necrosis virus is carried internally by the zoospores. Plasmodiophorids produce resting spores and the viruses are carried through these spores.

11.4 Specificity

The fungi are specific too in the transmission of plant viruses. To demonstrate specificity, the virion of CNV binds to zoospores (Kishore et al., 2003). Tobacco necrosis virus is highly specific to zoospores of *O. brassicae* while others, like Cucumber necrosis virus-CNV and Tobacco ring spot virus-TRSV are not adsorbed with zoospores. *O. brassicae* transmits TNV-A

(*Necrovirus*) but not CNV (*Tombusvirus*). Similarly, *O. bornovanus* isolate is a successful vector of Melon necrotic spot virus-MNSV and Cucumber leaf spot virus-CLSV but with differential efficiency. The transmission efficiency is variable in the case of *O. brassicae* as well. The specificity in CNV is afforded due to the glycoproteins. It has been demonstrated that, by treating the zoospores with periodate and trypsin, the binding was significantly reduced. Further, the binding of CNV zoospores is mediated with mannose and fructose, these contain oligosaccharides that are instrumental in bringing about change in zoospores (Kakani et al., 2003). CNV particles recognize the glycoprotein receptors on zoospores via virion axis that contains sugar receptor elements. Thus, the binding initiates swelling of the virion. The zoospores plasmalemma has specific reognition receptors. and these are known to afford specificity. The virus reccognition sites have two types of criteria, i.e., saturation and specificity. TNV-A is transmitted by *O. brassicae* but not by *O. bornovanus*. Furthermore, MNSV and CLSV are not vectored by *O. brassicae* but are transmissible by *O. bornovanus*. It can, therefore, be concluded that NV-A is not adsorbed to *O. bornovanus* at the site but it is more efficiently bound to zoospores of *O. brassicae*, and the role of proteins and glycoproteins has been highlighted (Kakani et al., 2003).

11.5 Virus Vector Relationships

The virus-vector relationship between fungal vector and virus is of two kinds, one in which the virus is carried on the surface of zoospores while in the other the virus is carried internally in zoospores. In the former, the virus is released from the host roots *in vitro* (soil water) and adsorbed by the plasma membrane of zoospores. It then enters into the protoplasm but not the thallus of a fungus. The process ends with the formation of resting spores. In the latter case, the fungus vector is in the roots *in vivo* condition in the infected host. The vector releases the virus from the host cell which carried in the protoplast of the host cell. It multiplies, occurs in the thallus as zoospores and ends up as resting spores. In the case of external transmission, the virus is shed on treating the zoospores with acid or triphosphate. Among chytrids, *Olpidium brassicae* (Tobacco necrosis virus), *O. radicale* (Melon necrotic virus) and *O. cucurbitacearum* (Cucumber necrosis virus) are vectors of isometric viruses. In addition, Tobacco stunt virus and Lettuce big vein viruses are rigid rods, transmissible through *O. brassicae.* Among the Plasmodiophorids, *Polymyxa graminis* (Barley yellow dwarf virus, Wheat spindle streak mosaic virus, Oat mosaic virus, Wheat yellow mosaic virus, Rice necrotic mosaic virus) transmits filamentous viruses and P. beta (Beet necrotic yellow vein virus) and *Spongospora subterranea* (Potato mop top virus) are vectors of rod shaped viruses. *Polymyxa graminis* also act as vectors of rod-shaped viruses (Soil-borne wheat mosaic virus, Oat golden streak virus, Peanut clump virus, Broadbean necrosis virus).

QUESTIONS (EXERCISE)

Q 1. Name the fungal vector species of the following:

a) Potato virus X
b) Soil-borne wheat mosaic virus
c) Peanut clump virus
d) Wheat necrotic yellow vein virus
e) Tobacco necrosis virus

Q 2. Describe the mechanism of spread of potato mop top virus through *Spongospora subterranea*.

Q 3. Name the genera of viruses transmissible through fungi and discuss virus transmission mechanisms taking carmovirus into consideration.

Q 4. Write a short note on the role of zoospores in the transmission of viruses.

Q 5. What is the level of specificity in fungi with respect to the transmission of viruses? Give your opinion.

CHAPTER 12

Phytotoxemia

12.1 Symptoms of Different Agents in Crop Plants

Crop diseases/disorders are the result of two distinct types of agents: living (biotic agents) and non-living (abiotic agents). Living agents include micro-organisms (viruses, viroids, *Phytoplasma*, bacteria, fungi, nematodes, protozoans, *Rickettsia-like-organisms*, etc.) and macro-organisms (insects, mites, snails, slugs, birds and small animals). Non-living agents include mechanical (human activity through farm operations), physical (environmental factors like temperature and light), chemical agents (pesticides, industrial smoke, fertilizers and soil conditions) and physiological factors. These agencies are known to cause specific damage symptoms and signs by which these maladies can be distinguished. Symptoms, such as canker, chlorosis, leaf rolling and curling, blight, etc., are an expression and function of the presence of a pathogenic organism, while a sign in the form of spores, mycelium or ooze is evidence that a pathogen is present in affected areas. Of the two categories, the damage caused by non-living factors is mainly confined to the leaves. The damage is also apparent in other plants in the same field as well as other species of plants adjacent to affected plants. The non-living agents, such as pollutants, are responsible for tip-burning of leaves; the damage could be due to the application of herbicides or fertilizers, for example. Wood or bark cracking is generally due to freezing temperatures. Above all, the damage caused by non-living agents appears suddenly; whereas, the gradual decline of diseased plants is always due to a living agency. These microorganisms inflict a different range of symptoms on their hosts, such as galls (outgrowth on leaf, stem or roots), canker (raised, depressed or sunken spots on various plant parts), blight (sudden/rapid burning of parts of plant), rots (decomposed spots) and necrotic lesions (dead spots). The bacterial spots or rots can be differentiated from the fungal spots due to conspicuous

12.2.2 Local lesions with secondary infection

The local lesions are also produced under this category due to feeding of toxicogenic insects. Instead of becoming necrotic, these lesions become rough on account of secondary infection of other pathogens. The abrasions could be in the form of scabs, cankers, fissures, etc. Besides these localized spots, the toxicogenic arthropods secrete a sweet sugary substance on which the other microorganisms develop. The sucking pests like aphids, whiteflies, mealybugs, etc., secrete honey dew on the upper surface of leaves while sucking the sap from the lower surface of leaves. *Capnodium* spp. of fungi develop on sooty mold and cover the entire leaf areas and create obstruction in the photosynthetic activities of the plant. As a result, the affected plants show poor growth. The premature shedding of leaves is a common indicator of the presence of aphids and whiteflies. The secondary infection by microorganisms is not directly on the point of feeding of toxicogenic insect, but on the sweet sugary material secreted by such category of insects.

12.2.3 Malformation/deformities in plants

Under this category of phytotoxemia, the plant parts or plants become misshapen or malformed due to the feeding of insects with toxic saliva. These symptoms are the outcome of feeding of hemipteran insects, thrips and mites. The toxic saliva of *Toxoptera citricida* induces rolling and twisting of citrus leaves. Likewise, the rolling and severe twisting of leaves in peaches (*Brachycaudis helichrysi*), cotton (*Aphis gossypii*), mustard (*Liphaphis erysimi*) and potato (*M. persicae*) is caused by aphid feeding. There are many more cases in which such symptoms are produced. The twisting of lower canopy leaves in chili is the result of feeding of *Hemitarsonemus latus* mite. As time passes, the attacked leaves become brittle in texture and easily break. The hemipteran insects, aphids, leafhoppers, planthoppers, psyllids, froghoppers, mealybugs, treehoppers, and true bugs, are responsible for these symptoms. Thysanopterous insects (thrips) also inject toxic saliva in cotton leaves in the seedling stage of crop. The young seedlings become misshapen, so much so that it becomes extremely difficult to recognize the cotton crop. Aphids are known for the production of phytotoxemia in plants. The cherry aphid *M. cerasi* causes leaf curl symptoms in cherry trees, on account of injection of toxic saliva. The symptoms bear a great resemblance to the leaf curl symptoms produced by a viral infection. In this case, the plant may recover from the symptoms once the aphids are controlled. However, the presence of insects confirms that the leaf curl was due to toxic saliva initially. The aphid grape *Phylloxera* (*Daktulosphaira vitifoliae*) caused the failure of the European grapevine industry in the nineties. It inflicted two kinds of damage, i.e., galicoles (galls on leaves) and radicoles (root deformities) as well as production of galls due to feeding beneath the bark, followed by death of trees. The production of symptoms has also been recorded as being due to the feeding of leafhoppers. Leafhopper *Empoasca fabae* is known to cause hopper burn in alfalfa as a result of injection of toxic saliva. The laceration of vascular tissues by *E. kraemeri* leafhopper is also responsible for the disruption of flow of water to the aerial parts of plant, thereby causing yellowing of leaves (hopper burn) and finally necrosis of leaves, as experienced in beans in South America. Similar damage symptoms are the outcome of feeding of another species, *E. tybica*, in North Africa. *Amrasca biguttula* is also known for the production of hopper burn in cotton and *Nilaparvata lugens* in rice. Another category of insects, like the squash silver leaf whitefly *Bemisia argentifolii*, also inflict phytotoxemia in squash. The affected plants show bronzing of leaves. The potato psyllid *Paratrioza cockerelli* is instrumental in producing phytotoxemia in potato. The production of galls is another category of symptoms that are the outcome of injection of toxic saliva. The galls are tumorous outgrowths of stem, leaf, petiole, flower, and root due to mitosis cell division and morphogenesis. They are of varying color, size and shape and are caused by insects, mites, parasitic plants (mistletoe) and microorganisms (fungi, bacteria and nematodes). These galls are not very destructive to plants but do reduce the aesthetic value of the crop. Some of the galls are useful to man as they serve as a home for symbionts in insects and be a source of food, they create suitable niches to live in, help in the process of pollination and act as guards against predators. They can also be a rich source of tannins, required to tan leather, and supply Gallic acid, garlic acid and many dyes used by various cultures. Besides these uses, the galls are mildly destructive to plants. The insects known for the production of galls include psylla (causes blister and nipple galls in hackberry), aphids (produce cone shaped galls in the leaves of elm and witch hazel of white spruce), phylloxerins hailing from the Family phylloxeridae (inducells on leaves of blue spruce) in Colorado (USA), jumping wasp (responsible for production of galls in a number of plants), midges belonging to family Cecidomyiidae (produce leaf galls in maple trees), gall flies (dipterous insects - Urophila gall flies) which are known for the production of stem galls, saw flies (*Phyllocolpa* species) that create leaf galls on willows and poplar and Pontania gall flies that are known for the production of globular galls. Petiole galls are the result of feeding of *Earisa* flies. In addition, lepidopteran larvae of moths (*Gnorimoschema*) create stem galls on golden rod and the mid rib moth causes leaf deformities in buckthorn plantations. The coleopterous beetles of family Buperistidae (*Agrilus ruficollis, Agrilus champlaini* and *Saperda* spp.) create galls in blackthorn, ironwood and poplar, respectively and gall weevils (*Podapion gallicola*) cause galls on twigs of pine trees. Besides insects, mites are also well known for causing galls in plants. In this context, aloc mite (Eriophyidae) has been found to inject toxin containing a growth regulator similar to 2,4-

compositions. The major function of saliva is lubrication of food due to the presence of water. The water is instrumental in the dissolution of sugar. The organic compounds of saliva contain enzymes (amylase, invertase, proteases, and lipases) that perform the digestion. In plant bugs, enzymes break down and liquefy the food particles and the food of this kind is sucked through the mouth. The saliva is of two categories viz. sheath saliva or watery saliva. The sheath forming saliva is known to contain polyphenoloxidas enzymes which oxidize defensive mechanisms and covert them into harmless material. The watery saliva, being enriched with enzymes, does the job of digestion. As referred above, the toxicogenic insects inflict localized symptoms or are responsible for the production of systemic symptoms due to diffusion of toxins in the whole plant. The symptoms are the result of growth hormones being triggered by chemical constituents of the toxic saliva. The remission of symptoms could be achieved by treating the plant with chemicals. For example, the toxic effect of tarnished bug feeding can be neutralized with the application of alpha naphthalene acetic acid. Besides these hormones, the saliva being toxic contains enzymes that are responsible for production of symptoms in plants.The toxic content of saliva is known in toxicogenic insects. These toxic substances that are picked up by the insects pass through the alimentary canal to get into the hemolymph and hit the salivary glands. From here, the toxic secretions are injected into the plant issues and produce disease symptoms. Some of these chemicals are inactivated in the salivary glands due to the presence of inhibitors. The saliva contains pectinase, as in aphids, and these enzymes are known to dissolve plant cell walls. *Miris dolabratus* are known to contain plant growth inhibiting chemicals in their saliva. Leafhoppers have saliva enriched with proteinases or amylases, irrespective of their feeding sites (phloem or mesophyll). Psylla species contains diastase enzyme responsible for disease production. Notably, the insects belonging to order Hemiptera and Thysanoptera, including mites, are known for the production of phytotoxemia. Amongst the hemipterans, heteropterous insects are known to contain toxic saliva. In these insects, the salivary fluid is taken into the chamber through the duct and ejected via the saliva canal formed by maxillary stylets. These stylets pierce plant tissues through the epidermis, after which, the stylets either follow the intercellular/intracellular course in the phloem or the intracellular course in the parenchyma.

12.6 Mechanism of Phytotoxemia

The phytotoxemia in plants causes hopper burn or gall formation. The mechanism of production of phytotoxemia has not been properly understood yet, however, some headway has been made with respect to the production of phytotoxic symptoms of hopperburn caused by the feeding of leafhoppers and planthoppers (Backus et al., 2005). According to these workers, the hopper burn is a non-contagious disease caused by feeding of leafhoppers and planthoppers and is characterized by tip yellowing in the early growth of plants, followed by complete yellowing of foliage coupled with the stunted growth of plants in most of the crops like cotton and rice. In alfalfa, however, triangular yellowing has been noticed. It is now known that phytotoxemia is not only the outcome of salivary secretions but also a result of the interaction of insect feeding stimuli and complex plant responses (hopper burn cascade) (Plate 12.1 Hopper burn in cotton due to cotton jassid). It is confined to *Empoasca* spp. (Cicadellidae: Typhlocybinae). It is measured using modern technology based on Electrical Penetration Graph (EPG). This is to measure probing and stylet penetration using AC and DC signals. Hopper burn was recorded in twenty-three species, mainly confined to division Auchenorrhyncha. Out of these species, nineteen were from the category of leafhoppers. The feeding behavior of two different types was noted in this category of insects, viz. sheath feeding and lacerate-flush feeding. *Empoasca* spp. is from both categories of stipplers and burners. Originally this important investigation dealing with stylet penetration was

Plate 12.1 Hopperbun in Cottondue to Cotton Jassid.

called "probing" (Tjallingii, 1978; Mclean and Weight, 1968; McLean and Kinsey, 1967; Mclean and Kinsey, 1964). The AC and DC major monitors of EPG have been used to quantify duration and number of probes. The feeding behavior was associated with hopper burn in two categories of feeding, viz. sheath feeding and lacerate-flush feeding (cell rupture feeding) (Smith and Poos, 1931)). The hopper burn is caused by sheath feeding and not cell rupture feeding. Another study on cellular feeding has demonstrated the ingesting of mesophyll content along with salivary sheath. *Empoasca* are a primary cell rupture species and do not make a true salivary sheath. These are clubbed into the category of stipplers and burners because of plasticity in their feeding. The stylet penetration of the same species on different hosts is usually different. There are two types of saliva, viz. watery and sheath-forming. The watery saliva of these insects contain enzymes (digestive, hydrolyzing, cell wall degradation) but their composition is variable in different species. The saliva also has carbohydrates (amylases, pectinases, cellulases, lipases and proteases, hydrolases and alkaline phosphatase). In addition, hydrolytic enzymes have been detected in watery saliva, along with oxidative enzymes (such as Catecholoxidase, polyphenol oxidase, peroxidase). On the other hand, the sheath saliva is found enriched with lipoproteins, phospholipids, and conjugated carbohydrates. Insects with watery saliva release saliva with every probe. Similarly, the sheath feeder also releases solid saliva with every probe. The durability is variable (thirty-four days in *Circulifer tenellus* and eighty days in *Homalodisca coagulates*) but the duration of cell rupture is for around one to four days. A combination of mechanical and salivary stimuli is necessary to trigger hopper burn. Another kind of symptom of toxic secretions is the formation of galls. The production of galls is also the outcome of oviposition. The injury inflicted by feeding or oviposition triggers the increase in the size of cells (Hypertrophy) or increase in the number of cells (Hyperplasia) and the resulting malady is the formation of galls. The galls could be on leaf stem, bud petiole or flower, and gall formation is generally caused by psyllids on Hackberry (nipple or blister galls), aphids on poplar (stem or petiole galls), wasps on oak (jumping, bullet galls) and rose (stem galls), midges on willow pine (cone galls, gouty galls, vein gall) and on grapevine (Filbert galls) and Eriophyid mites on maple (ash galls, bladder galls spindle galls, velvet galls, and bud galls) (Wawrzynski et al., 2005).

References

Backus EA, Serrano MS and Ranger CM (2005). Mechanism of hopper burn: An overview of insect taxonomy, behavior, and physiology. Annual Review of Entomology, 50: 125–151.

McLean DL and Kinsey MG (1967). Probing behavior of pea aphid *Acyrthosiphon pisum*. 1 Definitive correlation of electronically recorded waveform with aphid probing activities. Annals of Entomological Society of America, 60: 400–405.

McLean DL and Kinsey MG (1964). A technique for electronically recording aphid feeding and salivation. Nature, 202: 1358–1359.
McLean DL Jr and Weight WA Jr (1968). An electronic measuring system to record aphid salivation and ingestion. Annals of Entomological Society of America, 61: 180–185.
Purcell AH (2009). Phytotoxicity: Phytotoxemia. Encyclopedia of Insects (Second Edition) 2009, pp. 800–802.
Smith FF and Poos FW (1931). The feeding habits of some leafhoppers of genus *Empoasca*. Journal of Agricultural Research, 43: 267–285.
Taliaferro CM, Leuck DB and Stimmann MV (1969). Tolerance of Cynodon clones to phytotoxemia caused by two lined spittlebug. Crop Science, 9: 765–760.
Tjallingii WF (1978). Electronic recording of penetration behavior by aphids. Entomologia Experimentalcs et Applicata, 24: 721–730.
Wawrzynski R, Hahn J and Ascerno M (2005). Insect and mite gall insects. University of Minnesota Extension.

QUESTIONS (EXERCISE)

Q 1. What is meant by phytotoxemia? List the types of phytotoxemia and describe the malformation with suitable examples.

Q 2. Describe the mechanism of phytotoxemia in detail.

Q 3. How can the toxic effect of insect feeding be differentiated from the damage of fungal pathogens?

Q 4. What is hopper burn? Clarify its mechanism with examples.

Q 5. What are galls? Name the arthropods responsible for creating galls. Give a brief description of the physiology of gall formation.

CHAPTER 13

Plant Pathogens and Electron Microscope

13.1 Plant Pathogens

Pathogens are known to cause diseases both in plants and animals. The plant pathogens responsible for inflicting diseases are briefly discussed here. The pathogens belong to different categories including viruses, bacteria, fungi, protozoa, Mollicutes (*Mycoplasma, Phytoplasma* and *Spiroplasma*), nematodes, trypanosomes and *Rickettsia-Like-Organisms*. All the plant pathogens spread through insects, mites, nematodes, and fungi are included in this write-up.

13.1.1 Plant viruses

The viruses are sub-microscopic infective entities, capable of reproducing only in living cells. They possess one type of nucleic acid (RNA/DNA), make use of ribosomes of plant cells, lack lipman system of energy release, are potentially pathogenic and are obligate parasites of plants. The viruses are different from other organisms in characteristic features such as the presence of only one kind of nucleic acid (either RNA or DNA) in viruses, compared to other organisms that have both. Viruses reproduce from their nucleic acids only, while other organisms reproduce through cell division. Virus particles are unable to grow and undergo binary fission. Viruses are devoid of genetic information pertaining to production of energy, therefore, they lack the lipman system of energy release. These organisms are obligate parasites and make use of ribosomes of host cells (Lwoff and Tournier, 1966). Over 300 species of viruses are known, of which the chemistry is known in barely 10% of cases. With regard to the shape of viruses, half of the identified viruses are cubical/icosahedral/isometric/polyhedral (e.g., Tobacco necrosis virus-TNV) while the remaining half of these viruses are of either elongated/helical symmetry or they are flexuous threads (e.g., Potato virus Y) or rigid rods (e.g., Tobacco mosaic virus-TMV) and a few are complex symmetry cylindrical bacilliform rods (Cocoa swollen shoot virus-CSSV) (Caspar and Klug, 1963). In addition, geminate (e.g., Maize streak virus) viruses are also identified. The particle size of these viruses is 26 nm, 20–25 x 300 nm, 12 x 740 nm, 30–300 nm and 30 x 18 nm in TNV, TMV, Potato virus Y-PVY, CSSV and MSV, respectively (Plate 13.1 Electron microscope). The icosahedral symmetry is a polyhedron with 20 faces, in which protein sub-units are arranged in such a manner so that the nucleic acid remains embedded on the protein sub-units. The diameter of such viruses is between seventeen to sixty nm. The important and well-worked viruses are Turnip yellow mosaic virus-TYMV and Wound tumor virus-WTV. Whereas, the other category with a long length (300 nm) and short breadth (fifteen nm) have been named "helical viruses". This category is further divided into rigid (TMV) and flexuous rods (Citrus tristeza virus-CTV). Of these two kinds, flexuous rods are slightly longer than the rigid rods. The third category contains bullet shaped viruses, known as Bacilliform viruses (complex category). These viruses of the complex category are normally isometric, helical or both but their capsid covers the ends, as well as the sides of the virion with the envelope of protein sub-units and lipoprotein

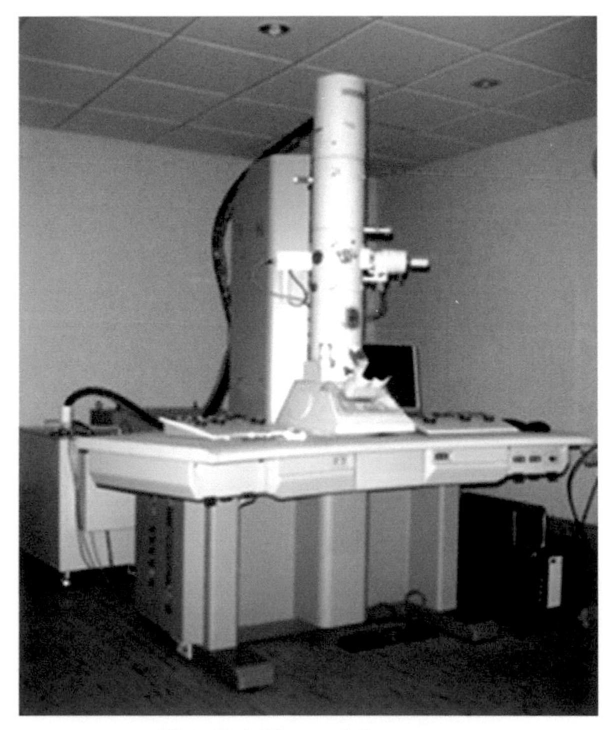

Plate 13.1 Electron Microscope.

layers. The best-known example of this shape is the Alfalfa mosaic virus-AMV (28–58 x 18 nm). The viruses consist of infective identity, the nucleic acid (5–40%) and the protein coat (60–95%). Depending on the species, the nucleic acid may be RNA (ssRNA/dsRNA) or DNA (ssDNA/ds DNA). The nucleus acid is enclosed in a protein coat called capsid. The protein coat is divided into sub-units and each sub-unit is called capsomere. The capsid enclosing the nucleic acid is referred to as nucleocapsid. The complete virus particle is called a virion, which could either be a nucleocapsid or may have other components, such as lipid envelope/enzyme/structural proteins. The whole component of nucleic acid is known as the genome. The virus genome has either two components (tobraviruses and comoviruses), three parts (cucumoviruses and hordeiviruses) or twelve parts (phytoreoviruses). The genome of viruses is of five types:

i) Monopartite: *Dianthovirus; Caulimovirus; Phytoreovirus; Carmovirus; Luteovirus; Potexvirus; Capillovirus; Carlavirus; Potyvirus; Closterovirus* (ten genera);
ii) Bipartite: *Comovirus; Alphacryptovirus; Betacryptovirus; Tobravirus; Furavirus* (five genera);
iii) Tripartite: *Cucumovirus; Ilarvirus; Pomovirus; Begomovirus; Varicosavirus* (five genera);
iv) Quadruplicate: *Alfamovirus; Ormiavirus; Benyvirus* (three genera); and
v) Pentapartite: *Tenuivirus* (only genus).

The nucleic acid of plant viruses exists in two forms, RNA and DNA, and it is either single stranded or double stranded and occurs in rings. The ssRNA may be positive or negative. If the nucleotide sequence to code protein of RNA functions in a similar manner as messenger RNA (mRNA), it is a positive genome but if the nucleotide sequence for coding protein is complementary to mRNA, it is considered as a negative genome. Since the RNA viruses are labile to mutation, changes in pathogenicity, host range or transmission manner can occur at any stage. The protein coat protects the genome from degradation, controls movements within the cell and transmission of the pathogen from one host plant to another plant. Besides nucleic acids and proteins, viruses also contain a small amount of carbon, nitrogen, sulphur, phosphorus and ash (metals). In addition to nucleic acid and protein, the complex particles also contain 20% lipid. Likewise, traces of enzymes, polyamines and metal ions can also be found in viruses. There are marked variations in the contents of nucleic acids in different plant viruses but the role of these differential contents is yet to be ascertained. Highly diverse contents have been recorded in bacilliform viruses of genera *Alfamovirus* (16%), *Nucleorhabdovirus* (1–5%), *Cytorrhabdovirus* (5%), and *Ourmiavirus* (15–25%). Besides the nucleic acids, the diversity has also been noted in contents of lipid; 15–37% in nucleorhabdoviruses and

25% in *Cytorrhabdoviruses*. While the lipid content is altogether missing in *Alfamoviruses* and *Ourmiaviruses*. The isometric viruses also differ in NA content, fluctuating between 14–24 per cent. Higher content has been found in viruses belonging to genera *Tombusvirus* (28%), *Fabavirus* (35%), *Comovirus* (38%), *Tymovirus* (39%), *Sequivirus* (40%), *Oryzavirus* (42%) and *Nepovirus* (46%) (ICTVdB, 2006; Fauquet et al., 2005; Hull, 2002; Brunt et al., 1996).

Nomenclature and classification of viruses: The viruses are known to produce different symptoms on different host plants. The diseases are named taking into consideration the host and symptoms; TMV contains the name of the host (tobacco) and mosaic a kind of symptom. Soon it was found that a plant is attacked by 2 or more viruses such as TMV, Tobacco leaf curl, and TRV and under such a situation it became an arduous task to name them. It was further refined by taking into account the modes of transmission in the naming of viruses. It was FO Holmes (Holmes, 1939) who proposed a traditional binomial system of naming these organisms which takes into consideration the symptomatology. It was only in 1962 that a provisional committee was constituted in Moscow, called the International Committee on Nomenclature of Viruses, and assigned the job taking into consideration the morphological and biophysical properties. The committee earlier proposed 16 groups (Gurr et al., 2015) which later on increased to 20 in 1975 (Shepherd et al., 1975). The concept of virus cryptogram also came into being, taking into consideration the four pairs of characteristics (nucleic acid/strands; molecular weight of NA/per cent content; particle shape-spherical/elongated/elongated with round ends/complex type; host-plant/fungus/invertebrate/vertebrate) (Gibbs et al., 1966). Accordingly, cryptogram of Tobacco mosaic virus (R/1:2/5:E/E:S/O,Se) in which R stands for NA (RNA/DNA), 1 for RNA strand (single or double stranded), 2 for molecular weight of NA, 5 for NA (%), E for shape of particle (spherical/elongated), another E for category of particle (Isometric/bacilliform/helical/complex), Similarly, cryptogram for AMV-R/1:.1/16 + 0.8/16 + 0.3/16:U/U: S/Ap (U for elongated particle with round ends, S for seed plant, Ap for aphid) (*Alfamovirus*), BrMV-R/1:1.2/22 + 1.0/22 + 0.8 + 0.3/22: S/S: S/Cl (S for spherical: S for seed plant, Cl for beetle) (*Bromovirus*), CMV-R/1:1.1/18 + 1.0/18 + 0.7 + 0.3/18: S/S: S/Ap (*Cucumovirus*), CaMV-D/2:4/16:S/S:S/Ap,/Cl (*Caulimovirus*), CPMV Carnation ring spot virus-CRSV R/1:1.5 + 0.5/20.5: S/S: S/Ne (Ne for nematodes) (*Dianthusvirus*), Pea enation mosaic virus-PeMV R/1:1.7/29 + 1.4/29: S/S: S/Ap (*Enamovirus*), Barley stripe mosaic virus-BSV R/1:1.4/4 + 1.2/4 + 1.1/4:E/E:S/O (O for no vector) (*Hordeivirus*), Tobacco ring spot virus-TRSV R/1:2.4/42 + 1.4–2.2/27–40 + 2 x 1.4/46:S/S:S/Ne (*Nepovirus*), TNV-R/1:1.5/19:S/S:S/Fu (Fu for fungus) (*Necrovirus*), Potato virus X-PVX-R/1:2.1/6:E/E:S/O (*Potexvirus*), Potato virus Y-PVY-R/1:3.5/5:E/E:S/Ap (*Potyvirus*), Southern Bean mosaic virus-SBMV-R/1:1.4/21:S/S::S/Cl

(*Sobemovirus*), Tomato bushy stunt virus-TBSV R/1:1.5/17:S/S:S/* means no information regarding transmission (*Tombusvirus*), Tobacco rattle virus TRV-R/1:2.3/5: + 0.6–1.3/5:E/E: S/Ne (*Tobravirus*), Clover wound tumor virus-CWTV-R/1:16–19/11–22:S/S:1/I/Au (Reovirus).

Based on this classification, twenty-four groups have been named (Matthews, 1979; Fenner, 1976). The classification is the act of naming, describing and placing viruses into taxa. Accordingly, the committee classed these viruses under the phylum Vira, which was divided into two sub-phyla, based on type of nucleic acid viz Deoxyvira (containing DNA viruses) and Ribovira (containing RNA viruses). The sub-phylum Deoxyvira contained three classes, namely Deoxybinata (order: Urovirales – family: Rhabdoviridae), Deoxhelica (order: Chitovirale – family: Poxyviridae) and Deoxycubica (order: Piplovirales – family: Herpesviridae; order: Haplovirales – families: Iridoviridae, Adenoviridae, Papiloviridae, Paroviridae and Microviridae). Ribovira is divided into two classes namely Ribocubica (order: Togovirales – family: Arboviridae; order: Limvirales – families: Nepoviridae and Reoviridae). Similarly, Ribohelica is divided into two orders namely Sagovirales (Stomatoviridae, Paremyxoviridae, and Myxoviridae) and Rhabdovirales (Mesoviridae and Peptoviridae in sub-order Flexviridales, and Pachyviridae, Protoviridae and Polichoviridae in sub-order Rigidovirales). Another classification based on the nature of the genome and the relationship of genome and mRNA was given by David Baltimore in 1971. According to this classification, the viruses are grouped into the following categories: double stranded DNA, single stranded DNA, reverse Transcripting viruses, double stranded RNA, negative sense single stranded RNA and positive sense single stranded RNA. To classify viruses, a committee was constituted in 1992 called the International Committee on Taxonomy of Viruses (ICTV). It is a Committee of the Virology Division of International Union of Microbiology Societies and consists of President, Vice-President, Secretaries, national members and life members. The sole objective of the committee is the indexing, naming and placement of viruses according to their taxonomic positions. Up to July 2016, the committee classified viruses into 7 orders, 111 families, 27 sub families, 609 genera and 2388 species. The earlier classification of viruses was based on the morphology of the virion. The earlier systems of classification of plant viruses were considered totally obsolete. The recent classification system has a taxonomic group that contains a set of characters. The levels in the system include order (-virales), family (-viridae), sub family (-virinae) and genus (virus). The species are separated by genome properties, structure, physiochemical and serological characteristics. The names are vernacular and based on the host plant and symptom of the disease (like a tomato as host and leaf curl as symptom and virus as a pathogen). These are abbreviated to ToLCV (Tomato leaf curl virus) as per norms prescribed by ICTV. The variants of virus species with unique properties (vector, symptoms host range) are designated as strains which are depicted by letter (Cucumber mosaic virus–CMV-S and CMV-M) or number (Pea enation mosaic virus-PeMV-1 and PeMV-2). The viruses with similar genome characteristics and morphology of virion are grouped in genera. The genus name like *Bromovirus* of Brome mosaic virus—BMV was given. Before the classification of viruses, it was essential to describe the structure and architecture. To describe the virions of viruses, the electron microscope was used to study their structure. After a complete examination, the characteristic features were recorded and taken into account for classification of viruses. The features include particle morphology (spherical/isometric, bacilliform and elongated-rigid/flexuous threads), genome properties (nature of genome, number of components, number of genes, translation strategy genome relatedness), biological properties (families and hosts and mode of spread), and serological properties (antigen based). The particle morphology is of several kinds, viz. elongated, bacilliform and spherical. The elongated viruses are either rigid or flexuous threads in which the width is very small and length is extremely long. The rigid elongated viruses are TMV (15 x 300 nm in size) and Barley stripe virus-BSV (20 x 10 nm in size) while elongated flexuous threads are PVX (10–13 nm x 480 nm) and CTV (10–14 x 2000 nm in size). The bacilliform is rod shaped, in which the length is 3–5 times of the width of the particle, as in Potato yellow dwarf virus-PYDV (75 x 380 nm), Wheat striate virus (65 x 270 nm) and Lettuce necrotic yellows virus (52 x 300 nm). Whereas, the viruses with isometric shape are Tobacco necrosis satellite virus (17–60 nm in diameter), WTV (60 nm in diameter) and Tomato spotted wilt virus-TSWV (70 nm in diameter). The plant infecting viruses are either RNA (with 0.3–28 kb size) or DNA (with 3–10 kb size). A few viruses are enveloped ones with slightly complex morphology/divided genome. The viruses attacking prokaryotes are either of RNA (5–9 kb size) or DNA (10–200 kb). Similarly, the viruses infecting fungi are known to contain RNA (2.3–23 kb), they are also enveloped, lack capsid and have a complex morphology and divided genome. To study the viruses, it is mandatory to isolate and purify them. The procedures to isolate and purify are briefly described. The isolation of viral RNA is the process through which the RNA is separated from protein coat, protein and lipid through the use of phenols. This chemical acts as a denaturant for proteins and inhibits nuclease. The system based on phenol-ethanol-water is used to release the RNA. In addition, the use of Sodium dodecyl sulfate (1%)/heat in the presence of sodium chloride is also employed in order to separate the NA and protein components of a plant virus. Besides these methods, guanidine hydrochloride, urea, and acetic acid calcium chloride can also be used to separate the virus components. The purification is the next step and is done through fractionation using a sucrose density gradient (equilibrium density gradient sedimentation). Subsequently, the purified material (RNA and DNA) with low infectivity due to inactivation with nuclease can be tested. The viruses replicate and pass through several stages, viz. adsorption, penetration, un-coating, viral genome replication, maturation and finally release. In adsorption, the virus attaches to the cell through glycoprotein spikes/coat protein fibers and specific adsorption is carried out by cellular receptors glycoprotein, phospholipids or glycolipids. After adsorption, there is a penetration of virion through receptor mediated endocytosis. The virus gets endocytosed, acidified and destroys the arrangement of coat protein components. The next step is un-coating of the virus particle to expose the hydrophilic sites. Further, it causes extrusion of the viral core into crystal and leads to the replication of the virus. In virulent viruses, the DNA/RNA stops the protein synthesis from disaggregating cellular polyribosomes. In naked viruses, the pre-assembled capsomeres are joined together to form empty capsids which are a precursor of virions. The assembled capsomeres form procapsid on account of reorganization. Under the enveloped ones, nucleocapsids are formed from the viral protein and nucleic acid by binding to membrane/ envelope. After this, the proteins are synthesized on cytoplasmic polysomes and readily come together to form capsid component. The proteins move to appropriate the cell membrane by displacing host proteins. Glycoprotein formation leads to the formation of the polypeptide. Finally, the maturation of highly organized viruses takes place in the cytoplasm. Tobacco mosaic virus is a well-worked rod shaped virus with helical symmetry. In the helical pattern, RNA and protein are arranged in a helix, in which nucleic acid is completely enclosed by protein units in a capsid with 3 nucleotides. The TMV particle has 6400 nucleotides (2130 units) in RNA that forms the 300 nm length of the particle. In isometric symmetry, the diameter is 20–70 nm and the sub units are icosahedral in shape (12 vertices and 20 triangular faces). Cauliflower mosaic virus-CaMV is a well-worked, typical example of this category of viruses. In isometric shape, there is one sub-unit (CaMV) but may have 2 or more in a capsid, i.e., 7 in non-enveloped phytoreoviruses and 4 in enveloped tospoviruses. In Bacillus rods, the envelope is either absent or present.

Structural morphology and chemistry of plant viruses. After purification, the viruses are defined by different structural variations. The plant viruses are grouped based on their shapes and sizes. The important categories are rod-shaped/helical shape (rigid rods/flexuous threads), spherical/cubical symmetry (isometric/polyhedral) and bacilliform/ complex symmetry; they are made up of of nucleic acid and proteins. The rod-shaped viruses (TMV and PVX/PVY) are separated into rigid rods

(300 x 15 nm) and flexuous threads (480–2000 x 10–13 nm). The rigid rods are of two kinds, i.e., rigid helical rods and flexuous threads. In these viruses, the central region is core consists of nucleic acid (5%) and it is enclosed on all sides by protein (95%) in typical TMV. In flexuous, elongated rods, the protein in the form of sub-units in the outer coat and the nucleic acids are arranged in a spiral, as in PVY. The rod-shaped viruses generally have a low content of NA and high content of protein; sub-units of protein are helically arranged with ssRNA. The length is generally long and variable among different viruses, the longest (2500 nm) is CTV. The other category is of spherical or icosahedral symmetry.

The viruses are identified by the symptoms. These symptoms resemble deficiency of nutrients or herbicide phytotoxicity. On account of this property, the distinction between the symptoms of virus and herbicide toxicity/nutrient deficiency cannot be made easily. Under such conditions, grafting or sap inoculations are done in order to identify the malady. Furthermore, methods based on serology and electron microscope are used to identify the plant viruses. The detection of viral pathogens is also done through the indicator plants. The important indicator plants are *Chenopodium amranticolor, Datura stramonium* (TMV-produces local necrotic lesions on both), *Nicotiana glutinosa* (local necrotic lesions of TMV) and *Nicotiana tabacum* (systemic infection of TMV). Similarly, PVX produces local necrotic lesions on *C. amranticolor, Gomphrena globosa*, and *D. stramonium* in addition to systemic infection in the latter. Besides, PVX is known to cause systemic mosaic and necrosis on *N. glutinosa* and local ring spot lesions, systemic ring spot, and necrosis on *N. tabacum.*

The virus symptoms are expressed as a result of infection by a virus in cells and cellular organelles, subsequently extending to tissues. There are viruses which do not express symptoms in plants (symptomless carriers) and the infectivity is known only after inoculation of a susceptible host of the virus. On replication of the virus in cells, symptoms are expressed as yellow chlorotic/black/brown lesions at or near the point of infection on account of a hypersensitive reaction. The improper infection can be known using staining with iodine that extends further to adjoining cells and ultimately to the phloem. The mechanism in operation leads to several effects on the structure and pigmentation of chloroplasts during replication. In addition, reduced production of vesicles, grana, chloroplasts ribosomes, ribulose biphosphate carboxylase (responsible for the conversion of carbon dioxide into organic carbon), starch accumulation/transport, alteration in photosynthetic pathway, aggregation/fragmentation of chloroplasts and changes in chlorophyll and carotene content is common. The symptoms start with vein clearing and are followed by vein yellowing, vein banding, and mosaic. It further leads to the development of vesicles in mitochondria and disintegration of the nucleus. All these effects are responsible for increased respiration, reduced carbon fixation, and net assimilation rate. The changes at the cellular level set thing in motion for changes in the tissues. These changes may be in the form of necrosis, hyperplasia, and hypoplasia and the symptoms are expressed as yellowing, reddening, striations, mottling, ring spots, interveinal chlorosis, rolling/distortion of leaf lamina, galls/enations, color break in flowers, stem pitting and grooving, fruit mottling/small fruits. In phloem infection, the pathogen destroys sieve tubes and the productivity is reduced. The reduction in productivity is caused by reduced photosynthesis, translocation of starch, stomatal opening, nitrogen fixation and nutrient uptake. The reduction in the aforementioned processes leads to epinasty, leaf rolling, sinesense, premature shedding of foliage and gall formation. Lethal effects like quick decline due to CTV in citrus and foliar decayvirus-CFDV in coconuts are also noted. In addition, mixed infection of PVX and PVY in potato is responsible for severe damage to the crop. The recent details of known families and genera are presented (Fauquet et al., 2005). The families include:

i) BROMOVIRIDAE

Alfamovirus: Alfalfa mosaic virus – AMV/ssRNA:

Ilarvirus: Tobacco streak virus-TSV/ssRNA

Bromovirus: Brome mosaic virus–BMV/ssRNA

Cucumovirus: Cucumber mosaic virus-CMV/ssRNA

Oleavirus: Olive latent virus–2–OLV-2/ssRNA

ii) BUNYAVIRIDAE

Tospovirus: Tomato spotted wilt virus-TSWV/ssRNA

iii) GEMINIVIRIDAE

Begomovirus: African cassava mosaic virus–ACMV/ssDNA

Curtovirus: Beet curly top virus–BCTV/ssDNA

Mastrevirus: Maize streak virus–MSV/ssDNA

Topocuvirus: Tomato pseudo curly top virus–TPCTV/ssDNA

iv) CLOSTEROVIRIDAE

Crinivirus: Lettuce infectious yellows virus–ssRNA/LIYV

Closterovirus: Beet yellows virus–ssRNA/BYV

Ampelovirus: Grapevine leaf roll virus–3–ssRNA/GLRaV-3

v) CAULIMOVIRIDAE

Badnavirus: Banana streak virus–dsDNA/BSV

Soymovirus: Soybean chlorotic mottle virus–dsDNA/SbCMV

Caulimovirus: Cauliflower mosaic virus–dsDNA/CaMV

Cavemovirus: Cassava vein mosaic virus–dsDNA/CaVMV

Petuvirus: Petunia vein clearing virus-dsDNA/PVCV
Tungrovirus: Rice tungro bacilliform virus–dsDNA/RTBV

vi) COMOVIRIDAE

Comovirus: Cowpea mosaic virus–ssRNA/CPMV
Nepovirus: Tobacco ring spot virus–1–ssRNA/TRSV
Fabavirus: Broad bean wilt virus–1–ssRNA/BBWV-1

vii) FLEXIVIRIDAE

Carlavirus: Carnation latent virus–ssRNA/CLV
Capillovirus: Apple stem grooving virus–ssRNA/ASGV
Foveavirus: Apple stem pitting virus-ssRNA/ASPV
Potexvirus: Potato virus X—ssRNA/PVX
Trichovirus: Apple chlorotic leaf spot virus–ssRNA/ACLSV
Allexivirus: Shallot virus X–ssRNA/ShVX
Vitivirus: Grapevine virus–AGVA
Mandarivirus: Indian citrus ring spot virus–ssRNA/ICRSPV

viii) LUTEOVIRIDAE

Luteovirus: Barley yellow dwarf virus–MAV-ssRNA/BYDV-MAV
Polerovirus: Potato leaf roll virus–ssRNA/PLRV
Enamovirus: Pea enation mosaic virus–1–ssRNA/PEMV-1

ix) METAVIRIDAE

Metavirus: Arabidopsis thaliana athila virus–ssRNAs/AthAthV

x) NANOVIRIDAE

Babuvirus: Banana bunchy top virus–ssDNA/BBTV
Nanovirus: Faba beans necrotic yellows virus–ssDNA/FBNYV

xi) POTYVIRIDAE

Potyvirus: Potato virus Y–PVY/ssRNA
Ipomovirus: Sweet potato mild mosaic virus–ssRNA/SpMMV
Rymovirus: Rye grass mosaic virus–ssRNA/RGMV
Bymovirus: Barley mild mosaic virus–ssRNA/BaMMV
Tritimovirus: Wheat streak mosaic virus–ssRNA/WSMV

xii) PARTITIVIRIDAE

Alphacryptovirus: White clover cryptic virus–1–dsRNA/(WCCV-1)
Betacryptovirus: White clover cryptic virus–2–dsRNA/(WCCV-2)
Endomavirus: Oryza rufipogon endomavirus dsRNA/(ORV)

Table 13.1 The detail of kind and characteristic features of plant pathogens vectored through different taxa.

Sl. No	Pathogen	Characteristic features
1	Virus (Kingdom: Vira; Order: Virales; Family: Viridae; Genus: Virus)	• Sub-microscopic entities • Contain only one type of nucleic acid (RNA/DNA) • Make use of own ribosomes and lack lipman system of energy release • Obligate parasites and plant pathogenic (e.g., Tobacco mosaic virus; Potato virus Y; Tomato spotted wilt virus)
2	Satellite virus (Sub-viral agent)	• Defective viruses • Dependence on other virus for replication • Contain only one nucleic acid • Particle contains lipid layer (e.g., Maize white line mosaic satellite virus, Tobacco mosaic satellite virus)
3	Viroid (Sub-viral agent)	• Virus without protein coat • Replication in nucleus and chloroplasts of cells • Transmission through seed, pollen and insect vectors (e.g., Potato spindle tuber viroid, Chrysanthemum stunt viroid, Coconut cadang-cadang viroid)
4	Virusoids (Sub-viral agent)	• Also sometimes called satellite viruses • Contain single stranded, circular RNA • Dependence on other viruses for replication and encapsidation • Large number of nucleotides in genome • No encoding of protein (e.g., Maize white line satellite virus, Tobacco necrosis satellite virus, Tobacco mosaic satellite virus)
5	Virophage (Sub-viral agent)	• Great resemblance to satellite viruses • It is a double stranded virus • Encodes it's on protein (e.g., Sputnik virophage, Zamilon virophage, Mavirus virophage)
6	Prions/Slow viruses (Sub-viral agent)	• Infective proteinaceous particles • Absence of nucleic acid • Great resemblance to virus but is much smaller (680–1600 kb) in size • Infectious like viruses (e.g., causes diseases in human being such as Creutzfeldt–vCJD)
7	Bacteria	• Single celled prokaryotes • Size in the range of 0.5–5.0 micron • Absence of crisp nucleus, mitochondria, and chloroplasts • Reproduction is through binary fission • Single circular DNA chromosome in cytoplasm (e.g., *Erwinia, Pseudomonas, Agrobacterium, Pantoea*)
8	Fungi	• A unicellular/multicellular Eukaryote • Parasitic/saprophytic in nature • Spore formers • Chitin is the main constituent of cell wall (e.g., *Fusarium, Verticillium, Sclerotinia, Rhizoctonia*)

Table 13.1 contd. ...

...Table 13.1 contd.

Sl. No	Pathogen	Characteristic features
9	Protozoa	• Eukaryote, single celled, non-photosynthetic protists • Vascular nucleus is enclosed in a membrane • Surrounded by outer layer • Parasite/predator on bacteria algae and plants (e.g., *Entamoeba, Bilantidium*)
10	Nematodes	• Animals like threads and taper from both ends of body • Animals with resemblance to insects as both shed exuviae (e.g., Root knot nematode)
11	Mycoplasma (Kingdom: Bacteria; division: Firmicutes; Class: Mollicutes; Order: Mycoplasmateles; Family: Mycoplasmataceae; genus: *Mycoplasma*)	• Members of a class Mollicutes and order Mycoplasmatales • Gram negative, unicellular, non-motile prokaryotes • Bacteria lacking cell wall • Cells coverage with rigid triple layer enriched with protein and sterol • DNA is located in cytoplasm • Parasites of plants • Characteristic disease symptoms are yellowish leaves, phyllody inflorescence and witches broom type plants (e.g., Mulberry dwarf; Sesamum phyllody; Little leaf of brinjal; Sandal spike)
12	*Phytoplasma* (Genus of Class: Mollicutes)	• A genus of class Mollicutes • Nonculturable • Phloem-borne • Candidatus (Ca) as prefix present before the name, e.g., *Candidatus Phytoplasma allocasuarinae* (e.g., Aster yellows; Rubus stunt; Little cherry disease of cherry; Elm disease; Black currant reversion)
13	*Spiroplasma* (Genus of Class: Mollicutes)	• A genus of a class mollicutes • Helical/spiral shape organisms • Movement of organism is in cork-screw mode • Present in the gut and hemolymph of insect vector • Growth at 30°C is optimum (e.g., *Spiroplasma citri* (citrus stubborn disease), *Spiroplasma kunkelii* (Corn stunt disease))
14	*Rickettsia* (Order: Rickettsiales)	• Bacteria belonging to order Rickettsiales • Gram negative, non-motile, spore-forming bacteria • Non-culturable on artificial media • Sensitive to antibiotics (e.g., Papaya bunchy tops, Clover club leaf, Pierce disease of grapevine, Beet latent rosette disease)

Sub group-3. (Circular satellite RNA, s or Virusoids) has Arabis mosaic virus small satellite RNA; Cereal dwarf virus RPV satellite RNA; Chicory yellow mottle virus satellite; Lucerne transient streak virus satellite;

(CCD) in the form of a photograph that can be viewed on a computer. For this purpose, thin sections of specimens are embedded in polymer resin to be viewed under a microscope.

13.3.2 Scanning electron microscope (SEM)

In this microscope, the electric beam from the electric gun is used to scan the specimen from three directions and produce the image on a gray background. Images of different background colour can also be achieved with the attachment of special software that is now widely available. The colour-coded signal is obtained with the use of energy dispersion X-ray spectrophotoscopy (EDXS) and cathode luminescence microscope (CM).

13.3.3 Reflection electron microscope (REM)

In this kind of microscope, a beam of scattered electrons, coupled with reflection of high electron diffraction and reflection of high energy loss spectroscopy, is captured.

13.3.4 Scanning transmission electron microscope (STEM)

It is an electron microscope, in which the action is focused prior to the bombardment of specimens with electrons and after the transmission of the electron microscope. The information generated is recorded in series rather than in a parallel pattern. The scanning is often done with the transmission electron microscope so as to obtain results from both transmission electron microscope and scan transmission electron microscope.

13.4 Preparation of Specimen for Examination

In order to attain a clear image of the sample, it is important to prepare the sample properly. The samples are prepared for microscopy of pure particles and sections of tissues. The electron microscopy of a purified suspension of particles has several steps, viz. removal of small impurities and drying of virus particles for compatibility within the microscope (this can be done using critical point method and freeze-drying method). The freeze-drying method is extremely useful for the preservation of large specimens for low-resolution electron microscopy. However, for high-resolution microscopy of particles, electric contrast is introduced in the specimens. The contrast can be increased by staining the samples. The best method to increase the contrast is to provide shades with the metal of platinum or uranium; using this method produces three-dimensional shapes of specimens. Metals do not require any kind of preparation but non-metals require a proper sample. The non-metals need to be coated with special coating material called a "squatter coater" (e.g., gold). Gold is acquired through the use of an electrical field and argon gas. The electrical field dislodges an electron from the argon, resulting in a purely +vely charged ion attaching to gold foil (–vely charged). After settling on the gold, the argon gas expels gold atoms which fall onto the specimen, covering it with a thin coating. The steps followed to prepare the sample to view under the microscope are given in the next paragraphs. The water vapors obstruct the movement of electrons in the vacuum; therefore, it is essential to dehydrate the sample so as to remove water from it. Earlier scanning microscopes required a vacuum, but contemporary ones do not since they use energy dispersive x-ray spectroscopy (EDS).

Step 1. The fixing of the specimen in chemicals is done by the cross-linking of protein and lipids with aldehydes and osmium tetroxides, respectively.

Step 2. The mixing of specimens is done in transparent solutions, such as ammonium molybdate or phosphotungstic acid, for negative staining. After the thorough drying, the use of blotting paper is essential.

Step 3. The cryofixation is done by freezing the specimen in liquid ethane maintained in liquid nitrogen and at liquid helium temperature.

Step 4. After freezing, the dehydration is done using ethanol or acetone solvents followed by drying and infiltrations in embedded resins.

Step 5. After completing this step, the specimens are embedded directly into water-miscible resin, such as acrylic, and cut into sections with a microtome.

Step 6. After preparation of thin sections, the embedded specimens are put to ultrafine abrasives for polishing with metal.

Step 7. Surface shadowing is done with metals, such as platinum and a mixture of Carron-platinum. This is followed by removal of the specimen and washing with acid using enzymes and mechanical separator. The specimen is now ready for examination.

Step 8. The specimens can be stained using uranyl acetate followed by lead citrate. These specimens require immediate freezing in cryofixation.

Step 9. It is advisable to prepare thin sections of the sample using microtome until the sample looks transparent.

Step 10. It is also important to apply an ultra-thin coating so as to prevent static electric field concentration at the specimen.

Step 11. The earthing up is done using an electrically conductive adhesive to prevent charge concentration on a sample by connecting electrically to the metal.

...Table 14.1 contd.

S. No	Virus	Vectors	Effect(s)	Source
11	Maize mosaic virus (MMV)	Planthopper, *Peregrinus maidis*	Production of higher population of macropterous over brachypterous	Clesson et al., 2013
12	Rice stripe virus (RSV)	Planthopper, *Laodelphax striatellus*	Reduced egg development of some isolates (Ls-dorsal and Ls-CPO)	Li et al., 2015
13	Rice stripe virus (RSV)	*Laodelphax striatellus*	Affected fecundity & development	Wan et al., 2015
14	Tomato spotted wilt virus (TSWV)	*Frankliniella occidentalis; Thrips tabaci*	Higher death rate of *T. tabaci* and lower of *F. occidentalis* on healthy plants	Stumpf and Kennedy, 2007; Inoue and Sakurai, 2006
15	Tomato spotted wilt virus (TSWV)	*Frankliniella fusca*	Enhances fecundity of vector	Sherestha et al., 2012
16	Western X disease (strain–peach yellow leaf roll)/(Mollicute)	*Colladonus montanus, leafhopper*	Reduction in longevity of vector	Jensen, 1959
17	*Candidatus liberibacter solanacearum* (*Ca L. solanacearum*)	Potato psylla, *Bactericera cockerelli*	Enhances preference towards healthy plants	Prager et al., 2015
18	Citrus greening (*Candidatus liberibacter asiaticus*)	Citrus psylla, *Diaphorina citri*	Modification in dispersal behavior, flight activity & sexual attraction	Martini et al., 2015
19	Nematode (Entomogenous) *F. occidentalis*	*Thrips tabaci* (Vector of TSWV)	Alter behavior of thrips	Sims et al., 2009
20	Rust fungi (*punctiformis punctiformis*)	*Aphis fabae* ssp. *cirsiiacanthoidis* and *Uroleucon cirsii; Cassida rubiginosa* (Beetle)	Enhances population of aphids but is detrimental to beetles	Kluth et al., 2002
21	Bacterial wilt of cucurbits (*Erwinia tracheiphila*)	*Maxican beetle, Epilachna varivestis, Acalyma vittatum*	Higher spread of viruses (Soybean mottle virus and Bean pod mottle virus) on diseased plants	Musser et al., 2003

22	Maize bushy stunt phytoplasma	*Dalbulus maidis*, leafhopper	Increases fecundity and survival on diseased host	Madden et al., 1984
23	*Flavescence doree phytoplasma*	*Scaphoideus titanus*	Preference for diseased plants	Bressan et al., 2005
24	*Spiroplasma citri*	*Circulifer tennellus*	Increases birth rate of vector on diseased	De Almeida et al., 2008
25	Corn stunt spiroplasma	*Dalbulus maidis*	No effect on *D. maidis*; Enhances the fecundity & survival of *D. elimatus* & *D. gelbus* vectors	Beanland et al., 2000
26	Tomato leaf curl virus	*Bemisia tabaci*	Reduction in fecundity and longevity of vector	Butter and Rataul, 1977

antagonistic to beetles as the diseased plants considerably reduced the development rate and increased the mortality of beetles (Kluth et al., 2002).

14.1.6 Hemiptera and bacteria

Zebra chip of potato is caused by a bacterium *Candidatus liberibacter solanacearum* and spread by potato psyllid *Bactericera cockerelli*. The studies were carried out to examine the effect of bacterium on its vector. It was found that bacterium significantly reduced the fecundity of female and survival of nymphs as compared to the non-vector population (Nachappa et al., 2012).

14.1.7 Coleoptera and bacteria

The *Cucurbita pepo* is attacked by the plant pathogen *Erwinia tracheiphila*, causing bacterial wilt, vectored by Mexican bean beetle *Epilachna varivestis* (Musser et al., 2003) and striped beetle *Acalymma vittatum*. Zucchini suffers from Zucchini yellow mosaic virus spread by aphids. Similarly, *Phaseolus vulgaris* is a host of Soybean mottle virus (SoMoV) and Bean pod mottle virus (BPMV) transmitted by Mexican bean beetle. This beetle spreads the viruses (SbMV and BPMV) more efficiently once fed on plants suffering from bacterial wilt disease. The bacterium is instrumental in bringing about changes in the behavior of Mexican bean beetle. The effects of these two pathogens were analysed, with a focus on their spread in nature. The beetle-infested plants with bacterial wilt disease were triggered to emit vegetative and floral odors. These odors enhanced the attraction of the beetles towards wilted plants of cucurbits. The growth rate of beetle larvae was much higher in the case of Mexican bean beetle and the larvae were heavier on plants suffering from bacterial wilt disease, compared to those on healthy cucurbits. As a result, the spread of wilt disease was higher through beetles (Shapiro et al., 2012). However, these beetles did not prefer Zucchini plants infected with viral disease.

14.1.8 Hemiptera and Mollicutes

The hemipteran vectors are also influenced while carrying *Phytoplasma/ Spiroplasma* (Mollicutes) in their bodies like the plant viruses. The Maize bushy stunt *Phytoplasma*, vectored by leafhopper *D. maidis*, significantly increased its survival and fecundity but the effect was, however, less pronounced on *D. elimatus* and *D. gelbus* at a low temperature of 20–29°C. At higher temperature, the least effect of these parameters was recorded (Madden et al., 1984). While in the case of Corn stunt *Spiroplasma*, there was no effect on *Dalbulus maids* but the survival and fecundity of the other two species (*D. elimatus* and *D. gelbus*) was greatly enhanced. Thus, the plants of maize infected by mollicute *S. kunkelii* had a favorable effect on leafhopper vector (*D. maidis*) of Mollicutes (Beanland et al., 2000; Ebbert and Nault, 1994; Moya-Raygoza and Nault, 1998). Similarly, the carrot plants infected with Aster Yellows Phytoplasma were found to be preferred by leafhopper vector over healthy carrot plants but the birth rate of *Scaphoideus titanus* was better on a diseased substratum of *Phytoplasmic* etiology (Flavescence doree *Phytoplasma*), compared to healthy plants (Bressan et al., 2005). Similar results were obtained in the case of *Circulifer tenellus,* a vector of *Spiroplasma citri* (pathogen responsible for Pseudo curly top disease of sugar beet), where the birth rate of the leafhopper was better on diseased sugar beet (de Almeida et al., 2008).

14.2 Climate/Weather Parameters and Virus-Vector and Host Interaction

In nature, four components of weather (the pathogen, insect vector, plant and the environment) are interacting with each other constantly. The environment is bound to affect the interaction of host plant, vector, and pathogen. There has been a big change in the global climate in recent years. Among the parameters of climate, temperature is the dominant factor. The climate is changing and is getting warmer at a faster rate. The average temperature has risen by 0.6°C over the last 100 years. If global warming continues at this pace, the rise in temperature would be around 6.3°C by 2100. Global warming is bound to affect both immobile flora and mobile fauna on this earth. Among these, crop growth is bound to be affected first and this will subsequently influence the herbivores that are totally dependent on crops for food. It is a fact that the major component of global warming is the rise in temperature due to increase in the concentration of CO2. The rise in temperature boosts the growth of crops harboring insect fauna, particularly the vectors of plant pathogens. Of the insect fauna, aphids are likely to be influenced the most and these are major vectors of a large number of plant viruses. In addition, the effects of weather on other insects from the sucking category, like whiteflies, leafhoppers, planthoppers, and thrips have also been recorded. As a result of global warming, the change in the biology of vectors, expressed in terms of the size of population, generations, longevity, dispersal pattern and feeding behavior is evident. Plants in the system determine the level of susceptibility of the pathogen, multiple infections of different strains, etc. The influence of various components of the environment (temperature, moisture, air currents and cultural practices) has been noted on other partners. Transmission of the pathogen from one host to other is an important link in the life cycle of the pathogen. The viruses also influence the physiology and behavior of vector. The pathogen affects the vector by altering feeding and behavior and the vector stresses on the plant show differential transmission of the pathogen. Besides, these microbes develop in the body of vector and

Trebicki P, Vandegeer RK, Bosque-Perez NA, Powell KS, Dader B, Freeman AJ, Yen AL, Fitzgerald GJ and Luck JE (2016). Virus infection mediates the effects of elevated Co2 on plants and vectors. Scientific Reports, 6: article no 22785(doi; 10. 1038/strep 22785 (2016).

Tu Z, Ling B, Xu D, Zhang M and Zhou G (2013). Effects of Southern rice black streaked dwarf virus on the development and longevity of its vector, *Sogatella furcifera*. Virology Journal, 10: 145.

Wan G, Jiang S, Wang W, Li G, Tao X, Pan W, Sword GA and Chen F (2015). Rice stripe virus counters reduced fecundity in its insect vector by modifying insect physiology primary endosymbionts and feeding behavior. Scientific Reports, 5: 12527.

Whitfield AE, Rotenberg D, Aritua V and Hogenhout SA (2011). Analysis of expressed sequence tags from Maize mosaic Rhabdovirus infected gut tissues of *Peregrinus maidis* reveals the presence of key components of insect innate immunity. Insect Molecular Biology, 20: 225–242.

Xu HX, He XC, Zheng XS, Yang YJ and Lu ZX (2014). Influence of Rice black streaked dwarf virus on the ecological fitness of non-vector planthopper, *Nilaparvata lugens* (Stal) (Hemiptera: Delphacidae). Insect Science, 21: 507–514.

Zheng X, Jhang J, Chen Y, Dong J and Jhang Z (2014). Effects of Tomato zunate spot virus infection on the development and reproduction of its vector, *Frankliniella occidentalis* (Thysanoptera:Thripidae). Florida Entomologist, 97: 549–554.

QUESTIONS (EXERCISE)

Q 1. What are the favorable and adverse effects of plant pathogens on aphids and leafhopper vectors?

Q 2. Does global warming affect the vectors of plant pathogens? Give your opinion.

Q 3. Discuss the effects of plant pathogens on their beetle vectors.

Q 4. Write an essay on the beneficial effects of plant pathogens on their vectors.

Q 5. Write short notes on:

a) Effect of pathogens on bio agents of host
b) Effect of *Phytoplasma* on their vectors

Q 6. List the pathogens affecting their host vector and discuss the effects of nematodes on thrip vectors, if any.

CHAPTER 15

Vector-Virus Management

15.1 Vector Management Strategies

The insects are distributed among 32 insect orders in phylum Arthropoda (Raccah and Fereres, 2009). The insects are known to exert both beneficial and harmful effects on mankind. The beneficial ones include production of honey by honey bees (*Apis malleifera*), lac by lac insect (*Lacifera laca*) and silk by mulberry silk worm (*Bombyx mori*). Besides, the insects also serve as an important source of food and various dyes. Another category of insects called 'parasitoids' (parasites and predators) are deployed as biological control agents to contain harmful insect pests. The parasitoids cover both parasites (*Trichogramma chilonis, Apanteles ruficrus, Bracon* sp., etc.) and predators (lady bird beetles-Coccinellids, praying mantis, green lacewing-*Chrysoperla carnea,* flies, etc.) and their use is largely as one of the components in integrated pest management. Therefore, it can be said that not all insect species are injurious to the interests of human beings. The phytophagous species of insects are the only category of insects that are against the interests of man and are, therefore, regarded as pests. The insects inflict direct damage by feeding on plants and destroying them completely. Under such a situation, it becomes almost difficult to realize an economical yield of crops under domestication by the farmer. Above all, there are species of insects known to act as vectors of dreaded plant pathogens in cultivated/ domesticated crops, such as viruses, bacteria, fungi, *Phytoplasma*, *Rickettsia*, nematodes, and protozoa. The insects are known vectors of several diseases of human beings and animals but these have not been discussed here as the topic does not fall under the purview of this book. The management of vectors of plant pathogens is difficult since a small population of the vector can spread the pathogen in throughout an entire field, particularly the plant viruses. Thus, insect vectors are much more dangerous and greater limiting factors in the successful raising of crops. To contain the damage

Insects and Pests Act of 1914 to deal with pests and diseases, (ii) Plant Parts and Seed Order of 1984 (introduced in 1985) for seed introduction and (iii) National Bureau of Plant Genetic Resources (NBPGR), New Delhi for introduction of germplasm. The basic duty of these organizations is to regulate the introduction of plant material, seed and germplasm to avoid the pests and diseases. At all these centers, the consignment is rigorously inspected and treated to ensure that it is free of any pests and diseases. The material is indexed for the presence of pests and diseases. If pest and disease-causing pathogens are detected in a consignment, it is treated with chemicals or given hot air/hot water treatment, depending on the nature of malady, in order to get rid of pathogen. Heat treatment in quarantine stations is used against nematodes, insects, mites, viruses and fungi. The temperature ranges are identified for flower bulbs (44°C for 240 minutes), Chrysanthemum (48°C for twenty-five minutes) and potato tubers (45°C for five minutes) to make consignment free from nematodes, mites and insect vectors and viral and fungal pathogens. Similarly, the temperature range for insect pests is already identified for Narcissus bulbs (44°C for 180 minutes) and strawberry runners (46°C for ten minutes). Temperatures are also known to eliminate viruses from grapevine (45°C for 180 minutes), sugarcane setts (50°C for 120 minutes) and potato tubers (50°C for seventeen minutes). The hot water treatment of celery seed (50°C for twenty-five minutes) and wheat seed (52–54°C for ten minutes) to get rid of fungi is also available. As a result of strict regulation and inspection, plant viruses Fig mosaic virus-FMV (fig plants from North America), Grapevine fan leaf virus-GFLV (grapevine plants from South Pacific Asia), Jasmine mosaic virus-JMV (Jasmine plants from Asia/Europe), Pea enation mosaic virus-PeMV (pea seed from South Pacific Asia), Orchid virus-OV (Orchids plants from Asia), Chrysanthemum yellows virus CYV (Chrysanthemum sp. plants from Europe) and Tobacco mosaic virus-TMV (Dahlia tubers from Europe and North America) have been detected. The external seed-borne infection of Lettuce mosaic virus-LMV in lettuce seed or TMV in tomato is eliminated by treating seed with hydrochloric acid or trisodium phosphate or sodium hypochlorite. Domestic regulations are enforced to check the movement of disease from areas of its presence to the new areas where the disease is absent. To check the spread of Banana bunchy tops from Kerala to other states in India, the restrictions are already enforced. Therefore, the disease is kept under control through the application of domestic regulations by preventing the movement of diseased suckers. The importance of domestic regulations to contain the menace of plant viruses has been highlighted

15.3 Use of Virus Free Seed/Elimination of Virus Sources

Mitigating the menace of virus diseases by avoiding the use of virus infected seed is paramount. There are many diseases transmissible through seed

(Cucumber mosaic virus-CMV, Cowpea mosaic virus-CPMV, etc.) and vegetative propagation (Potato virus Y-PVY, Potato leaf roll virus-PLRV, Citrus tristeza virus-CTV, Sugarcane ratoon stunt virus-SRSV, etc.). The sowing of virus-free seed eliminates the primary source of seed-borne pathogens. With the sowing of pathogen-free seed, the chances of spread of pathogens become nil, the only option left is the introduction of pathogens through the incoming vectors. To avoid the pathogen in a given area is to plant the crop at the time when the inoculum of pathogens is low. In this context, the sanitation takes care of the pathogen. To ensure the sowing of virus-free seed, indexing of seed lots is mandatory. Another measure is to have a time gap between harvesting of old crop and planting of new crop or to keep a physical gap or space between crops.

15.3.1 Virus indexing

It is the testing of seed or planting material for the presence of virus. This technique is meant to ensure the healthy state of seed. There are two options, viz. virus-free seed or virus-infected seed. The virus-free seed can be put to use while the infected seed/planting material can be treated to eliminate virus. The meristem or callus or callus-derived plantlets are tested before using them as a mother plant to produce virus-free stock. For indexing, sap testing is done to detect the presence or absence of virus. Sap inoculations are done on indicator plants to detect plant virus. Other serological methods such as Enzyme Linked Immunosorbent Assay (ELISA), Dot Immuno Binding Assay (DIBA), Tissue Blot Immuno Assay (TBIA) and Immuno Electron Assay Microscopy (IEBA) are used to detect viruses. If the indicator plants show the presence of virus, the lot is discarded; otherwise, the lot is taken up for further multiplication. The use of virus-free seed is the best remedy to manage plant viruses. In case the seed is infected with virus, the heat treatment (35–40°C) of seed or planting material is done or the infected seed can be discarded. The above temperature range inhibits the replication of most plant viruses. It is a rather successful proposal particularly in seed production programmers. It is being widely used to detect the plant viruses in seeds and vegetatively propagated plants. Besides seed-borne viruses (CPMV), regular indexing in Potato for viruses (PLRV, PVY, PVS, etc.) and fruit plants/citrus for viruses (Citrus greening, BBTV) is done. The reduced replication of viruses of potato like PVA, PVY, and Potato spindle tuber viroid is achieved by storing potato at 5–15°C, followed by meristem tip culture. The viruses are systemic in nature and all the plant parts of vegetative propagative material may carry viruses. All countries have their own certification procedures to check potato seed for the presence of potato viruses. In this context, the International seed testing association is important. Additionally, the European Plant Protection Organization (EPPO), Mediterranean Plant Protection Organization (MPPO) and North

American Plant Protection Organization (NAPPO) are also important organizations linked to indexing of viruses.

15.4 Elimination of Pathogens through Tissue Culture/ Chemicals

The best practice is to use virus-free seed or planting material so that there is no virus source available to the aerial vector to spread the pathogen further. This is generally achieved through different means. For example, various therapies (thermotherapy, chemotherapy, cryotherapy and electrotherapy) are in use. The apical meristems are usually free from virus, thus are exploited to exclude the virus infection. The tips of plants are always free from virus as the movement of virus is slower than the growth of plant; this can be used to produce virus-free seed through meristem tissue culture techniques. In this technique, the tip of plant is excised (0.1–0.5 mm in size) in aseptic condition on artificial media and incubated for 3 days at 24C± 1. It is, therefore, a technique of producing plants from cultured tissues obtained from meristems, cultured tissues, protoplast culture and chemically (cytokines for inhibiting virus multiplication) treated media. This technique can be made effective with the use of pre-treatment (heat, chemical, etc.) to eliminate viruses. It was first tried in Dahlia, in 1952, and then used in tomato (Tomato aspermy virus-TAV), lily (Lily symptomless virus-LSV), carnation (CLV and cucumber (CMV) to manage plant viruses. Aside from this, work to produce virus-free plants through tissue culture technique in citrus has been in progress since 2002.

15.4.1 Thermotherapy

It is a therapy where heat is utilized to eliminate the pathogens. Heat treatment is a method extensively used in our quarantine stations to eliminate the viral pathogens. The temperature range between 35–40°C is followed by a varying period to eliminate different plant viruses. The virus-susceptible cultivar Bebecou of apricot has been under cultivation in Greece since long ago; it was covering the areas to the tune of 90 per cent. It was highly susceptible to Plum pox virus. Thermotherapy was applied to this cultivar and proved highly useful to apricot growers. Two methods of elimination of virus were compared and, of these, the method in which the plants were kept at 30–35°C temperature for eight weeks eliminated the virus completely. By employing this method, 104 plants free of the virus were produced in four months, *in vitro*, as compared to another method where only eighteen plants were produced in six months of *in vivo* treatment (Koubouris et al., 2007). The latest method was more efficient and less time-consuming. These plants, after heat therapy, were developed from shoot cultures *in vitro*. Thus, the method involving heat treatment followed by *in vitro* shoot culture proved highly effective in Greece. Heat therapy (at 40°C temperature) is effective against Carnation latent virus-CLV in carnation (Mangal et al., 2004). Likewise, meristem culture of Chrysanthemum following heat therapy at 38–40°C for three weeks proved successful in Chrysanthemum in Indonesia for eliminating CMV (Budiarto, 2011). Furthermore, Prunus necrotic ring spot virus-PNRSV was eliminated from commonly cultivated early blue cultivar using heat therapy (38°C and 35°C for ten and twenty-nine days, respectively). CTV was a limiting factor in the successful cultivation of citrus in Pakistan. It was tackled through the use of heat therapy (keeping plants at 35/30°C for 2 weeks followed by 45/30°C for another week and then incubated at 50/40°C for one week) in order to produce virus-free plants of sweet orange, sour orange and eureka lemon (Arif et al., 2005). It eliminated the virus completely. Overall, thermotherapy has been found to be successful against viruses belonging to thirteen plant virus families (Panattoni et al., 2013). The heat treatment is commonly practiced against Chrysanthemum B virus (30–40°C), CarRSV (35–40°C), CMV in Banana (35–43°C), Gooseberry vein banding virus-GbVBV (35°C), PVX, PVY, and PVS (33–38°C). Thermotherapy through hot air treatment (37°C for three to six weeks) is also used to eliminate Alfalfa mosaic virus, Potato leaf roll virus and Tomato black ring viruses completely from potato (Kaiser, 1980). Likewise, using thermotherapy (37°C for forty-eight days) on grapevine *in vitro* completely eradicated Fan leaf grapevine virus (Panattoni and Triolo, 2010). Another disease of cassava, Cassava mosaic virus disease, is eliminated with thermotherapy (35–37°C) (Acheremu et al., 2015). Besides viruses, the diseases caused by MLO have also been managed with the use of heat therapy. Peach yellows caused by MLO has been managed through the use of heat therapy by keeping trees and root stocks (immersed in hot water) at a temperature of 35°C for two to three weeks.

15.4.2 Chemotherapy

Another method to eliminate viruses is through the use of chemicals that prevent their replication. Besides, there are chemicals, like antibiotics, that prevent translation or transcription in infected cells and analogues act as a barrier in replication of virus through the replacement of false nucleotides. Chemotherapy is achieved through the use of antiviral agents (Ribavirin, Virazole, and DTH), growth promoting chemicals (Cytokines) and antimetabolites (Azaguanine and Thiouracil). A polypeptide, an inhibitor of the virus, has been isolated from tobacco plants infected with CMV and is used on infected plants to reduce the multiplication of virus (Loebenstein, 2015). Ribavirin (Virazole) @25–100 mg/liter successfully eliminated Apple chlorotic leaf spot virus-ApCMV from Myrobalan and Prunus necrotic ring spot virus-PNRSV from express plum shoots (Cieslinska, 2007). It also proved successful in the elimination of Sugarcane mosaic virus and

subsequently changed to predict the invasion of *M. persicae, Macrosiphum euphorbiae* and *Aulacorthum solani* vectors of potato viruses (PVY; PLRV). In all the above cases, the overwintering population of aphids depends on the duration and harshness of winter conditions.

15.5.3.4 Repellents. It is on account of a property of the chemicals that the insect vectors do not feed on the substratum sprayed with chemical repellent. One such chemical (polygodial) isolated from water pepper (*Polygonum hydropiper*) is known as sesquiterpenoid and it is highly effective against Barley yellow dwarf virus transmitted by *Rhopalosiphum padi* (Griffiths et al., 1989).

15.5.3.5 Pesticides of plant origin. Many pesticides of plant origin are in use to control plant viruses. Azadirachtin was the first product of Neem tree registered with Environmental Protection Agency of the USA. It was tried against vectors (*R. padi* and *Sitobion avenae*) of plant viruses at a concentration of 500 ppm and showed promising results (West and Mordue, 1992). The conventional pesticides used against viral diseases, being slow-acting, failed to check these diseases satisfactorily. Instead, insecticides of plant origin, like Pyrethrum, with quick knock-down were relied upon. On account of some limitations, these pesticides too could not become popular. These pesticides were highly unstable under light and required frequent applications. The frequent application caused an increase in the cost of control operations. Therefore, these plant origin pesticides did not gain popularity. Subsequently, considering their level of effectiveness, new synthetic insecticides (pyrethroids) were synthesized. These chemicals were photo-stable and could, therefore, retain effectiveness for a long time without losing knock-down effectiveness. This new group of pesticides, in combination with other tactics, showed promise in their use against insect vectors.

15.5.3.6 Behavior manipulating chemicals (Pheromones). These are the chemicals secreted by an insect species to elicit a response in receiving individuals of the same species. The response is for foraging activities, mating or for defense purposes. The alarm pheromone (Beta farnesene) is produced by aphids *Myzus persicae* to indicate the entry of predators (coccinellid beetles) in an area. The pheromone is generally released in order to avoid the attack of predators in the area. All the aphids, on perceiving the pheromone, become active and fall down on the ground so as to save themselves from predator attack. The increased activity of aphid vectors in the area significantly increases the chances of spread of non-persistent type viruses. It is also pertinent to mention that the release of such chemicals makes the predators more active and that activity is responsible for effective management of viral diseases. The use of predators along with carbamate insecticide provided better control of aphid vectors of Barley yellow dwarf virus in Japan.

15.5.3.7 Baits. Baits have been in use since long ago to attract and kill mainly fruit flies using methyl eugenol; these have not been used so commonly against vectors of plant pathogens. Recently, the baits containing Manuka oil (obtained from *Leptospermum solariums*) and Phoebe oil (obtained from *Phoebe porosa*) have been found effective in monitoring the population of *X. glabratus*, the Redbay Ambrosia beetle. This beetle is instrumental in the spread of laurel lethal wilt of red bay.

15.6 Management of Vector/Invading Population of Vector

The vectors are not sessile and they cover long distances and spread plant viruses, particularly the persistent category of viruses. These viruses are not immediately inoculated by vectors as these pathogens have a latent period. The latent period lasts hours to days, depending on vector species. The vectors are carried through air currents and appear suddenly in a field. Repeated applications of pesticides are required to check the invading population of vectors. To manage invading arthropod vectors, previously planned and immediate strategies are mandatory to check vectors of plant viruses. The strategies have been mentioned and discussed in following paragraphs.

15.6.1 Resistant varieties/cultivars

The plant resistance is the relative amount of heritable qualities possessed by the plant that ultimately decides the degree of damage done by insect vectors that would occur in their absence. The plant resistance has different components, viz. non-preference, antibiosis and tolerance. These parameters are under the genetic control. Besides this, there is a type of resistance called ecological resistance. It is normally the susceptible plants which, under a certain set of conditions, behave like resistant ones. The ecological resistance is also known as "induced resistance" and has been exploited in the management of vector population. Induced resistance has been utilized through cross protection and the development of transgenic plants to mitigate the problems of vectors of plant pathogens. The host plant resistance is the most important, economical and effective means to check plant viruses. This method is a base on which any kind of method can be used to realize great success. This method is highly compatible with almost all the available tactics of pest management without any ill effects. It is one of the most important tactics by which the population of the vector can be drastically reduced on vector-resistant cultivar and the pathogen-resistant/ immune cultivar prevents the availability of virus source from the field. To develop such a strategy, the first step is to locate/identify bases of resistance to insect vector. To quote an example, the basis of resistance to whitefly in cotton crop has been identified. These bases of resistance identified include morphological (Butter and Vir, 1989), biochemical (Butter et al., 1992a) and nutritional bases (Butter et al., 1996). Taking in to account all these factors, the resistant varieties can be developed to mitigate the problems of cotton whitefly, the major emphasis should be given to developing pathogen resistant cultivars rather than to vectors. The development of variety involves the modification of physiological characters making them unfavorable for the vector species for feeding, oviposition or shelter. The resistance could be antibiosic, affecting the life processes of the insect vector through the ingestion of toxic chemicals. It has been seen that some lines of tobacco did not suffer from the infection of CMV, TMV, and TNV because of thickened leaf lamina or nature of glandular trichrome of leaf and there was no replication of viruses in those lines. A genetically modified cultivar of papaya has been developed in order to manage Papaya ring spot virus.

15.6.2 Cross protection

In the phenomenon of cross protection, the introduction of mild strain/isolate/closely related virus protects the plant against severe strain/isolate/challenge virus. The phenomenon of cross protection was first investigated as early as 1929, in the case of Tobacco mosaic virus, but now it has been extended to several crops/viruses such as Potato virus X, Potato leaf roll virus in potato and Citrus tristeza in citrus; also many more RNA and DNA viruses and viroids (Pennazio et al., 2001; Gal-On and Shaboleh, 2006). The mechanism involved until 1970 was based on the hypothesis that the mild strain produces the antibodies, erodes the essential metabolites, occupies all the required sites for multiplication, and specifically adsorbs new cell compounds against the entry of severe strain (Zhou and Zhou, 2012; Pennazio et al., 2001). With the advancement of science, the basis of cross protection was further investigated. This consisted of the prevention of the replication of severe strain or challenge virus through blockade of the initial translation of the invading viral RNA, preventing the transcription of viral nucleic acid of the incoming virus and halting the cell-to-cell movement. In the phenomenon of cross protection, two isolates (type P and type W) were identified. Of these two isolates, type P infected papaya and other plants of family Cucurbitaceae, while type W did not infect papaya but could infect other members of cucurbits. Cross protection is successful in Apple mosaic virus in apple (New Zealand), Passion fruit woodiness in passion fruit (Australia), Zucchini yellow mosaic virus-ZYMV in squash/melon/watermelon (Israel), Cocoa swollen shoot in cocoa (West Africa), Tomato mosaic virus-ToMV in tomato and pepper (China) as well as in Papaya ring spot virus in papaya (USA, Taiwan, Thailand). It is a new method and must be used with extra care as it has limitations too. The introduction of mild strain itself is known to reduce some yield of the crop, serve as virus source in the area and both strains may produce more deadly disease.

15.6.3 Use of transgenics

The plant viruses are a limiting factor in the successful cultivation of agricultural crops; their spread in nature is through various vectors. It is a laborious task to manage vector population in the field. To mitigate the vector problem, it is advisable to tackle such problems using various tactics. These tactics are applied to eradicate the diseased plants so as to check secondary spread through aerial vectors, to control vector behaviour and to breed plants for resistance against plant pathogens. The production of genetically modified plants is the only solution to mitigate the problems of such category of pests. The genetic sources are not available in all plant species. Moreover, the resistance gets broken due the high rate of mutation. To overcome these problems, RNA silencing is the mechanism which was exploited to develop transgenics (Baulcombe, 1996). Furthermore, the genetically modified plants are now developed based on virus genes of coat protein and movement proteins to inhibit the replication of viruses. The transgenic plants are now available in several countries to tackle plant-pathogenic viruses. The important ones includes tomato in The Netherlands and UK (against TMV) and tomato/pepper in Japan, France, USA, China (against TMV), pepper in China (against CMV), cocoa in West Africa (against Swollen shoot of cocoa), squash melon/water melon in Israel (against ZYMV), apple in New Zealand (against Apple mosaic virus-ApMV), passion fruit in Australia (against Passion fruit woodiness virus-PFWV), papaya in Taiwan/Hawaii/Thailand/Mexico/Florida (against Papaya ring spot virus-PRSV) and citrus in South Africa/Australia/Peru (against CTV).

15.6.4 Use of para-transgenics

The arthropods carrying genetically modified bacteria to prevent the transmission of plant pathogens are referred to as para-transgenic. This technique came into existence only on account of limitations of chemical control and this method was exploited to manage *Phytoplasmas* transmitted through arthropods (Weintraub, 2007). The para-transgenic bacteria are successful if they occupy the same organs as pathogens and replicate and spread rapidly throughout the body. Two bacteria were identified, cultured, modified and reintroduced into leafhopper vectors *Alcaligenes xylosoxidans*, sub-species *denitrificans* (a modified symbiont in *H. coagulata* sharpshooter) (Bextine et al., 2005), and *Cardinium hertigii* (a modified bacteria in leafhopper *Sogatella titanus)* (Bigliardi et al., 2006) to manage *Xylella fastidiosa* and FD pathogen, respectively. The *Phytoplasma* vectors

15.7.5 Inter cropping/mixed crops

To cover the risk of adversity of nature, two or more crops are planted and mixed with the main crop. This practice is quite common in minimizing the incidence of non-persistent plant viruses. As the viruliferous vector, particularly aphids, feed for a little longer on the non-host inter-crop, they lose the virus since it is of the non-persistent category. The inter-cropping of pigeon pea with gingelly in the ratios of 1:6 is a practice to contain Phyllody virus of pigeon pea. The intercropping of tomato with cucumber in Jordan is followed to mitigate the Tomato yellow leaf curl disease. The cucumber, being the preferred host of *Bemisia tabaci* and non-host of Tomato yellow leaf curl virus, is planted about one month earlier so that whitefly should remain confined to cucumber only where these whiteflies are controlled through the use of pesticides. The mixed crops (3 rows of maize/sorghum/pearl millet) around/between the main crop (black gram/green gram) is another important method to keep the main crop free from Yellow mosaic virus diseases. Likewise, planting 3 rows of barley in cauliflower and sugar beet is a good practice to contain two important non-persistent plant virus diseases of cauliflower (Cucumber mosaic virus) and sugar beet (Beet yellows virus), respectively. The trap crop of tomato in onion or garlic in Egypt (Afifi and Hayder, 1990) and mixed cropping of carrot and onion in England (Uvah and Coaker, 1984) are instrumental in reducing the population of thrips to the tune of 80% and 20%, respectively.

15.7.6 Use of trap crops

The yellow, sticky polythene sheets erected vertically, windward side of red pepper, are instrumental in checking the diseases Potato virus Y and Cucumber mosaic virus in Israel. In addition, to protect the potato crop from Potato leaf roll virus (aphid-borne) and green gram/black gram from Yellow mosaic virus (whitefly-borne) diseases, the use of sticky traps is also quite common. The mulches of non-hosts are used to keep the viral diseases under control. Aluminum strips/gray or white plastic sheets are used extensively in Israel to get rid of aphid vectors and control CMV and PVY in red pepper. These mulches are also used in Imperial Valley of California to check Water melon mosaic virus-WMMV in summer squash. The reflection of these mulches is instrumental in the killing of whiteflies due to heating effect. The only drawback in the use of these mulches is that after some time the yellow colour becomes faded and fails to attract the whiteflies. In this situation, the mulches need to be replaced, but this can be costly. Therefore, straw mulches are preferred over these mulches. The use of straw mulches is recommended to control Tomato yellow leaf curl virus-TYLCV transmissible by whitefly in tomato in Israel. The landing can be prevented with camouflaging nets and these are being commercially used to manage Papaya ring spot virus-PaRSV in Taiwan. The white nets with 2–8 mesh are placed on the crop grown under them and the aphids get camouflaged as these aphids are unable to see beyond 50 cm. As a result, the incidence of the virus in pepper was reduced to 2% from 95% after 77 days. The use of kaolinite and montmorillonite (a mixture of kaolin and bentonite) on citrus lime crop is also known to reduce the population of *Aphis citricola* from 13.5% to 2/3% (Bar-joseph and Frenkel, 1983).

15.7.7 Barrier crops

Any kind of protective crop planted around the primary crop to save it from the vector-borne viruses is known as a barrier crop. It is a cultural measure used to mitigate the severity of plant viruses through the interference with vector actions. The barrier crops are also known as protective crops. The mechanism of virus reduction is not being fully investigated so far but the significant reduction in the incidence of non-persistent aphid-borne viruses is amply clear (Fereres, 2000). A possible explanation put forth by various workers is that the barrier crops affect appropriate/inappropriate landing due to the insects' attraction towards visual stimuli to locate the host in the background of soil. The barrier crop acts as sink tank as they land on the borders and spend some time on the protective crop, in the meantime, they lose the non-persistent plant virus. Besides acting as virus sinks, these barrier crop plants block most vectors by acting as physical barriers. The aphids do not accept the host without making exploratory probes and by doing so the vectors lose the stylet-borne viruses. These barrier crops provide substratum for providing floral nectar to parasitoids, the biological agents (Hooks et al., 2007; Hooks and Fereres, 2006). These barrier crops are mostly non-hosts of both the viruses or vectors and due to their attractiveness to vectors, these are, therefore, planted around the field. In addition, these plants are generally taller than the main crop. These barrier plants interfere in the landing of alate aphids by blocking their landing areas. It is being done to manage PVY in pepper in USA (Simons, 1957). In addition, the planting of wheat, sunflower or mustard as barrier plants is a practice to mitigate the incidence of Potato leaf roll virus (Mannan, 2003). The cultivation of oat as a barrier crop in lupins is instrumental in reducing the incidence of Barley yellow dwarf virus–N (strain) in Australia (Jones, 2005). Besides, these plants act as sink tank for non-persistent aphid-borne viruses likely to infest the main crop as the viruliferous vectors lose the virus in minutes while feeding. Once the vectors become non-viruliferous, they may enter into the main crop but would not be able to carry virus and inoculate the main crop. The use of barrier crops (maize, sorghum) as sink tanks for CMV and PVY in pepper is a practice used in Spain (Fereres, 2000). In India, the benefit of cultivation of barrier crops (sunflower, sorghum, maize, pearl millet) in chili to reduce the incidence of CMV is apparent of such insects becomes very high in the crop and the management becomes mandatory through pesticides. These pesticides may kill natural enemies of vector species of insect. In the absence of natural enemies, the vector population will build up and cause economic damage to the crop. In spite of such limitations, the biocontrol is successful. The good control of CTV was obtained with the use of biocontrol approach through the introduction of the mild strain of the virus. A good control of aphid vector *Pentalonia nigronervosa* and virus, BBTV (semi-persistent category) was recorded in Japan. However, the biocontrol measures have not proved effective in the case of Papaya ring spot virus.

15.12 Integrated Pest Management

Hitherto, these chemicals have been used in isolation. The efforts have been made to exploit other tactics of pest management (physical control, biological control, and cultural control, etc.) but these tactics also failed to check the vector population as these had been used singly or in isolation. The control achieved through these tactics took little longer. To be successful, these tactics should be carefully combined to combat these maladies (Fig. 15.1). There is a need to intelligently combine the tactics to be utilized in a system that is known as integrated pest management. It is defined by the Food and Agricultural Organization (FAO) as a system which, in consideration with plant environment and pest population dynamics, integrates all the sustainable techniques of pest management possible and maintains a population of pests below the economic injury level so as to avoid economic damage. To contain vectors, such as aphids, an integration of tactics is to be done intelligently. Besides, it is essential to understand the steps for successful transmission of plant viruses through aphid vectors. The suitable strategy/tactic to avoid landing, prevent brief probing, check feeding and avoid settling of aphids is to be selected for a given agroecosystem. The required strategies are available but the efficient control will depend on the sequence followed to manage plant viruses. In line with this, the invasion of aphids can be prevented through the installation of reflective devices/can be kept away from the crops grown in enclosures. In situations where the landing is completed, the aphids can be prevented from short probing through the use of repellent/antifeedant chemicals. If the aphids are in the process of settling, the management can be done with the chemical tactic to avoid congregation of aphids in a crop ecosystem. Based on a highly suitable combination of various tactics of integrated pest management, the virus management strategy through vector control in Zucchini squash (*Cucurbita pepo*) is found to be highly useful in controlling the Papaya ring spot virus in cucurbits in Australia (Pinese et al., 1994). It is an important case of implementation of integrated pest management which can be replicated in other crops as well. Another kind of integration

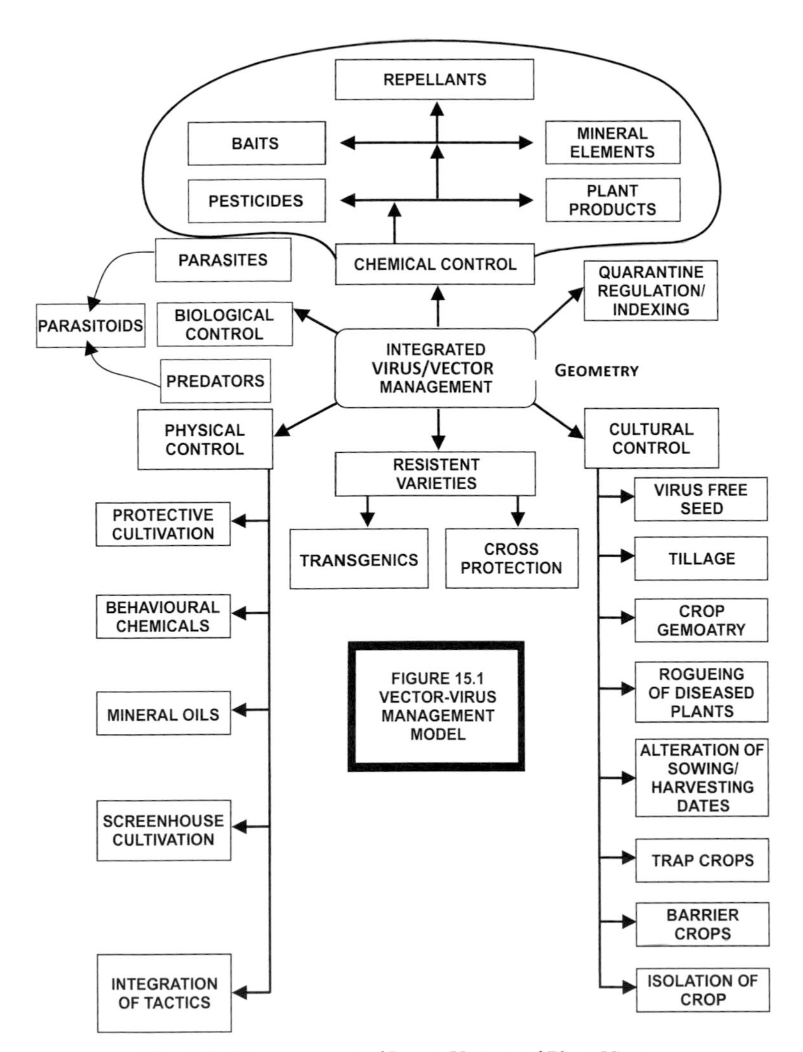

Fig. 15.1 Management of Insect Vectors of Plant Viruses.

involving the combined use of mineral oil and pyrethroid is also identified to contain viral diseases (Raccah, 1986). The integrated approach based on pre-planting and post-planting among the several tactics for the management of whitefly vector in cucurbits is advocated. Pre-planting practices include virus-free transplants of vegetables from the USA, avoidance of yellow clothes while transplanting seedlings, use of whitefly resistant screen houses (0.25–0.8 mm mesh) with roof cover of UV absorbing films and UV reflective mulches (Aluminum) on the ground around screen house. The post-planting measures include removal of diseased plants, weeds hosts, and non-application of neonicotinoids within seven days of transplanting.

Summers CG, Mitchell JP and Stapleton JJ (2004). Management of aphid-borne viruses and *Bemisia argentifolii* (Homoptera: Aleyrodidae) in Zucchini squash by using UV reflective plastic and wheat straw mulches. Environmental Entomology, 33: 1447–1457.
Uvah III and Coaker TH (1984). Effect of mixed cropping on some insect pests of carrots and onion. Entomologia Experimentalis et Applicata, 36: 159–167.
Weintraub PG (2007). Insect vectors of *Phytoplasmas* and their control, an update. Bulletin. Insectology, 60: 169–173.
West AJ and Mordue AJ (1992). The influence of azadirachtin on feeding behavior of cereal aphids and slugs. Entomologia Experimentalis et Applicata, 62: 75–79.
Wintermantel WM (2004). Emergence of greenhouse whitefly and *Trialeurodes vaporariorum* transmitted criniviruses as threats to vegetable and fruit production in North America. APSnetFeatures.Online:Doi;10 1094/APSnetFeature-2004-0604.
Wrobel S (2009). The retention of Potato virus Y in the stylet of *Myzus persicae* Sulz after the application of mineral oil on a potato plant. Plant Breeding and Seed Science, 60: 3–12.
Zhou C and Zhou Y (2012). Strategies for viral cross protection in plants. Methods in Molecular Biology, 894: 69–81.

QUESTIONS (EXERCISE)

Q 1. Describe the use of mineral oils in the management of non-persistent whitefly-borne viruses.

Q 2. What is meant by cultural measures? Discuss different measures which are successful in the management of invading populations of insect vectors.

Q 3. Why is the host plant resistance considered as an important method of vector management? Discuss with suitable examples.

Q 4. How is the physical method of control used to manage insect vectors? Give your views.

Q 5. Chemical and biocontrol are hesitantly used for vector control of plant viruses. Why?

Q 6. Write short notes on the use of the following in vector management:

a) Cross protection
b) Transgenic
c) Behavior modifying agents
d) Quarantine measures
e) Thermotherapy

Glossary

Acquisition: The act of acquiring/picking up the virus from the diseased plant by insect vector

Acquisition threshold: The minimum time required by insect vector to become viruliferous after an acquisition access on virus source, or

Acquisition threshold: It is the minimum time required by the vector to pick up the virus from the plant tissues

Acrostyle: A point of fusion of food and salivary canals in hemipterans

Alate: The winged form of aphids

Androcyclic: A type of reproduction in which parthenogenesis occurs at the very end of the season to produce only male offspring

Angstrom (A0): A unit of wavelength and its length is equal to 0.0001 micron

Arthropod: Organism with jointed appendages from phylum Arthropoda

Auchenorrhyncha: A division of order Hemiptera; contains leafhoppers, planthoppers, froghoppers

Autoecious life cycle: The completion of life cycle on one host only

Bacteria: These are microscopic, 0.5 to 5.0 um in length, single-cell prokaryotic organisms without nucleus, mitochondria, chloroplasts and reproduce through binary fission and contain a single DNA chromosome in cytoplasm

Bacteriophage: A virus that attacks bacteria

Biotype: A sub-division of a species producing genetically identical organisms

Bromosomes: Waxy material of waxy glands produced by leafhoppers to repel water from their bodies

Capsid: The protein coat encloses the nucleic acid

Cauda: A triangular shaped last abdominal segment covers the anus that secrets honeydew in aphids

Cerci: The forceps like two structures at the end of last abdominal segment meant to hold the prey

Circulative: It is a category of viruses that circulate in the body of insect vector, are picked up by stylets, passed through gut, enter into the hemolymph and finally the salivary glands, from where the virus is ejected into the healthy plants

Complete metamorphosis: A metamorphosis in which all the four stages viz. egg, larva, pupa and adult are present

Cornicles: A pair of the tube-like structures present on the dorsal side of last abdominal segment meant to secrete glycerols/honey; it is a synonymous with siphunculi

Crawler: The first instar nymph of whitefly

Ecdysis/molting: The act of shedding of old skin/exuviae during metamorphosis

Elytra: The forewings of hardened consistency in beetles

Endocytosis: A process of cellular ingestion by which the plasma membrane folds inward to bring substance into a cell

Envelope: A host-cell derived membrane with virus specific antigens that are acquired during virus maturation

Eriophyid: The mites hailing from the family Eriophyidae.

Exocytosis: The process of vesicles fusing with the plasma membrane and releasing the contents to out side the cell

Filter chamber: A part of the alimentary canal located at a junction of anterior ends of midgut and hindgut

Fungi: These are eukaryotic unicellular/multicellular, saprophytic spore-formers; they are net-like, filamentous, lack chlorophyll and are classed under the fungi kingdom and their cell wall is of chitin

Galls: An outgrowth on plant parts resulting from toxin injected into the plant by toxeniferous insects

Genome: A set of genes

GroEL: A synonym of protein produced by symbiotic bacteria in insect vectors required for safe transport of virus in the body of vector

Gustatory: An act of testing the host plant by tasting/chewing a small amount of food to continue the feeding by an insect vector using sensory organs

Halteres: A pair of dumbbell-shaped hind wings of dipteran insects

Hemocoel: A body cavity containing blood in which the organs of the insect float

Heteroecious life cycle: The completion of life cycle on two or more hosts

Heteropterous: The insects included in the sub-order Heteroptera, under order Hemiptera

Horizontal transmission: A mechanism of spread of plant pathogens from diseased plants to healthy ones through an aerial vector, vegetative propagation or grafting

Icosahedral: A geometric figure with 12, 20 and 30 vertices, triangular faces and edges, respectively

Incomplete metamorphosis: A metamorphosis in which three stages viz. egg, nymph, and adult are present

Incubation period: It is the time taken by the plant to show symptoms after inoculation of virus by a viruliferous vector

Inoculation: The act of inoculating/injecting the virus in plants by viruliferous vector

Inoculation threshold: The minimum time required by the vector to inoculate the virus in healthy plants

Instar: The stage between two molts during metamorphosis

Internal transmission: The mechanism of internally carrying the pathogen from diseased to healthy plant by the vector

Kataplastic: The new tissues of overgrowth are almost alike

kb (kilo bars): A unit of pressure equal to 1000 bars

Labium: Lower lip of insect mouthparts formed from a pair of maxillae

Labrum: Upper lip of insect mouthparts formed by a pair of mandibles

Latent/incubation period: The time that elapses between inoculation of pathogen and the appearance of symptoms

Longevity: Duration of survival in insects

Maggot: A legless larva of dipterous flies

Malpighian tubules: The long tube-like structures at the junction of posterior end of midgut and anterior end of hindgut, serving the excretory function

Mandibles: The pincer-like outer part of the mouth of insects that facilitates penetration into plant tissues

Maxillae: A pair of appendages; part of mouth of insects

Metamorphosis: Sum total of all changes taking place from egg laying to adult formation in insects

Meiosis: A type of cell division that reduces the chromosome number by half, forming four haploid cells each distinct from the parent cell

Micron (u): 1/1000th of a millimeter

Transovarial transmission: Transmission of plant viruses from one generation to the next through eggs

Transstadial transmission: This occurs when a pathogen remains with a vector from one instar to the next

Trichome: A glandular hair on the leaf surface

True bugs: Members of sub-order Heteroptera in the order Hemiptera

Trypanosoma (protozoan): A unicellular animal lacking true organs

Vascular tissues: The tissues responsible for transport of water and nutrients from roots to the aerial parts and for the supply of ready made food material from the leaves to other parts; xylem and phloem, respectively

Vector: The living organisms transmitting the disease-causing pathogen from diseased to healthy plant, e.g., insects

Vegetative reproduction: Asexual mode of reproduction

Vertical transmission: A mechanism in which the pathogen passes from parents to offspring, or transmission of seed-borne pathogen (virus) from seed to its progeny

Virion: A virus particle

Viroid: A virus without protein coat

Virology: The science deals with the study of viruses

Viruliferous: An insect that is carrying a virus

Virus: A sub-microscopic infective entity, capable of reproducing in living cells. Viruses contain only one kind of nucleic acid, are potentially pathogenic, lack lipman system of energy release and make use of ribosomes of plant cells

Xylem: The vascular tissue in plants which conducts water and dissolved nutrients upwards from the root and also helps to form the woody element in the stem

Zygote: A single cell formed by merging of an egg cell and a sperm cell

Acronyms *

**All abbreviations may not be according to the list approved by the ICTV due to the addition of more new viruses*

ACLSV	:	Apple chlorotic leaf spot virus
ACMV	:	African cassava mosaic virus
AgMV	:	Agropyron mosaic virus
AILV	:	Artichoke Italian latent virus
AIVV	:	Abutilon infectious variegation virus
AIVV	:	Artichoke infectious variegation virus
AlpMV	:	Alpinia mosaic virus
BYSV	:	Bean yellow stipple virus
BYV	:	Beet yellow virus
BYVaV	:	Blackberry yellow vein associated virus
BYVMV	:	Bhindi yellow vein mosaic virus
BYVV	:	Black bean yellow vein virus
CABYV	:	Cucurbit aphid-borne yellows virus
CaMV	:	Cauliflower mosaic virus
CapLCV	:	Cape gooseberry leaf curl virus
CapMV	:	Capsicum mosaic virus
CarMV	:	Carnation mottle virus
CaYMV	:	Canna yellow mottle virus
CBDV	:	Colocasia bobone disease virus
CBSV	:	Cassava brown streak virus
CcDV	:	Cadang cadang disease virus(now viroid)
CCDV	:	Citrus chlorotic dwarfing virus
CCMoV	:	Cucumber chlorotic mottle virus
CCMV	:	Cowpea chlorotic mottle virus
CaCV	:	Capsicum chlorosis virus
CCV	:	Cardamom chirke virus
CCYV	:	Cucurbit chlorotic yellows virus
CEDV	:	Cohern enation disease virus

CP	:	Coat protein
CpCDV	:	Cowpea chlorotic dwarf virus
CPMV	:	Cowpea mosaic virus
CPSMV	:	Cowpea severe mosaic virus
CpYMV	:	Cowpea yellow mosaic virus
CRDV	:	Cherry rosette disease virus
CRLV	:	Cherry rasp leaf virus
CRSV	:	Carnation ring spot virus
CRV	:	Coconut rosette virus
CSMV	:	Chlorosis striate mosaic virus
CSNV	:	Chrysanthemum stem necrosis virus
CSSV	:	Cacao swollen shoot virus
CTDV	:	Cereal tilling disease virus
CTV	:	Citrus tristeza virus
CVCV	:	Cucumber vein clearing virus
CVV	:	Citrus variegation virus
CVYV	:	Cucumber vein yellowing virus
CWMV	:	Chinese wheat mosaic virus
CWTV	:	Clover wound tumor virus
CYDV	:	Cereal yellow dwarf virus
CYMBV	:	Citrus yellow mosaic badnavirus virus
CymMV	:	Cymbidium mosaic virus
CymRSV	:	Cymbidium ring spot virus
CYMV	:	Citrus yellow mosaic virus
CYSDV	:	Cucurbit yellow stunting disorder virus
CYSV	:	Cucurbit yellow stunting virus
CYV	:	Cucumber yellows virus
CYVMV	:	Croton yellow vein mosaic virus
DDV	:	Docking disorder virus
DIBA	:	Dot immune binding assay
DiSV	:	Digitaria striate virus
DNA	:	Deoxyribonucleic acid
DoYMV	:	Dolichos yellow mosaic virus
dsRNA	:	Double stranded RNA
DVCV	:	Diodia vein chlorosis virus
ELISA	:	Enzyme linked immunosorbent assay
EPMV	:	Eggplant mosaic virus
EPPO	:	European plant protection organization
EpYVV	:	Eupatorium yellow vein virus
ERSV	:	Echinocloea ragged stunt virus
EuMV	:	Euphorbia mosaic virus
EWSV	:	European wheat streak virus
EYVMV	:	Eclypha yellow vein mosaic virus

FBNSV	:	Faba bean necrotic stunt virus
FDV	:	Fiji disease virus
FLNV	:	Fereesia leaf necrosis virus
FMMV	:	Finger millet mosaic virus
FMV	:	Fig mosaic virus
GARSV	:	Grapevine antolian ring spot virus
GBMV	:	Golden bean mosaic virus
GBNV	:	Groundnut bud necrosis virus
GbVBV	:	Gooseberry vein banding virus
GFkV	:	Grapevine fleck virus
GFLV	:	Grapevine fan leaf virus
GLRaV	:	Grapevine leaf roll associated viruses
GLRaV-3	:	Grapevine leaf roll associated virus-3
GRSV	:	Groundnut ring spot virus
GV-A	:	Grapevine virus–A to F
GYMV	:	Grapevine yellow mosaic virus
HC	:	Helper component
HgYMV	:	Horsegram yellow mosaic virus
HLCV	:	Hibiscus leaf curl virus
HoYMV	:	Hollyhock yellow mosaic virus
ICTV	:	International Committee on Taxonomy of Viruses
ILCV	:	Ipula leaf curl virus
INSV	:	Impatient necrotic spot virus
ITP	:	Insect Transmission Protein
IYSV	:	Irish yellow spot virus
IYVV	:	Ipomea yellow vein virus
JacYMV	:	Jacquemontia yellow mosaic virus
JCRSV	:	Jasmine chlorotic ring spot virus
JGMV	:	Johanson grass mosaic virus
JLCV	:	Jatropha leaf curl virus
JuYMV	:	Jute yellow mosaic virus
Kb	:	Kilo bite
kDa	:	Kilodalton
LBVaV	:	Lettuce big vein associated virus
LBVV	:	Lettuce big vein virus
LClV	:	Lettuce chlorosis virus
LCV	:	Little cherry virus
Li NV	:	Lisianthus necrosis virus
LiYV	:	Lettuce infectious yellows virus
LLCV	:	Lupin leaf curl virus
LMV	:	Lettuce mosaic virus
LNYV	:	Lettuce necrotic yellows virus
LonWBV	:	Longan witches broom virus
LP	:	Latent period

Annexure

Virus Familes, Genera and Important Type Species of Plant Viruses

1. Bromoviridae

Alfamovirus: (Alfalfa mosaic virus-AMV/ssRNA).

Ilarvirus: (Tobacco streak virus (TSV/ssRNA); American plum line pattern virus (APLPV); Apple mosaic virus (ApMV); Asparagus virus-2 (AV-2); Blueberry shock virus (BlShV); Citrus leaf rugose virus (CiLRV); Citrus variegation virus (CVV); Ilm mottle virus (IMV); Fragaria chiloensis latent virus (FCiLV); Humulus japonicus latent virus (HJLV); Parietaria mottle virus (PMoV); Prune dwarf virus (PDV); Prunus necrotic ring spot virus (PNRSV); Spinach latent virus (SpLV); Tobacco streak virus (TSV); Tulare apple mosaic virus (TAMV); Lilac ring mottle virus (LiRMV).

Bromovirus: (Brome mosaic virus–BMV/ssRNA); Broad bean mottle virus (BBMV); Cassia yellow blotch virus (CYBV); Cowpea chlorotic mottle virus (CCMV); Melandrium yellow fleck virus (MYFV); Spring beauty latent virus (SBLV).

Cucumovirus: (Cucumber mosaic virus–CMV/ssRNA); Peanut stunt virus (PSV); Tomato aspermy virus (TAV).

Oleavirus: (Olive latent virus–2-OLV-2/ssRNA).

virus (GaMV); Japanese iris necrotic ring virus (JINRV); Melon necrotic spot virus (MNSV); Pea stem mosaic virus (PSMV); Pelargonium flower break virus (PFBV); Saguaro cactus virus (SaCV); Turnip crinkle virus (TCV); Weddel water-borne virus (WWBV).

***Necrovirus*:** Tobacco necrosis virus–A–ssRNA (TNV–A); Ahlam water-borne virus (AWBV); Beet black scorch virus (BBSV); Chenopodium necrosis virus (ChNV); Leek white stripe virus (LWSV); Olive latent virus 1 (OLV1); Tobacco necrosis virus D (TNVD).

***Dianthovirus*:** Carnation ringspot virus–ssRNA CRSV); Red clover necrotic mosaic virus (RCNMV); Sweet clover necrotic mosaic virus (SCNMV).

***Machlomovirus*:** Maize chlorotic mottle virus–ssRNA (MCMV).

***Avenavirus*:** Oat chlorotic stunt virus–ssRNA (OCSV).

***Panicovirus*:** Panicum mosaic virus–ssRNA (PMV).

***Aureusvirus*:** Cucumber leaf spot virus–ssRNA (CLSV); Oat chlorotic stunt virus (OCSV); pothos latent virus (PoLV).

18. Tymoviridae

***Tymovirus*:** Turnip yellow mosaic virus–ssRNA (TYMV); Andean potato latent virus (APLV); Belladonna mottle virus (BeMV); Cacao yellow mosaic virus (CaYMV); Calopogonium yellow vein virus (CalYVV); Clitoria yellow vein virus (CYVV); Desmodium yellow mottle virus (DYMoV); Dulcamara mottle virus (DuMV); Eggplant mosaic virus (EMV); Erysimum latent virus (ErLV); Kennedya yellow mosaic virus (KYMV); Melon rugose mosaic virus (MRMV); Okra mosaic virus (OkMV); Onion yellow mosaic virus (OYMV); Passion fruit yellow mosaic virus (PFYMV); Peanut yellow mosaic virus (PeYMV); Petunia vein banding virus (PetVBV); Physalis mottle virus (PhyMV); Plantago mottle virus (PlMoV); ScrMV mottle virus (ScrMV); Voandzeia necrotic mosaic virus (VNMV); Wild cucumber mosaic virus (WCMV).

***Marafivirus*:** Maize rayado fino virus–ssRNA (MRFV); Bermuda grass etched line virus (BELV); Oat blue dwarf virus (OBDV).

***Maculavirus*:** Grapevine fleck virus–ssRNA GFkV).

Without Family Allocation

***Hordeivirus*:** Barley stripe mosaic virus–ssRNA (BSMV); Anthoxanthum latent blanching virus (ALBV); Lychnis ringspot virus (LRSV); Poasemi latent virus (PSLV).

Cheravirus: Chery rasp leaf virus–ssRNA (CRLV); Apple latent spherical virus (ALSV).

Benyvirus: Beet necrotic yellow vein virus-ssRNA (BNYVV); Beet soil-borne mosaic virus (BSBMV).

Pomovirus: Potato mop top virus–ssRNA (PMTV); Beet soil-borne virus (BSBV); Beet virus Q (BVQ); Broad bean necrosis virus (BBNV); Broad bean necrosis virus (BBNV).

Sobemovirus: Subterranean clover mottle virus–ssRNA (SCMoV); Blueberry shoestring virus (BSSV); Cocksfoot mottle virus (CfMV); Lucerne transient streak virus (LTSV); Rice yellow mottle virus (RYMV); Rye grass mottle virus (RGMoV); Sesbania mosaic virus (SeMV); Solanum modiflorum mottle virus (SNMoV); Southern bean mosaic virus (SBMV); Southern cowpea mosaic virus (SCMV); Sowbane mosaic virus (SoMV); Turnip rosette virus (TroV); Velvet tobacco mottle virus (VTMoV); Blueberry shoestring virus (BSSV); Ryegrass mottle virus (RGMoV).

Umbravirus: Pea enation mosaic virus–ssRNA (PEMV); Carrot mottle mimic virus (CmoMV); Carrot mottle virus (CMoV); Groundnut rosette virus (GRV); Lettuce speckles mottle virus (LSMV); Pea enation mosaic virus–2 (PEMV2); Tobacco bushy top virus (TBTV); Tobacco mottle virus (TmoV); Groundnut rosette virus (GRV).

Ourmiavirus: Ourmia melon virus–ssRNA (OuMV); Cassava virus c (CaVC); Epirus cherry virus (EpCV); Ourmia melon virus (OuMV).

Soil-borne wheat mosaic virus–ssRNA (SBWMV); Chinese wheat mosaic virus (CWMV); Oat golden stripe virus (OGSV); Sorghum chlorotic spot virus (SrCSV); Soil-borne cereal mosaic virus (SBCMV).

Tobamovirus: Tobacco mosaic virus ssRNA (TMV); Cucumber fruit mottle mosaic virus (CFMMV); Cucumber green mottle mosaic virus (CGMMV); Frangipani mosaic virus (FrMV); Hibiscus latent fort pierce virus (HLFPV); Hibiscus latent Singapore virus (HLSV); Kyuri green mottle mosaic virus (KGMMV); Obuda pepper virus (ObPV); Odontoglossum ring spot virus (ORSV); Papika mild mottle virus (PaMMV); Pepper mild mottle virus (PMMoV); Rye grass mosaic virus (RGMV); Sammon,s opuntia virus (SOV); Sunhemp mosaic virus (SHMV); Tobacco latent virus (TLV); Tobacco mild green mosaic virus (TMGMV); Tobacco mosaic virus (TMV); Tomato mosaic virus (ToMV); Turnip vein clearing virus (TVCV); Ullucus mild mottle virus (UMMV); Wassabi mottle virus (WMoV); youcai mosaic virus (YoMV); Zucchini green mottle mosaic virus (ZGMMV); Frangipani mosaic virus (FrMV); Ribgrass mosaic virus (RMV); Simmons,s opuntia virus (SOV).

Tenuivirus: Maize stripe virus–(ssRNA MSpV); Echinocloa hoja blanca virus (EHBV); Rice grassy stunt virus (RGSV); Rice hoja blanka virus (RHBV); Rice stripe virus (RSV); Urochloa hoja blanca virus (UHBV).

Pecluvirus: Peanut clump virus–ssRNA (PCV); Indian peanut clump virus (IPCV).

Varicosavirus: Lettuce big vein associated virus–ssRNA (LBVaV).

Ophiovirus: Citrus psorosis virus–ssRNA (CPsV); Lettuce ring necrosis virus (LRNV); Mirafiori lettuce virus (MiLV); Citrus psorosus virus (CPsV); Ranunculus white mottle virus (RWMV); Tulip mild mottle mosaic virus (TMMMV).

Tobravirus: Tobacco rattles virus (TRV); Pea early browning virus (PEBV); Pepper ring spot virus (PepRSV).

Idaevirus: Raspberry bushy stunt virus (RBSV); Tobacco rattle virus (TRV).

Sadwavirus: Satsuma dwarf virus–ssRNA (SDV); Strawberry latent ringspot v (SLRSV); Strawberry mottle virus (SMoV).

Macluravirus: Maclura mosaic virus (MacMV); Narcissus latent virus (NLV).

VIRUSES WITHOUT GENUS

1. **Rhabdoviridae** (ssRNA)
   Carrot latent virus (CtLV); Euonymus fascination virus (EFV); Lupin yellow vein virus (LYVV); Malva sylvestris virus (MaSV); Saintpaulia leaf necrosis virus (SLNV); Sambucus vein clearing virus (SVCV); Sarracenia purpurea virus (SPV); Atropa belladonna virus (AtBV); Beet leaf curl bacilliform virus (CBV); Carrot latent virus (CtLV); Cynara virus (CraV); Dendrobium leaf streak virus (DLSV); Digitaria striate virus (DiSV); Euonymus fascination virus (EFV); Finger millet mosaic virus (FMMV); Gerbera symptomless virus (GeSLV); Gomphrena virus (GoV); Hoicus lanatus yellowing virus (HLYV); Iris gemanica leaf stripe virus (IGLSV); Ivy vein clearing virus (IVCV); Laelia red leaf spot virus (LRLV); Launea arborescens stunt virus (LArSV); Lemon-scented thyme leaf chlorosis virus (LSTCV); Lolium rye grass virus (LoRV); Lolium rye grass virus (LoRV);Lotus stem necrosis virus (LoSNV); Lucerne enation virus (LEV); Maize fine streak virus (MFSV); Maize mosaic virus (MMV); Maize sterile stunt virus (MSSV); Melilotus latent virus (MeLV); Melon variegation virus (MVV); Oat striate mosaic virus (OSMV); Parsley virus (PaV); Phalaenopsis chlorotic spot virus (PhCSV); Pigeon pea proliferation virus (PPPV); Pineapple chlorotic leaf streak virus (PCLSV); Pisum virus (PisV); Plantain mottle virus

(PlMV); Ranunculus repens symptomless virus (RaRSV); Raphanus virus (RaV); Raspberry vein chlorosis virus (RVCV); Saintpaula leaf necrosis virus (SLNV); Sambucus vein clearing virus (SVCV); Sorghum virus (SrV); Soursoup yellow blotch virus (SYBV); Triticum aestivum chlorotic spot virus (TACSV); Vigna sinensis mosaic virus (VSMV); Wheat chlorotic streak virus (WCSV); Wheat rosette stunt virus (WRSV); Winter wheat Russian mosaic virus (WWRMV); Zea mays virus (ZMV).

2. **Luteoviridae** (ssRNA)
   Indonesian soybean dwarf virus (ISDV); Groundnut rosette assister virus (GRAV); Sweet potato leaf speckling virus (SPLSV).

3. **Tymoviridae**
   Poinsettia mosaic virus (PnMV).

4. **Closteroviridae** (ssRNA)
   Alligatorweed stunting virus (AWSV); Burdog yellow virus (BuYV); Grapevine leaf roll associated virus 7 (GLRaV–7); little cherry virus 1 (LChV–1); Megakepasma mosaic virus (MegMV); Olive leaf yellowing associated virus (OLYaV); Orchid fleck virus (OFV).

5. **Flexiviridae**
   Banana mild mosaic virus (BanMMV).

6. **Luteoviridae:** Barley yellow dwarf virus GPV (BYDV–GPV); Barley yellow dwarf virus–RMP (BYDV–RMP); Barley yellow dwarf virus SGV (BYDV–SGV); Potato virus T (PVT); Sugarcane striate mosaic associated virus (SCSMaV).

7. **Bromoviridae** (ssRNA)
   Pelargonium zonate spot virus (PZSV).

8. **Pseudovirdae**
   Phaseolus vulgaris pv2–6 virus (PvupvV).

9. **Potyviridae** (ssRNA)
   Spartina mottle virus (SpMV); Sugarcane streak mosaic virus (SSMV); Tomato mild mottle virus (TomMMoV).

Viruses Without Family and Genus

Raspberry necrosis virus (BRNV); Brachypodium yellow streak virus (BraYSV); Cassava Ivorian bacilliform virus (CsIBV); Chara australis virus (CAV); Flame chlorosis virus (FlCV); Hart's tongue fern mottle virus (HTFMoV); Hawaiian rubus leaf curl virus (HRLCV); Maize white line mosaic virus (MWLMV); Nicotiana velutina mosaic virus (NVMV); Parsley latent virus (PaLV); Pigeon pea sterility mosaic virus (PPSMV); Poinsettia latent virus (PnLV); Tulip streak virus (TuSV); Watercress yellow spot virus (WYSV); White clover virus 1 (WClV1); Brachypodium yellow streak virus (BraYSV); Flame chlorosis virus (FlCV); Hart's tongue fern mottle virus (HTFMoV).

Subject Index

Q

R

S

T

U

V

About the Author

Professor Nachhattar Singh Butter was born on October 13, 1948, in the small village of Bathinda district in the Punjab state of India. After completion of schooling from a rural village in 1967, joined Punjab Agricultural University, Ludhiana, and did BSc (Ag) in 1971, MSc (Entomology) in 1973 and PhD (Entomology) in 1976. Being a meritorious student, Dr Butter was awarded merit scholarship both in Bachelor and Masters programs and Senior Fellowship, CSIR during PhD. The fellowship was later on converted into Post Doctorate Fellowship on the completion of a degree in Nov. 1976 to work in the field of insect vectors. It continued up to Feb 1977 on joining of the post of Assistant professor. He continued to work till his promotion as Entomologist in July 1987. In March 1994 he joined as a professor. After serving as professor up to June 2006, he took over as Head, Department of Entomology and continued to discharge administrative duties till superannuation in Oct 2008. He developed Integrated Pest Management system for cotton crop in Punjab and Indian Society for Cotton Improvement, Mumbai, honored him with Hexamar Award, 1992 for outstanding research contributions in cotton. He guided nine post graduate students in Entomology and taught both under- and post-graduate courses in Entomology. He has published 77 research articles in journals of repute besides 200 popular articles, 22 presented papers in seminars/workshops. He has four books to his credit in addition to contributed chapters in books and laboratory manuals. Dr Butter has delivered 44 Radio/TV talks and 200 lectures to farmers on important topics. He is instrumental in establishing plant clinic in PAU, the model of which was emulated throughout the country. Dr Butter visited erstwhile USSR in 1987 to study Integrated Pest Management in cotton as a member of the 2-men delegation. For the outstanding contributions in research teaching and extension in agriculture, the State Govt awarded him Punjab Sarkar Parman Patra, 2002. Dr Butter was honoured with Life Time Achievement Award by Cotton Research and Development Association at the International conference on cotton and other fibre crops held at Umiam, Meghalaya, India on Feb 20, 2018, for outstanding research contributions in cotton entomology. Dr Butter is a fellow of Entomological Society of India and The Indian Society for the Advancement of Insect Science.